ADVANCES IN SOLUTION CHEMISTRY

ADVANCES IN SOLUTION CHEMISTRY

Edited by
I. Bertini
University of Florence
Florence, Italy

L. Lunazzi
University of Bologna
Bologna, Italy

and
A. Dei
University of Florence
Florence, Italy

PLENUM PRESS · NEW YORK AND LONDON

Library of Congress Cataloging in Publication Data

International Symposium on Solute–Solute–Solvent Interactions, 5th, Florence, Italy, 1980.
Advances in solution chemistry.

Includes index.
1. Solution (Chemistry) — Congresses. I. Bertini, Ivano. II. Lunazzi, L. III. Dei, A. IV. Title. [DNLM: 1. Solutions—Congresses. 2. Solvents—Congresses. QD 540 A244 1980]
QD540.I57 1980   541.3'4   80-28783
ISBN 0-306-40638-1

A part of the proceedings of the Fifth International Symposium on
Solute–Solute–Solvent Interactions, held June 2–6, 1980, in Florence, Italy

© 1981 Plenum Press, New York
A Division of Plenum Publishing Corporation
233 Spring Street, New York, N. Y. 10013

All rights reserved

No part of this book may be reproduced, stored in a retrieval system, or transmitted, in any form or by any means, electronic, mechanical, photocopying, microfilming, recording, or otherwise, without written permission from the publisher

Printed in the United States of America

PREFACE

From June 2nd to 5th 1980, the Vth International Symposium on Solute-Solute-Solvent Interactions was held in Florence, Italy. Owing to the large range of interests included in the program and to their interdisciplinary nature, a number of microsymposia on specific subjects were organized, in addition to plenary lectures, session lectures and poster sessions. The abstracts of the Conference were published in Inorganica Chimica Acta as a special publication and as an appendix to the issue of June 1980.

The plenary lectures have been published, as customary, in the Journal of Pure and Applied Chemistry (October 1980) and the results of the following microsymposia are being published in specialized journals:

1) Electronic Rearrangements Induced by Solute-Solvent Interactions (Local Editor J. H. Ammeter) - N. J. Phys. Chimie.
2) Anion Activation in Quaternary Salts, Crown Ethers, Cryptates and Related Systems (Local Editor F. Montanari) - J. Mol. Catalysis.
3) Solvent Effects in Homogeneous Catalysis By Metal Complexes (Local Editor R. Ugo) - J. Mol. Catalysis.
4) Theoretical Models in Biochemical and Related Systems (Local Editor E. Clementi) - J. Computational Chemistry.
5) Thermodynamic Approach in Coordinative Interactions in Solution (Local Editor P. Paoletti) - Advances in Molecular Relaxation and Interaction Processes.

The present book contains most of the session lectures as well as the contributions to the microsymposium on "Evaluation of Solvation Energies of Reagents and Transition States". The title "Advances in Solution Chemistry", although rather general and not pursuing a single definite goal, has been chosen in order to fit the frame of the title of the Symposium. The choice of the lectures has been based on the scientific prestige of the contributors in the various areas needed to be discussed in the meeting and, although to a lesser extent, following a geographical criterion.

The contributions to the microsymposium on "Evaluation on Solvation Energy of Reagents and Transition States" (E. Buncel, M. H. Abraham and G. Scorrano) have been included because we thought they fit well into the overall scheme of the reactivity in solution. We acknowledge the judgement of Professor Giorgio Modena, convener of the microsymposium, for the excellent choice of speakers.

The selection of the session lectures is a result of the joint efforts of the organizing committee: U. Belluco, E. Clementi, A. Dondoni, C. Furlani, F. Montanari, L. Sacconi, R. Ugo, and the editors.

By organizing the meeting we have tried to outline the most recent aspects and trends (physical, organic, inorganic, theoretical, biological, pharmaceutical) of the chemistry of solutions. This book witnesses in part this effort.

We take the opportunity of thanking all the colleagues, friends, and students of the University of Florence, who helped in the organization of the Conference. A special thanks is due to Ms. Simonetta Donzellini who, besides delighting the participants to the meeting with her presence, patiently and skillfully retyped all the manuscripts: of all the misprints which have been left in the final draft the editors have the sole responsibility. Indeed, we decided not to return the retyped texts to the authors in order to speed up the publication.

The Editors

Firenze, 9th October 1980

CONTENTS

Solute-Solvent Interactions as Required for
    the Existence of a Liquid   1
G. Resch and V. Gutmann

Thermodynamics of Aqueous Dilute Solutions
    of Non-Charged Molecules   13
S. Cabani

Classical Ionic Fluids in the Mean
    Spherical Approximation   41
R. Triolo and A.M. Floriano

A Comparison between Structures of Aqua and
    Ammine Complexes in Solution as
    Studied by an X-Ray Diffraction
    Method   67
H. Ohtaki

Copper(II) Chelate Complexes-Solute and/or
    Solvent Interactions   81
N.D. Yordanov

Photo-Induced Ligand Solvent Interchange
    in Transition Metal Complexes   91
L.G. Vanquickenborne and A. Ceulemans

Mechanism of Octahedral Substitutions on
    Transition Metal Complexes. Attempts
    to Distinguish between D and I_d
    Mechanisms 105
S. Ašperger

Use of Electron Paramagnetic Resonance
    Spectroscopy to Study the Interaction
    between Cobalt Schiff Base Complexes
    and Phosphines or Phosphites in Solution 115
J.B. Raynor and G. Labauze

An NMR Study of Solvent Interactions in a
    Paramagnetic System 129
  R.M. Golding, R.O. Pascual, and
  C. Suvanprakorn

Protonation and Complexation Equilibria of
    Macromolecular Bioligands in Aqueous
    and Mixed Solvent Solutions. The
    Solvent Effect 139
  K. Burger and B. Noszal

Intramolecular Hydrophobic and Aromatic-Ring
    Stacking Interactions in Ternary
    Complexes in Solution 149
  H. Sigel

Study on Some Dioxygen Carriers - New Models
    of Natural Systems 161
  B. Jeżowska-Trzebiatowska and A.A. Vogt

Direct NMR Studies of Ionic Solvation 175
  P. Laszlo, A. Cornélis, A. Delville,
  C. Detellier, A. Gerstmans, and
  A. Stockis

NMR Studies of Calcium and Magnesium in
    Biological Systems 191
  S. Forsén, T. Andersson, T. Drakenberg,
  E. Thulin, and T. Wieloch

Synthetic Molecular Membranes and Their
    Functions 209
  T. Kunitake and Y. Okahata

New Insights into the Host-Guest Solvent
    Interaction of Some Inclusion Complexes.
    Reaction Path Control in Cyclodextrin
    Inclusion as Lyase Model: Solvolysis
    of β-Bromoethyl-1-Naphthalene 221
  I. Tabushi

The Composite Physical and Chemical Approach
    to the Solution Spatial Structure of
    Polypeptide Neurotoxins 231
  V.F. Bystrov, V.T. Ivanov, V.V. Okanov,
  A.I. Miroshnikov, A.S. Arseniev,
  V.I. Tsetlin, V.S. Pashkov, and
  E. Karlsson

CONTENTS

Metalloenzymes and Model Systems. Carbonic
    Anhydrase: Solvent and Buffer Partici-
    pation, Isotope Effects, Activation
    Parameters and Anionic Inhibition   253
  Y. Pocker, T.L. Deits, and N. Tanaka

Pyridinium-N-Phenoxide Betaine Dyes as
    Solvent Polarity Indicators.
    Some New Findings   275
  C. Reichardt, E. Harbusch, and R. Müller

Some Applications of Liquid Crystals in
    Organic Chemistry   295
  G. Gottarelli and B. Samorì

Reactions and Behaviour of Organic Anions
    in Two-Phase Systems   309
  M. Makosza

Solute-Solvent Interactions in Ring Formation   319
  C. Galli, G. Illuminati, L. Mandolini,
    and B. Masci

Hydrogen Acidities and Brønsted Relations   331
  A. Streitwieser, Jr.

Solvent Effects on Some Nucleophilic
    Substitutions   341
  M.H. Abraham

Initial State and Transition State Solvent
    Effects: Reactions in Protic and
    Dipolar Aprotic Media   355
  E. Buncel and E.A. Symons

Solvation Energies in Acid Catalyzed
    Processes   373
  G. Scorrano

Index   385

SOLUTE-SOLVENT INTERACTIONS AS REQUIRED FOR THE EXISTENCE OF A

LIQUID

Gerhard Resch and Viktor Gutmann

Institute of Inorganic Chemistry
Technical University of Vienna
Vienna, Austria

Abstract - It is proposed that the existence of a liquid would be impossible in the complete absence of impurities, voids or ions. The structures and the dynamics of the system are decisively influenced by their presence. In liquid water and various other liquids ions are produced by "self-ionization". In so-called "non aqueous" liquids water is an unavoidable impurity. Solutes and voids are structure-modified and modifying centres ("SMM-centres") which are under control of the forces in the higher hierarchic levels, namely the proposed "channel-network" and the boundary areas.

1. INTRODUCTION

Any substance is known to contain impurities, which cannot be completely removed. In addition there is always "empty" space within a system and this is frequently referred to as "voids" or "holes". As is well-known for the solid state, their unavoidable presence suggests that they do constitute integral parts of the phase(1). The number of molecules immediately surrounding "empty" space is usually small as compared to the total number of molecules constituting the liquid system, and hence the former do not contribute significantly to the statistical results. Likewise, the number of molecules at the surface is too small to be statistically significant.

In approaching the liquid state one of the major obstacles is the absence of experimental techniques for the direct observation or for the measurement of the individual dynamic features. Most of the experimental results, such as the thermodynamic properties of water, its heat capacity, compressibility X-ray diffraction or

NMR-data are considered with regard to the so-called "diffusionally averaged structure"(2), corresponding to relaxation times of 10^{-11} s.

However, statistical results are unsuitable to recognize the individual properties of the constituent parts and their interrelationships within the complex system under consideration(3). They may lead to idealizations and these may serve as illustrations of certain aspects, but they do not lead to an understanding of the real object. It would be an illusion to expect to gain an understanding of nature by consideration from one single point of view. From any single point of view shifts and distortions in the projection are unavoidable, so that not even a certain part of the system can be observed fully. It is therefore necessary to consider each phenomenon in all respects from various points of view(4,5). All of the pieces of evidence should be related to the whole system with appropriate consideration of all of the environmental effects, which are so easily neglected in scientific investigations.

It seems obvious to choose reality in its apparent complexity rather than the idealized models as a point of departure. This starting point seems to offer no exit in cognition but only in the real thing itself(4). Acquisition of scientific knowledge from this starting point is difficult and vulnerable. However, this step appears unavoidable, if an understanding of matter is sought. Due to the mutual penetrations of the constituent parts and their continuous changes within a liquid it is unlikely to expect to find constituents of precisely the same properties. In applying the familiar abstract description by means of atoms and molecules it is important to emphasize the difference in properties of the constituent molecules and ions. The next step is to evaluate their relative significance within the complex and ordered relationships of the whole liquid system under consideration(3). The constituent parts may be ordered with respect both to their temporal local significance <u>within</u> the system and to their significance <u>for</u> the whole system(3-5). Regularities are expected to be found which have not been envisaged in the light of the existing approaches and additional knowledge to be found as reflected in the properties of matter and their changes as they are actually observed and not by the requirements of existing theories or model assumptions.

2. THE TEMPORAL LOCAL SIGNIFICANCE OF THE CONSTITUENT PARTS.

In choosing liquid water as an example, this liquid would be incapable of existence, if its constituent molecules were not differentiated(6). Even in the simple, idealized system of an isolated dimeric water molecule, the two consituent water molecules have assumed rather different structural aspects as has been demonstrated by the results of ab initio SCF-calculations(7):

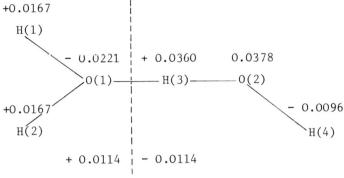

In this idealized state (without molecular environment) only two atoms are equivalent, namely H(1) and H(2), which are more strongly acidic than those in monomeric water; H(3) has obtained an even greater positive net charge, whereas the net charge of H(4) has become more negative, i.e. its acidity is lower than that of the hydrogen atoms in a monomeric water molecule. The bonds H(1) - O(1), H(2) - O(1) and H(3) - O(2) are longer whereas the bond O(2) - H(4) is shorter than in an isolated water molecule. The reported values refer to the fiction of a rigid dimeric water molecule, where the distortions from the symmetrical positions of H(1) and H(2) due to oscillations cancel in the statistically calculated values. In a real dimer each of the mentioned hydrogen atoms will always be in a slightly different molecular environment and hence none of the atoms are actually equivalent.

The formation of higher aggregates (with formation of "hydrogen-bonds") is therefore bound to lead to unsymmetrical structures. The changes may be illustrated by applying the bond length variation rules of the extended donor-acceptor approach(8). The cooperative effects lead to further lengthening of the bonds already lengthened and to shortening of O(1) - H(3). The cooperative effects caused by a certain number of water molecules may lead to lengthening of one of the former bonds to that extent that heterolysis is to take place with the formation of a hydrated proton and a hydrate ion(9).

Ionic species are known to be in higher states of energy and under more constrained conditions than solvent molecules(3).

In considering the solvent effects on the static aspects of structure and properties, the effects due to the presence of the other ions and solute molecules cannot be neglected. For example, the ^{23}Na chemical NMR-shifts of sodium tetraphenylborate of the same concentration in different solvents are linearly related to the donor number of the solvent: the greater the solvent-solute interaction, the more is the net positive charge at the sodium ion decreased. However, for sodium iodide no such relationship is found, because the stronger donor properties of the iodide ion as compared to those of the tetraphenylborate are strongly influenced by the nature of the solvent(10,11).

The ions are not only modified by the solvent, but at the same time they function as centers for the modification of the liquid structure. This fact is commonly expressed by considering such ions as "structure breakers", as they interact with the water molecules more strongly than do the water molecules between each other. IR- and NMR-spectra indicate changes which simulate the effects of a decrease in temperature of the pure water spectrum(12). What is termed "thermal vibrations" is actually a complicated dynamic pattern with a character all of its own that reflects the energetic influence by the environment including that of temperature as well as these of pressure, irradiation, fields, mechanical forces, drop size etc.

Bond lengths and bond angles around an atomic hydrogen ion (which in fact cannot be recognized as such within the complex liquid structure) are highly influenced by the number of water molecules attached and by nature and number of other solute particles. For example, in crystalline $HClO_4 \cdot H_2O$ the three hydrogen atoms originating from the central oxygen atom are coordinated to perchlorate groups, but the O·········O distances are slightly different, varying in mean values between 263 and 271 pm(13) (Figure 1). The structure of $HClO_4 \cdot 2 H_2O$ may be derived from that of the monohydrate, in that one of the weakly coordinating perchlorate groups is replaced by a strongly donating water molecule that is hydrogen-bonded by one of the three hydrogen atoms of the H_3O-group, to give the $H_5O_2^+$-ion. Thus, the O·········O distance between the two water

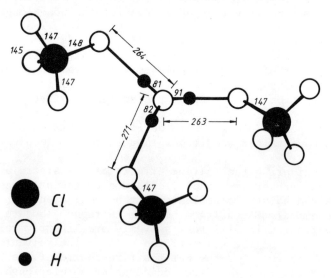

Fig. 1. Crystal Structure of $HClO_4 \cdot H_2O$

units is considerably shorter, namely 242 pm and the hydrogen bond nearly symmetrical. On the other hand the O····O distances connecting oxygen atoms of the H_5O_2-unit to perchlorate oxygen atoms are somewhat longer, 278 and 279 pm respectively(14) (Figure 2).

In an idealized way the hydration structure around a cation may also be illustrated by applying the bond length variation rules(8) (Figure 3) the greater the number of hydration layers, the shorter is the innersphere coordinate ion-oxygen (water) bond and the smaller the net positive charge residing at the cation as resulting from the modification of the ion by the environment(3). The smaller the distance from the ion, the shorter are the O·······O bonds and hence the greater is the local density. The greater the distance from the ion, the greater both the asymmetry and the deviation from linearity of the so-called hydrogen bonds between the solvent molecules. Each solvent molecule has therefore a slightly different environment and hence different properties. At any time the static aspects may be considered by a characteristic pattern of inhomogeneities that is established throughout the liquid system, influenced locally by the so-called solute-solute interactions, mediated through the solvent molecules.

Each ion is therefore modified by and modifying the solvent structure. The ions are therefore analogous to interstitial positions in solids(4,5), which may be termed structure modified and modifying centres or abbreviated "SMM-centres". The hydrogen and hydroxide ions which are in thermodynamic equilibrium in water, may be considered to be produced by such solvent-solvent interactions which produce the SMM-centres as they are required for the existence of the liquid.

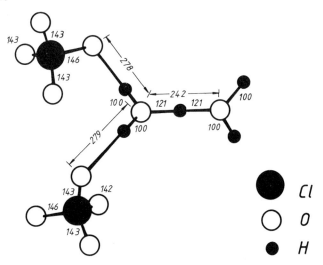

Fig. 2. Crystal Structure of $HClO_4 \cdot 2\ H_2O$

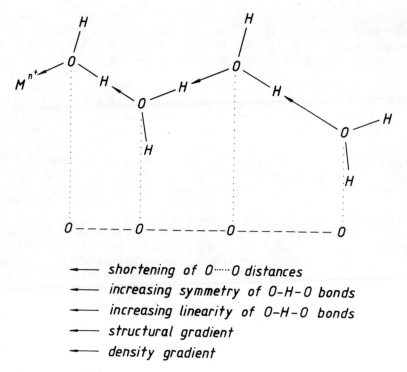

Fig. 3. Idealized illustration of a hydration structure around an ion M^{n+}

The long range effects by an ion on the water structure are supported by various facts. Likhtenstein(15) concluded from the linear relationships between activation energy and entropy of activation and the deviations from them, that the solvent transmits the charges in the solvent molecules caused by their interactions with solutes, to adjacent solvent molecules and from there to further molecules and so on. The macroscopic properties of a solution such as vapor pressure, surface tension or density are known to be changed even by small external changes, such as temperature or pressure and hence they must extend through the whole solution and in particular they must reach the bondary areas. Resolution of data into solvation and excess properties reveals that the latter are still present at an ionic concentration of 10^{-6} mol dm^{-3} (16). This corresponds to mean distances of about 100 nm or about 400 water layers. An indication for the extension of such effects up to 3 μm (corresponding to 10 000 water layers) is the solubility of quartz in water in "poly-water"(17).

In addition to the presence of ions in water, there is always "empty space" in a liquid. Water, purified by distillation contains more than 10^{-3} mol dm^{-3} of gases dissolved from the atmosphere. This corresponds to a mean sphere of influence of each gas molecules of about 17 water layers. Water, that has been degassed still contains about 10^{-6} mol dm^{-3} of gas, which apparently cannot be removed. There are always large areas of "empty space" and hence inner surface areas surrounding such "holes". In analogy to vacancies in crystal(4), such holes are also SMM-centres(3).

Unoccupied holes have been found in various clathrates(18,19). The clathrate structures and hence the holes are stable although there are no chemical bonds involved between the solvent and the host molecules with an exceptional degree of rotational mobility for the latter(20). It has been shown that the mean O······O distance of the water framework enclosing a cavity is shorter both for unoccupied and for occupied holes(21,22). The far-reaching effects on the water structure have been illustrated by applying the bond length variation rules and they are represented in an idealized, two-dimensional way(1) in Figure 4.

Interactions between SMM-centres are mediated through solvent molecules and hence nature and number of the ions will also be reflected in the "hole structure" which therefore contains information characteristic of the whole system.

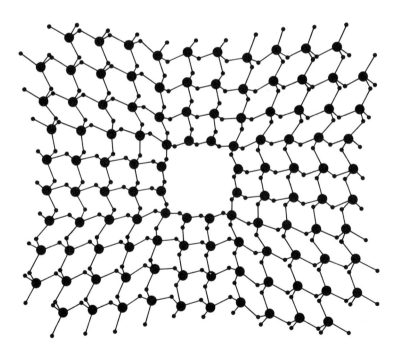

Fig. 4. Two-dimensional idealized illustration of water structure around a hole

All of the SMM-centres are known to migrate within the liquid system. As the macroscopic properties of the liquid under given conditions are not changed by the local migrations, a stationary pattern that is characteristic for a given system must be dynamically maintained.

Another point to mention is that extrapolation of data to infinite dilution leads to values that depend on the properties of the solution. In other words, by the fictious process of dilution to $c \to 0$, a rest memory appears to remain to some of the features that have been originally present in the solution.

Self ionization equilibria are known to exist in various other solvents and it may be proposed that their presence should, in principle, be admitted for any liquid system even if such ions were not detectable. The extent of the self-ionization equilibrium appears related to the strength of solvent-solvent interactions and hence to their amphoteric properties in order to mediate the cooperative effects, to the ease of heterolysis of bonds within solvent molecules and to the value of the dielectric constant. In a liquid with unfavourable conditions for the production of SMM-centres by self-ionization such as carbon tetrachloride, SMM-centres must be made available in other ways in order to provide for the existence of the molecular liquid.

Any liquid is known to contain at least trace amounts of water and it is virtually impossible to remove completely water molecules from any molecular liquid. The so-called "anhydrous" solvents are known to contain water in the ppm-region. These water malecules are strongly bonded and under strong strain within the liquid system, i.e. they are integrated in the liquid and cannot be considered any longer as "water" in a chemical sense. It may therefore be proposed that these "water-molecules" are SMM-centres for the structural arrangement of solvent molecules as required for the liquid under consideration.

Whereas the holes are considered as fairly small and closed entities within the system, the existence of a network of channels may be proposed which may contain even "sheet-like" areas(3). In analogy to the presence of dislocations in a solid material, channels should not end somewhere within the phase, but rather reach the surface at both ends or form a closed loop(3). The molecules surrounding channels are expected to be under great strain with appropriate bond contractions at the inner-surface areas.

Under even greater strain are the molecules at the phase boundaries. The bond lengths between surface molecules are particularly short and the energy per part rather high. Although the surface of a liquid is macroscopically smooth, it is microscopically highly differentiated. The strongly curved regions are those of greatest local surface energies. These may be described as "funnels" constituting points of emergencies of channels at the surface.

With regard to their temporal and local significance within the liquid, various groups of molecular species constituting the liquid may be distinguished namely those at boundaries, channels, SMM-centres and "normal" molecules. The latter represent the overwhelming majority of constituents, which determine the statistical results obtained for the macroscopic system.

3. THE SIGNIFICANCE OF THE CONSTITUENT PARTS FOR THE WHOLE LIQUID.

We may now approach the problem in what ways the temporary, local features are integrated within the complex system and how the ordered relationships between the different parts are established. A liquid has a great adaptability to respond to changes in the environment and a high capability to store energy and information. Building units which have been distinguished due to their temporal local significance within the system are found to serve the whole liquid system in different ways, and to influence each other in different intensities. According to their mutual interrelationships differences in domination are apparent, as they are typical for a so-called "hierarchic" order, i.e. the various parts serve on successively graded levels. A level is hierarchically higher, the greater its significance for the whole system. Forces in a higher level control and regulate the properties of the parts serving the lower levels.

The surface is of greatest significance for the whole liquid; it is the "first line of defense" by showing greatest adaptability towards changes. The units are under great strain and they have high energies, allowing them to retain information to a relatively high extent in that extra energy is passed on to other units within the system or it is stored at the surface by appropriate motions of its constituents. An increase in tension due to increase in energy per unit may be compensated by expanding the surface area with increase in the number of surface atoms over which the energy may be distributed. Any energy change at the surface will cause a redistribution of energy within the whole system and this redistribution is controlled and regulated by the forces acting on the surface. These actions, as well as the exchange and redistribution of matter involve continuous and dynamic interactions between surface and the other parts of the system and hence require ordered relationships. Each point of the surface must contain informations about the static and dynamic aspects in other parts of the system, as "seen" from the point under consideration. Likewise, each point within the system contains information of the other parts as "seen" from the said point. This means also that the fluctuations within the system are reflected in the surface.

For the dynamic interactions between the various surface and bulk regions the channel-network is of greatest importance. The areas of highest local surface energy, which are suggested to be considered "funnels" are interconnected by a network of fluctuating channels, which may be considered as possible pipes for the transportation and distribution of matter, energy and information. The funnels are also interconnected by the outer-surface areas between them. Outer- and inner-surface areas may be considered as providing closed loops which are connected to each other by nodes and intersections. Matter, energy and information is directed and stored according to the temporal and local requirements.

Units along the inner-surface of channels are under less constrained conditions than surface units, and they are controlled and regulated by the forces acting on the surface. The channel-network should therefore be considered as hierarchically inferior to the free surface.

SMM-centres are under the decisive influence of the forces acting both at the surface and at the channel-network. The migrations of SMM-centres follow the gradients in chemical potential with simultaneous formation of new gradients of chemical potential. In the various microscopic areas both the analytical composition and the positions af the parts are continuously changing, even though the macroscopic properties of the system remain constant. This requires that a certain motion pattern is maintained by the ordered motions within the system. As soon as the regularities have been established we can speak af a dynamic order. Like a field, the dynamic order cannot be observed directly, but recognized by its reactions. We can describe how the dynamic order makes use of the various parts within the complex relationships of the system. The order is lost by the mental or physical dismemberment of the system. Because the dynamic order uses the parts and not the parts the dynamic order, it is not always necessary to learn more and more about smaller and smaller parts. Even the most precise knowledge of the properties of the isolated parts or of the parts on a lower level is entirely inadequate to obtain information on the properties after promotion to serve on a higher level.

All of the other units of the system are dominated by the dynamic order established by the properties of SMM-centres of the system under consideration. It is therefore the great majority of solvent molecules which is serving on the lowest hierarchic level of the system, by which the statistical results are determined for the system under consideration. Figure 5 illustrates various aspects of the hierarchic order in a liquid.

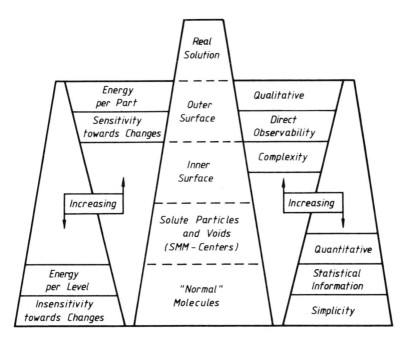

Fig. 5. Illustration of the Hierarchic Order in a Liquid.

4. CONCLUSION

The consideration of the hierarchic order is not in disagreement with statistical results. Quantum chemical approaches are still useful in aiming to calculate a system on statistical grounds, because the statistical results are approximated by considering the overwhelming majority of the parts, which serve the lowest level. However, it is impossible to represent the real system with its hierarchic structure by such methods, which - strictly speaking - are restricted by leading to ideal model systems and not to the real object as such.

According to the concept of hierarchic order SMM-centres are required for the existence of a liquid. These may be impurities (solute molecules or ions), voids and ions produced by self-ionization. In their absence a dynamic order between "equal" constituents would be impossible.

The presented approach may be extended by considering the distribution of energy between the various levels(3), their redistribution following any changes and tested by the changes in macroscopic properties as related to defined changes of energy, as due to changes in temperature, pressure, composition, irradiation, action of fields or mechanical forces.

REFERENCES

1. V. Gutmann and G. Resch, Z. Chem. 19:406 (1979).
2. D. Eisenberg and W. Kauzmann, "The Structure and Properties of Water", Oxford, Clarendon (1969).
3. V. Gutmann and G. Resch, Pure Appl. Chem. to be published.
4. G. Resch and V. Gutmann, Z. Physik. Chem. (n.F.), submitted for publication.
5. G. Resch and V. Gutmann, Abh. Braunschw.wiss. Ges. in the press.
6. V. Gutmann, E. Plattner, and G. Resch, Chimia, 31:431 (1977).
7. D. Hankins, J. W. Moskowitz, and F. E. Stillinger, J. Chem. Phys. 53:4544 (1970).
8. V. Gutmann, "The Donor-Acceptor Approach to Molecular Interactions", Plenum, New York (1978).
9. V. Gutmann and G. Resch, Acta Chim. Hung. in the press.
10. R. H. Erlich and A. I. Popov, J. Am. Chem. Soc. 93:5620 (1971).
11. M. Herlem and A. I. Popov, J. Am. Chem. Soc. 94:1431 (1972).
12. J. O. Bernal and R. H. Fowler, J. Chem. Phys. 1:515 (1933).
13. I. Olovsson, J. Chem. Phys. 49:1063 (1968).
14. J. O. Lundgren and I. Olovsson, Chapter 10 in "Hydrogen Bond-Recent Developments in Theory and Experiments", eds., P. Schuster et al., North Holland Publ., Co., Amsterdam (1976).
15. G. I. Likhtenstein, Russ. Journal Phys. Chem. (English transl.), 44:1079 (1970).
16. J. Barthel, Private Communication.
17. D. Schuller, Naturwiss. 60:145 (1973).
18. G. A. Jeffrey and R. K. McMullan, Progr. Inorg. Chem. 8:43 (1967).
19. M. V. Stackelberg and R. H. Müller, Z. Elektrochem. 58:251 (1954).
20. J. W. Tester, R. L. Birins, and C. C. Herrick, A.I.Ch. E. Journal, 18:1220 (1972).
21. D. W. Davidson, Can. J. Chem. 49:1224 (1971).
22. R. K. McMullan and G. A. Jeffrey, J. Chem. Phys. 42:2725 (1964).

THERMODYNAMICS OF AQUEOUS DILUTE SOLUTIONS OF NON-CHARGED MOLECULES

Sergio Cabani

Istituto di Chimica Fisica dell'Università, Pisa
Italy

Abstract - The thermodynamic aspects of the dilute aqueous
solutions of non-electrolytes have been largely investigated
in the last years and many data are now at disposal concerning
the thermodynamics functions ΔX_h° (X = G, H, S, C_p) for the
transfer from gas to water and the limiting partial molar
properties Φ_j° (j = v, C_p, C_v, E, k) in water on non-charged
organic molecules. The state of the present knowledge of
experimental data for classes of compounds is considered and
the empirical or semiempirical ways the ΔX_h° and the Φ_j° quantities can be correlated with the molecular structure are ana-
lyzed and discussed. A short survey is finally given of the
recent developments of the theories of the aqueous solutions
and of the results obtained by applying the Monte Carlo and
the molecular dynamics simulation methods to the study of
water and aqueous solutions.

1. INTRODUCTION

The thermodynamic state of a solute A in a dilute solution is
usually stated in terms of thermodynamic standard functions of transfer ΔX_t° (X = G, H, S, C_p) of A from a chosen standard state in a
thermodynamic medium α to an another chosen standard state in a
second thermodynamic medium β. The β phase may be constituted by a
pure liquid or by a mixture of substances, the α phase analogously
may be pure A (gas, liquid or solid) or A dissolved in another solvent C. When A is partitioned between the gas phase and a binary
dilute aqueous solution the thermodynamic standard functions of
transfer are called "thermodynamic standard functions of hydration"

and indicated with the symbol $\Delta X_h^{\circ}{}^{(a)}$. Butler(5) first realized the importance of the ΔX_h° functions for expressing the water-solute interactions and determined the values of the free energies, enthalpies and entropies of hydration for many organic compounds. The topic has since then been object of much interest and was revised in excellent review articles(4a,6) on aqueous solutions of non-electrolytes which cover the literature until 1972.

Here a brief survey will be made about the recent developments on the subject regarding the hydration functions and the apparent molar properties Φ_j° (j = v, c_p, E, k) of non-charged solutes in aqueous solution. In order we will examine the progresses achieved:

i) In the collection,
ii) In the organization,
iii) In the interpretation of the data until now known.

Only the progresses realized in the field in the years from 1973 to 1979 will be considered and limitedly to the dilute solutions.

2. THERMODYNAMIC FUNCTIONS OF HYDRATION, ΔH_h°, AND THERMODYNAMIC PARTIAL MOLAR PROPERTIES, Φ_j°, OF NON-CHARGED ORGANIC MOLECULES IN WATER.

In the above specified years, thanks also to apparatus recently get ready which allow to obtain a great deal of very high quality data in a short time, a noticeable amount of experimental data have been collected: the values of ΔG_h° of organic non-charged compounds have increased from 299 to 347, those of ΔH_h° from 127 to 196, of $\Phi_{c_p}^{\circ}$ from 85 to 243, of Φ_v° from 136 to 384, of Φ_E° from 18 to 38, of Φ_k° from 14 to 110. Values have been calculated of the partial molar heat capacities at constant volume $\Phi_{c_v}^{\circ}$ for 22 non-electrolytes too(7). These data refer to the temperature of 25°C and infinitely dilute solution, but numerous are also the organic substances

---

[a] In the ΔX_h° symbol the h index is selfexplaining, the superscript circle ° is instead confusing. In effect, by lacking a general agreement about the standard states for A in the gas phase and in aqueous dilute solution, it is necessary to indicate every time which standard states have been selected. Many authors(1-3) have considered the importance of the choice of proper standard states in expressing the standard thermodynamic functions of transfer, in particular Ben Naim(4), by using a statistical mechanic argument, proved that the most meaningful choice is that based on the molar concentration scale for both the phases in which A is partitioned. The superscript ° will be used for such a choice. The superscript asterisk * is instead reserved for the very commonly used choice from ideal gas at 1 atm. to hypothetical ideal solution having unit mole fraction of solute but properties as volume, heat capacity and enthalpy equal to those the solute exhibites in the infinitely dilute solution.

for which studies have been made regarding the dependence of gas solubilities in water on temperature(8,9) and of the partial molar heat capacities and volumes on temperature and concentration(10-13).

The most of new values of ΔG_h° have been calculated by using data of vapour pressure of pure substances together with solubility data in pure water and concern benzene and derivatives and polynuclear aromatic hydrocarbons(14-17). Other ΔG_h° values regarding alcohols(18a), various bifunctional compounds(18a,c) and pyridines (19, 20) have been obtained by static measurements of vapour pressures of the solutions together with the vapour pressures of pure compounds. The determination of the free energy of hydration of acetamide(21) by means of a dynamic technique and the use of radioactivity deserves then to be mentioned for the extremely low value of the absolute distribution coefficient (= 7.6×10^{-8}) from water to the vapour phase.

As far as the ΔH_h° quantities are concerned most of them regarding alcohols(18a,22), ketones(23), esters(24), pyridines(25), phenols(22,26a,b), fluoroalcohols(27), polyethers(18b,28), polyamines and aminoethers(18a,b), hydroxyethers(18b,28) and some hydrocarbon compounds(29,30) are the results of direct calorimetric measurements contrary to the values previously reported, in general obtained by differentiation of the ratio $\Delta G_h^\circ / T$ with respect to temperature.

Also Φ_{Cp}° values of alcohols(22,31,33), ethers(11b,34,35a), amines(34, 35b), ketones(11b,36), amides(11b,32,36,37a,b), pyridines and others heterocyclic compounds containing nitrogen(38), phenols (22,39), benzene and derivatives(11c,22b,40), esters(11b), polyethers (28,41), polyamines(35b,41), polyols(42-44), polycarboxylic acids (43), aminoalcohols(41), aminoethers(31,35b,41), hydroxyethers(11a, 28), xanthines(45), ureas(46) have been preferably obtained from direct measurements of the heat capacity of the solutions, but in some cases also by differentiation of the heats of solution with respect to temperature(22a,b,28,43,45) or as second derivatives with respect to temperature of solubility data(33,47).

For hydrocarbons and their halides derivatives, the ΔH_h° and Φ_{Cp}° values have been in general calculated from solubility data. The whole of the values collected by various authors has been submitted to a rigorous statistical analysis. The results of this analysis, made time ago by Alexander et alii(48) and by Wauchope and Haque(49) and recently by Wilhelm et alii(47) do no produce however, in some cases, ΔH_h° and $\Delta C_{p,h}^\circ$ values in satisfactory agreement; f.i. for the enthalpy and heat capacity of hydration of argon by using the same experimental data(50) values ranging from 11.7 kJ mol^{-1} to 12.0 kJ mol^{-1} and from 140.2 J mol^{-1} °K to 238.5 J mol^{-1} °K respectively are calculated according to the equation used for the dependence of the solubility on temperature(3).

The selection and weighting made on the same experimental data, together with the equation used, explain the above findings, but as a result, unfortunately, up to present the uncertainty in the ΔX°_h functions for hydrocarbons is high. The situation is better for the rare gases studied very accurately by Benson and Krause(51) and by Potter and Clynne(52) a short time ago.

The knowledge of reliable thermodynamic data of saturated and aromatic hydrocarbons in water is very useful in several practical and theoretical fields, f.i. it is essential for the developments of the concepts of the hydrophobic hydration and of the hydrophobic interaction largely used for the understanding of some important biological phenomena. So new experimental data of high quality are necessary, having the elaborations of the existing material reached their limits. In the last years instead new solubility data in pure water have been reported only for methane(10,53a) ethane(53), cyclopropane(54), n-butane(9,53a) and benzene(17). Enthalpies of hydration have been furthermore determined by direct calorimetric measurements only for benzene and some derivatives, cyclohexane, pentane and hexane(29).

As far as the volumetric properties are concerned, Φ°_v values at 25°C have been reported for alcohols(31,55-59a), ethers(11b,34, 58,60,61), amines(34,57,62-64), esters(11b,58), ketones(11b,36,58), amides(36,37,63,65), carboxylic acids(62a,b,67), hydroxycarboxylic acids(66c), polycarboxylic acids(66d), polyols(31,42;56,59b), pyridines and others unsaturated heterocyclic compounds containing nitrogen(38), ethylene glycol derivatives(11a,61,68-70), carbohydrates(72-74), urea and derivatives(44,74,75) polyamines(76,77a) aminoalcohols and aminoethers(77b), polyethers(6°,61,77c), alkylhalides (61), benzene and derivatives(11c,26c). Moreover the behaviour of the functions $\Phi^\circ_v = f(T)$ has been studied for alcohols(10,12,55,56) and polyols(12,55,56,68,78), ethers(12) and polyethers(11b,12,68), ketones and esters(11b), ethylene glycol derivatives(11a,68,71,78). Compressibility data at 25°C have been calculated from ultrasonic velocity measurements for ethers(79), amines(62), alcohols(79-82), polyols(79), ethylene glycol derivatives(68,71), polyethers(68,74, 79) urea and derivatives(74), carbohydrates(72,74). Finally some investigations about the influence of various types of non-electrolytes on the temperature of maximum density(82b,83) and the adiabatic compressibility minimum(84) of water are to be remembered.

3. EMPIRICAL RELATIONSHIPS BETWEEN ΔX°_h AND Φ°_j QUANTITIES AND THE MOLECULAR STRUCTURE OF THE SOLUTES.

The ΔX°_h functions, except some cases where specific chemical effects between functional groups and water arise, are the expression mainly of some modifications the liquid water structure suffers near the solute. On the contrary, the values of the Φ°_j properties are determined by both intermolecular and intrinsic factors. Taking that into account, it should not be expected that the values of $\phi_j^°$ and, more than over, of $\Delta X_h^°$ are correlated to the internal structure of the solute molecules in the way the thermodynamic properties of gaseous molecules are correlated to their constituting bonds or, better, groups(85).

Lacking a thermodynamic statistical support to the search of relationships between thermodynamic properties in solution and molecular structures, totally empirical ways to proceed are justified. In such a context three trends can be roughly identified:

a) Attempts to correlate the thermodynamic properties of the substances in solution with properties of the pure state;
b) Attempts to find reciprocal relationships between the thermodynamic properties in water themselves;
c) Attempts to find relationships between thermodynamic properties in water and the internal structures of the solutes expressed in some convenient way: f.i. atoms, bonds or groups.

As far as the first point is concerned, it has been confirmed (47) that the solubility of gases in water, expressed as $- \ln N_2$, is not correlated with solute properties like energy of vaporization at the normal boiling point, polarizability, hard sphere diameter, normal boiling temperature or critical temperature respectively, contrary to what happens when non polar solvents are involved.

The peculiarity of water and, of course, of aqueous solutions emerges also when enthalpies and entropies of hydration are compared (point b). In fact while for most solvents the enthalpy and entropy changes for the solution process are related by only one linear relationship, the Barclay-Butler relationship(86), it has been proved (3,18a,87,88) that plotting $\Delta H_h^°$ of organic solutes in water vs. $\Delta S_h^°$ a set of parallel straight lines are obtained whose slope ($\simeq 252°K$) is very different from the slope of the Barclay-Butler line ($= 660°K$) in non-aqueous solvents. To the same straight line belong substances having the same functional group. That allows to deduce that water is able to distinguish the different functional groups and it is so possible, for the monofunctional compounds, to consider independent the interactions the water has with the non-polar and with the polar part of the solutes.

Let us consider finally the $\Delta X_h^°$ functions or the $\phi_j^°$ properties have been correlated to the internal structure of the molecules (point c). Three endeavours may be recognized in this connection:

i) Methods based on observations of homologous series of compounds;
ii) Methods of group contributions;
iii) Semiempirical methods where some simplified models regarding the process of solution or the state of the solute molecules in solution assist in the attribution of the structural parameters.

3.1 Group contributions from trend of properties along homologous series.

In order to see whether a Y group carries a constant contribution to a thermodynamic property X the simplest but most significant way is to examine how the property changes along homologous series in which one member differs from each adjacent member for the Y group. This has been widely made for Y = CH_2 and independently of the homologous series or property considered, except particular structural situations, the CH_2 group was always found to carry a constant contribution amounting for saturated open chain compounds, to 90±4 J mol^{-1} K^{-1} for $\Phi^°_{C_p}(CH_2)$(28,31,32,35,43,89,90), to 16.0±0.1 ml mol^{-1} for $\Phi^°_V(CH_2)$(11c,31,55,56,57,66,81,91), to -3.6±0.4 K J mol^{-1} for $\Delta H^°_h(CH_2)$(28,29,89), to +0.69±0.05 K J mol^{-1} for $\Delta G^°_h(CH_2)$ (92,93). The methylene contribution in saturated cyclic compounds is always about 20% smaller(35,57,91,94a). When the methylene group is included between two hydrophilic centres as in $Z(CH_2)_n Z$ (Z = OH, NH_2, COOH; n<4)(28,43,56,66a,66d,77b) or it is next to an aromatic ring(11c,22a) its contribution is significantly different, smaller for $\Phi^°_{C_p}(CH_2)$ and $\Delta H^°_h(CH_2)$ and greater for $\Phi^°_V(CH_2)$, with respect to the above reported values.

The same operation which allows to assign the methylene contributions has been extended also to other central groups and these values have been reported: $\Phi^°_V(CH_2 CH_2 O)$ = 36.9 ml mol^{-1}(68,69,71), $\Phi^°_{C_p}(CH_2 CONH)$ = 90 J mol^{-1} K^{-1}(95), $\Phi^°_V(CH_2 CONH)$ = 38.0 ml mol^{-1}(95). A suggestion has been made(18b,41) to apply the procedure above described also to the terminal groups considering the open chain molecules as developed between two terminal hydrogen atoms H....H. In such a case to the terminal hydrogen atoms contributions have to be attributed and they result very high, namely: $\Phi^°_{C_p}(H)$ = 79±12 J mol^{-1} K^{-1}(11c,31,90). $\Phi^°_V(H)$ = 11.4±1 ml mol^{-1}(11c,31). Others prefer instead to calculate contributions for the terminal groups by considering in order molecules of the type $Z(CH_2)_n Z$ and $Z(CH_2)_n Z'$. In this case the X(Z) values result always extremely high when compared with similar groups occupying central positions, f.i. $\Phi^°_{C_p}(CH_3)$ = 167±12 J mol^{-1} K^{-1}(11c,90), $\Phi^°_V(CH_3)$ = 26.4±0.3 ml mol^{-1}(11c,55,56,66a), $\Phi^°_{C_p}(CH_2 OH)$ = 100 J mol^{-1} K^{-1}(22a), $\Phi^°_V(CH_2 OH)$ = 28.2 ml mol^{-1}(55,56). This does not mean that the terminal groups effectively carry a greater contribution to the thermodynamic properties, but is a result only of the different ways used to obtain their values with respect to the way used for the non-terminal groups Y.

The constancy of the Y contributions is an important, although indirect, test that in the saturated compounds the interactions solute-water is a sum of localized interactions between water and the groups constituting the molecule of the solute. This does not occur when the solute molecules possess more than one hydrophilic centres distant from each other less than four bond distances. The direct (i.e. internal) or indirect (i.e; through the solvent) interactions these groups exert on each other produce effects on the thermodynamic functions of hydration or Φ_j° quantities which result very evident when proper comparisons between selected molecules of solutes are made(18b,30,96).

The interaction between hydrophilic centres in saturated compounds usually does not produce important effect on the volumetric properties, but it is important in determining the values of the Φc_p°. When a hydrophilic group is introduced in hydrocarbons or in monofunctional compounds the Φc_p° decreases in both the cases but the decrease for the introduction of a second hydrophilic centre depends remarkably on the type of the hydrophilic group already present and on the distance between the two functional centres(42, 50,97). Effects of interactions between hydrophilic centres are also reflected in the different values the Φ_v° of many isomeric polyols and carbohydrates(42,72,73,98) show.

3.2 Group contributions by least square methods.

These completely empirical methods assume that atoms, bonds or groups constituting the molecule carry their own contribution to the property into consideration. Such a contribution is calculated by a least square method over a number of compounds containing the envisaged groups more than once. This procedure has been applied to the ΔG_h° functions by Hine and Mookerjie(96), to the Φc_p° properties by Guthrie(99), recently Cabani et al.(100) have extended this method to ΔH_h°, $\Delta C_{p,h}^\circ$, Φ_v°, Φ_E° and Φ_k° quantities and have repeated the calculation also for the ΔG_h° and Φc_p° quantities. The usefulness of describing the thermodynamic properties of organic compounds in water by means of a relatively reduced number of structural parameters is evident, however it is not justifiable attributing physical meanings to these contributions. Their values is in effect dependent on the number of compounds considered and on the calculation procedure adopted.

The ability to reproduce thermodynamic quantities by means of the structural parameters is usually better than 5% for substances belonging, or similar, to those taken into consideration in the fitting of the data. It is however very difficult to make prediction about the ability of the structural parameters to reproduce satisfactory values for compounds which are not similar to those considered in the calculation of the group contributions. As an example the experimental value of the ΔG_h° of acetamide (= -40.6 K J mol^{-1})(21) resulted very different with respect to that expectable from Hine's bond contributions (= -31.4 K J mol^{-1})(3). That is not surprising in view of the characteristics the water has to discriminate the various type of solutes better than others solvents do, see f.i. the peculiarity that the Barclay-Butler's plot presents in water with respect to non-aqueous solvents.

3.3 Group contributions from models.

A different but still empirical approach to the problem of the connection between thermodynamic properties in water and molecular structure of the solutes considers the thermodynamic functions of hydration of the saturated organic compounds ΔX_h° as a sum of terms arising from the formation of a cavity where the molecule of the solute is lodged (ΔX_{cav}), the interaction water-hydrocarbon part of the solute ($\Delta X_h^\circ(R)$) and finally the interactions water-hydrophilic parts of the solute. These latter are considered, at their turn, subdivisible in terms of water-hydrophilic centres interactions ($\Delta X_h^\circ(y)$) and parts in excess ($\delta X_h^\circ(Y_1, Y_2, r_{Y_1-Y_2})$) when more than one centre is present:

$$\Delta X_h^\circ = \Delta X_{cav} + \Delta X_h^\circ(R) + \Sigma \Delta X_h^\circ(Y) + \delta X_h^\circ(Y_1, Y_2, r_{Y_1-Y_2}) \qquad (1)$$

The terms arising from the formation of the cavity and from the water-hydrocarbon interactions are evaluated using the value of the area of the hydrocarbon surface $A(R)$ of the molecule together with the values of the parameters χ_X and β_X which represent the intercept and the slope of the best linear equation:

$$\Delta X_h(R) = \chi_X + \beta_X A(R) \qquad (2)$$

which represents how the $\Delta X_h^\circ(R)$ experimental values of the hydrocarbon saturated compounds vary with the hydrocarbon surface $A(R)$. The $\Delta X_h^\circ(Y)$ and the $\delta X_h^\circ(Y_1, Y_2, r_{Y_1-Y_2})$ quantities are then evaluated as differences considering in order the mono- and the polyfunctional compounds.

Table I. Values of increments $\Delta H_h^\circ(R)$ ($X = H, S, G, C_p$) in the thermodynamic functions ΔH_h° of hydration produced by elementary hydrocarbon radicals R ($T = 298.16$ K)[a]

R	$A(R) \times 10^{-9}$ /cm^2 mol^{-1}	$\Delta H_h^\circ(R)$ /kJ mol^{-1}	$\Delta S_h^\circ(R)$ /J mol^{-1} K^{-1}	$\Delta G_h^\circ(R)$ /kJ mol^{-1}	$\Delta C_{p,h}(R)$ /J mol^{-1} K^{-1}
CH$_3$	2.12	−4.81	−20.7	1.36	83
CH$_2$	1.35	−3.03	−13.3	0.86	53
CH	0.57	−1.29	−5.6	0.36	22
C	0.0	0	0	0	0
CH$_3$(Y)	2.12	+1.8	−3.0	2.69	63

[a] $\Delta X_h^\circ(R) = \beta_X A(R)$. For the methyl group bonded to a hydrophilic centre Y, a correction has been introduced to assign the same value to the $\Delta X_h^\circ(Y)$ contribution for the Y group, whether Y is bonded to the methyl or to other saturated alkyl groups (see ref. 30).

Table II. Values of polar centre contributions $\Delta X^°_{h,c}(Y)$ (X = H, S, G, C_p) to the thermodynamic functions of hydration of saturated organic monofunctional compounds RY at 298.15 K[a].

Y	$\Delta H^°_h(Y)$ /kJ mol^{-1}	$\Delta S^°_h(Y)$ /J mol^{-1}K^{-1}	$\Delta G^°_h(Y)$ /kJ mol^{-1}	$\Delta C_{p,h}(Y)$ /J mol^{-1}K^{-1}	$\sigma(Y)$[b] /ml mol^{-1}
OH (primary alcohols)	−37.6±0.5	−28.9±1.5	−29.0±0.2	−58±14	−4.3
OH (secondary and tertiary alcohols)	−42.8	−42.7	—	−24	—
NH$_2$ (primary amines)	−36.5±0.2	−30.7±1.3	−27.4±0.3	−65±16	−4.8
NH (secondary amines)	−42.0	−42.3	−29.4	−41	−3.6
N (tertiary cyclic amines)	−38.5±0.4	−38.8±2.1	−26.9±0.2	−36±10	—
O (ethers)	−26.6	−16.5	−21.7	−96	—
COOH (carboxylic acids)	−38.4±2.5	—	—	−83±8	−8.3
COO (esthers)	−27.5	−17.7	−22.1	−88 2	—
C=O (ketones)	−23.8	2.6	−24.7	—	—
CO·NH (N-substituted amides)	−60.2	—	—	−75±15	−11
CO·NH$_2$ (amides)	—	—	−47.6	—	−9.3

(a) From Reference 30.

(b) Shrinkage in volume (ref. 102)

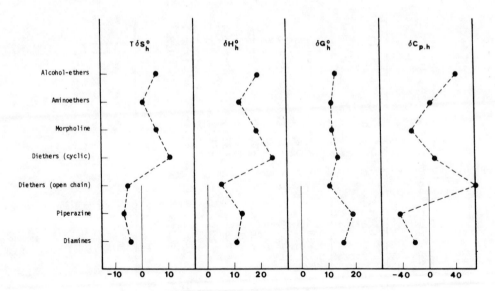

Fig. 1. Interaction parameters $\delta X_h^\circ(Y_1, Y_2, ry_1-y_2)$ between hydrophilic groups, in the relative 1,4 position, for some bifunctional saturated organic compounds at 25°C (X = G,H,TS in kJ mol^{-1}; X = C_p in J mol^{-1}K^{-1}).
(S. Cabani and P. Gianni, J. Chem. Soc. Faraday Trans. I, 75:1184(1979)).

Such a procedure, applied to the ΔG_h°, ΔH_h° and $\Delta C_{p,h}^\circ$ values of organic saturated compounds(30,97) allowed to obtain the data reported in Table I and II for the $\Delta X_h^\circ(R)$ and the $\Delta X_h^\circ(Y)$ contributions. As far as the interaction parameters between the hydrophilic centres are concerned, Figure 1 gives a picture of the situation.

It can be observed that the $\Delta H_h^\circ(Y)$ values are, on average, ten times larger than the $\Delta H_h^\circ(CH_2)$ value, while the $\Delta S_h^\circ(Y)$ values are about three times larger in magnitude than the $\Delta S_h^\circ(CH_2)$ value, except for ether oxygen whose contribution to the entropy of hydration is similar to the methylene contribution. Therefore the characteristic which distinguishes non-polar from polar groups is not the negative value of the entropy of hydration, but the different value of the ratio of enthalpy to entropy of hydration. It follows that the higher solubility in water, that monofunctional compounds exhibit with respect to the corresponding hydrocarbons, can be attributed to the enthalpic contribution of the hydrophilic centre, since the entropic contribution $\Delta S_h^\circ(Y)$ of the hydrophilic centre opposes solubility generally more than one non-polar group does.

As far as the $\delta X_h^o(Y_1,Y_2,r_{Y_1-Y_2})$ terms are concerned, the enthalpic term is always positive, while the entropic term is positive or negative according to the nature of the two hydrophilic groups. However the balance between enthalpy and entropy terms always produces a positive contribution to the free energy of hydration. Thus the interaction between hydrophilic centres lowers the thermodynamic stability of the aqueous solution of the bifunctional compound with respect to the stability the latter would have if the two hydrophilic groups acted as independent centres.

A procedure similar to that above described, i.e. the choice of hydrocarbon compounds as reference molecules in order to evaluate contemporaneously the effects connected to the formation of a cavity and to the hydrophobic hydration, was first applied to the partial molar volumes by Terasawa et al.(61). In this case, of course, in equation 2 the intrinsic volumes V_w, calculated according to Bondi (101), of the molecules are to be considered instead of surface areas and $\Delta X_h^o(R)$ identifies itself with Φ_v^o. Comparison of hydrophilic solutes with imaginary hydrocarbons with exactly the same dimension allows to deduce that the interaction water-hydrophilic centres produces a shrinkage $\sigma(Y)$ in the volume, f.i. the partial molar volume of solutes containing one hydroxyl group is lower by 4.3 ml mol^{-1} than that of the hypothetical hydrocarbon with the same intrinsic volume. Some other values of the shrinkage in volume due to some polar centres are (in ml mol^{-1}):

$\sigma(NH_2) = -4.8$, $\sigma(COOH) = -8.3$, $\sigma(O,ether) = 0$

$\sigma(CONH) = -11.0$, $\sigma(COHN_2) = -9.3$ (102)

While the direct comparison of the partial molal volumes of polar molecules with hypothetical hydrocarbons having the same intrinsic volume V_w produces $\sigma(Y)$ parameters satisfactorily constant for the various hydrophilic centres and that allows to reproduce with success the experimental values of the Φ_v^o for very large classes of compounds(102), this does not occur in another method, proposed by Edward and Farrell(103). In such a method by using the intrinsic and the partial molar volumes of the apolar molecules, supposed spherical, a constant parameter $\Delta (= 0.057$ nm) is obtained which, added to the radius r_w calculated from the intrinsic spherical volume of the solute, allows to calculate the volume the spherical cavity would have if no polar group were present in the guest molecule. The difference between the experimental Φ_v^o value and the value calculated for the cavity provides the value of the effect of the polar centre. A strong trend is however found when members of homologous series are examined: f.i. the shrinkage effect produced by -OH group varies from -7.4 to -5.0 ml mol^{-1} when n-alkanol from methanol to pentanol are considered and from -6.8 to 0.0 ml mol^{-1} when α,ω diols from ethyleneglycol to 1,10 decandiol are taken into consideration(102). The results are nevertheless better when compounds belonging to homologous cyclic series are examined(58,63).

4. HYDROPHOBIC HYDRATION

What was said above concerned mainly the evaluation of the effect of the hydrophilic groups as can be deduced by assuming the hydrocarbons as standard materials. We will examine now what has been realized, still on sole empirical basis, about the hydrophobic hydration.

At this purpose it is noteworthy the paper by Gill and Wadsö (104) who find convenient to attribute the hydrogen atoms the unusual thermodynamic properties of the aqueous solution of the hydrocarbon compounds. By expressing the variation in the free energy ΔG_s° and in the heat capacity $\Delta C_{p,s}^\circ$ in going from liquid hydrocarbons to the saturated aqueous solution in terms of the number n_H of hydrogen atoms, only two linear equations, one for the free energy (intercept = 6.4 kJ mol^{-1}, slope = 1.85 kJ mol^{-1}) and the other for the heat capacity changes (intercept = 0, slope = 35 J mol^{-1} K^{-1}) were found independently of whether the hydrogen atoms belong to cyclic or open chain, and to saturated or unsaturated hydrocarbons. By putting equal to zero the enthalpy change of solutions at room temperature, a general equation for the dependence of ΔG_s°, and therefore of the solubility of liquid hydrocarbons in water, on temperature and on n_H is proposed. From this equation the entropy of hydration at room temperature results to decrease by about 6 J K^{-1} for each additional hydrogen. That was interpreted as due to the highly localized interaction with water and a model was proposed in which the entropic effect would arise from the different ways a water molecule may form hydrogen bonds with other water molecules arranged around it in a tetrahedral manner following that a corner of the tetrahedron is occupied by another water molecule or by a C-H group.

Cramer(105) questioned the existence of some particular interaction between water and non-polar solutes as expressed in the hydrophobic models which postulate an ordering of water molecules immediately adjiacent to the non-polar surface of the inert solute. In his opinion the aqueous solutions of inert gases do not differ substantially from the solutions in other solvents like f.i. the 1-octanol, that is chosen as a typical solvent similar to lipids.

The feebleness of Cramer's argumentation has been however put in evidence by Abraham(93) who compared the free energies of transfer ΔG_s° of n-alkanes and rare gases from gas to solution in water or in a wide variety of non-aqueous solvents. Taking the benzene as reference solvent, the ΔG_s° of both these categories of solutes are represented satisfactorily by the same straight line:

$$\Delta G_s^\circ(\text{in solvent}) = m\Delta G_s^\circ(\text{in benzene}) + C \tag{3}$$

The values of the slopes m are feebly different in the various solvents ranging from 0.767 (in methanol) to 1.114 (in PhI). On the contrary the ΔG_h°(water) functions plotted against the ΔG_s°(benzene) functions give rise to two straight lines, in the first, with positive slope m = 0.5, the rare gases lie, while in the second, with negative slope m = -0.16, the n-alkanes are situated.

In order to justify this singular behaviour a particular effect, namely the hydrophobic effect, has to be invoked. Taking into account the experimental methylene contribution to the free energy of hydration, $\Delta G_h^\circ(CH_2) = 0.74$ k J mol^{-1}, and the contribution of -1.53 k J for mole of methylene calculated assuming that this group behaved like a rare gas, a value equal to 2.27 k J mol^{-1} is calculated for the free energy change due to the hydrophobic hydration of one methylene group. Therefore the experimental value for the free energy of the methylene i.e. $\Delta G_h^\circ(CH_2) = 0.74$ k J mol^{-1}, obtained considering the way the ΔG_h° of saturated aliphatic compounds change along homologous series of compounds, is the result of an unfavourable hydrophobic effect to 2.27 k J mol^{-1} and of a favourable normal solvent effect of -1.53 k J mol^{-1}. The hydrophobic effect of 2.27 k J for mole of methylene is much larger than previous estimates suggested. It will be interesting to see how the enthalpy and entropy terms concur to produce this effect. The analysis made by Abraham is interesting, however it is difficult to accept a model in which the rare gas acts on water in a way which is completely different with respect to the way a molecule like methane act, in spite of the fact that f.i. methane and argon or xenon have very similar values of the partial molar heat capacity and volume in water.

5. THERMODYNAMIC PROPERTIES OF DILUTE AQUEOUS SOLUTIONS OF NON-CHARGED MOLECULES IN THE FRAMEWORK OF SOME CURRENT THEORIES

After the pioneering papers of Barker and Watts(106) and Dashevsky and Sarkisov(107) on the Monte Carlo simulation of pure water and dilute aqueous solutions of non polar solutes respectively, much attention has been devoted in these last ten years to investigations on the structure of water and aqueous solutions and to their equilibrium and kinetic properties by means of simulation techniques. The criteria adopted by the various authors differ in the potential functions representative of water-water, water-solute and, sometimes, solute-solute intermolecular interactions, as well as in the number of water molecules in the base unit and in the choice of the type of ensemble.

Pairwise potential functions obtained by models(106,107,108) or by ab initio quantum-mechanical calculations(109) have been hitherto both employed with a similar degree of success(110). The number of water molecules in the base unit has been increased from 32 up to 256(111), a case is also reported where 343 water molecules per unit cell have been considered(109c). The NVT ensemble is in general chosen, but the NPT ensemble has been also considered(112).

The theory and the molecular models for pure water have been revised by Stillinger(113) with particular regard to the results obtained by means of the molecular dynamics simulation, Wood(114) has given a survey about the computer simulation of water and also of the aqueous solutions considering both the Monte Carlo and the molecular dynamic methods.

Recent papers on the topic, limitedly to aqueous solutions of non charged molecules, have considered the aqueous dilute solution of non polar molecules in order to recognize the structure of the water near a non-polar solute and to calculate the thermodynamic functions for the hydrophobic hydration(112b,115-119) and the hydrophobic interaction(119-121,139). As significant examples of aqueous solutions of a non polar molecule, the dilute aqueous solutions of methane(112b,115,117) neon(119) and argon(118) have been studied, very interesting are also the studies of the hydrophobic hydration around a pair of apolar species in water made by Pangali et al.(121b) and by Geiger, Rahaman and Stillinger(109). In spite of the different methods the various authors have employed(117) as far as the choice of the ensemble, the number of water molecules and the intermolecular potential functions are concerned, they all have arrived at very similar results, i.e.:

i) for the water molecules near to the solute there is an increase in the four coordination and stronger binding with respect to the bulk water;

ii) the local solution environment near to inert molecule (methane, neon or argon) is similar to that found in the clathrate cage, however there is no immobilization of the hydration water, but only a reduction in mobility of shell water against the bulk water;

iii) the differences between the calculated and the experimental values of the thermodynamic properties are considerable, the agreement is usually only in the sign: f.i. the calculated partial molar internal energy for methane is -91.2 ± 27.6 k J mol^{-1} (115) and -46 ± 62 kJ mol^{-1}(112b) against the experimental value $\Delta U = -10.9$ k J mol^{-1} likewise for the partial molar volume the value of 25 ± 34 cm^3 mol^{-1} was calculated(112b) for the methane in water against a 37 cm^3 mol^{-1} as experimental value.

For pure water molecular dynamics calculations carried out by using a model potential(108a, 108d) and the Monte Carlo simulation carried out by using intermolecular potential obtained for the water dimer from ab initio quantum-mechanical methods(109a-109c) both allow to see the water as constituted by a continuum distribution of structures rather than a mixture of free and hydrogen bonded molecules. The agreement between calculated and experimental data is still poor, f.i. for the heat capacity at constant volume the Raman - Stillinger dynamic simulation produces $C_v = 145.2$ J mol^{-1} K^{-1} at 0°C(108d) which is significantly larger with respect to $C_v = 75.7$ J mol^{-1} K^{-1} at 0°C for real water. In spite of the very important progresses made in the field, correct intermolecular potentials for water-water and water-solute interactions are not at present available(121,122). For this reason it is still valid to apply semiempirical theories which make use of the experimental behaviour of real water. Such approach is typical of the scaled particle theory (SPT) which develops on the basis of semiempirical correlations involving properties of the solute and the solvent. Such theory was devised for single sphere fluids(123) or mixture of hard spheres(124), but it was applied time ago with success by Pierotti(125) also to aqueous solutions of non polar solutes. A summary of the theory with specific reference to its applications to the dilute aqueous solutions may be found in two reviews articles by Lucas(126) and by Pierotti(127).

According to the SPT the variation of the free energy ΔG_h^* for the transfer of a solute from gas at fugacity f_2 to aqueous solution at X_2 mole fraction can be written in the form:

$$\Delta G_h^* = G_c + G_i + RT \ln RT/V_1^\circ \qquad (4)$$

where $\Delta G_h^* = RT \ln K_H$ is an experimental quantity ($K_H = f_2/X_2$ is the Henry's law constant), V_1° is the molar volume of the solvent, G_c is a quantity corresponding to the reversible work required in order to create in the solvent one mole of cavities having suitable size to receive one mole of hard sphere molecules of solute of σ_2 diameter, G_i is the free energy for mole arising from the interactions between one molecule entering one cavity and the surrounding water molecules.

The calculation of the partial molar Gibbs free energy for interactions G_i for a polarizable solute in water may be managed taking into account the inductive, dispersion and dipole-dipole energies. The knowledge of the dipole moments and the polarizabilities together with that of the Lennard-Jones parameters or, eventually, the molecular magnetic susceptibilitis of the solvent and the solute are however required(127).

The thermodynamic quantities for the formation of the cavity are obtained by using for the calculation of the G_c value the expression:

$$G_c/RT = -\ln(1-y) + 3yz/(1-y) + |3y/(1-y) + 9y^2/2(1-y)| z^2 +$$
$$+ (yz^3/e_1 kT)P \qquad (5)$$

where $y = \pi e_1 \sigma_1^3/6$. To obtain the G_c values the solvent hard-sphere diameter σ_1 ($\sigma_1 = 2.76$ Å for water), the solvent number density e_1 and the $z = \sigma_2/\sigma_1$ ratio are required together with the experimental pressure P. The pressure P is sometimes(128) substituted by the P_{HS} pressure value corresponding to the pressure of the hard-sphere fluid at the given density and temperature as it is calculated from some equation of state for the hard sphere fluids(129).

The G_c values for the formation of cavities of various sizes at various temperatures calculated for water and other polar and non polar solvents allow to justify the negative ΔG_t° values of transfer of non polar solutes from water to a non polar solvent as due mainly to the fact that the non polar solvent molecules have a large size than water molecules(130). The large negative entropies and the unusual high value of heat capacity observed for the transfer on non polar solutes from gas to water can be also justified largely from the process of formation of the cavity accompanied however by solvent structural effects too(127,128,131). De Voe(128) has shown that the SPT allows to demonstrate that non polar solute molecules cause large increases in the structural ordering of water while cause smaller changes, possibly also inverted, in the structural order of non polar solvents and to evaluate the relative magnitude of this effect on the entropy values for the transfer from gas to dilute solutions.

As far as the enthalpies of hydration are concerned, the computations lead instead at small and positive values of H_c, due to the small values of the coefficient of thermal expansion α_p for water, which do not agree with the negative and relatively large in magnitude enthalpy changes observed for the solution of non-polar gases in water(132). As an example Arakawa(132) has calculated for the methane (σ_2 = 3.59 Å) the following values of the thermodynamic functions for cavity formation in water at 25°C: G_c = 22.6 k J mol^{-1}, H_c = 3.8 k J mol^{-1}, S_c = -63 J mol^{-1} K^{-1}, C_c = 159 J mol^{-1} K^{-1}. The hydration functions for the transfer of methane from gas (1 atm) to hypothetical ideal aqueous solution (N_2 = 1) are: ΔG_h^* = 26.3 k J mol^{-1}, ΔH_h^* = -13.6 k J mol^{-1}, ΔS_h = -135 J mol^{-1} K^{-1}, $\Delta C_{p,h}$ = 209 J mol^{-1} K^{-1}(47a). When the solvent is CCl_4, the computed(132) thermodynamic functions for cavity functions are: G_c = 15.8 k J mol^{-1}, H_c = 15.5 k J mol^{-1}, S_c = -0.8 J mol^{-1} K^{-1}, C_c = 54 J mol^{-1} K^{-1}. The experimental values for the transfer of methane from gas (1 atm) to hypothetical ideal solution (N_2 = 1) in CCl_4 are(47b): ΔG_s^* = 14.5 k J mol^{-1}, ΔH_s^* = -3 k J mol^{-1} and ΔS_s^* = -58.7 J mol^{-1} K^{-1}.

Recent applications of the theory have been made for calculating the contribution the formation of a cavity gives to the free energy of transfer of a series of isomeric ketones(134), or various other solutes(132), from H_2O to D_2O, for the studies of solubility of many apolar gases in water and other polar solvents(135), for comparing the experimental thermodynamic data for the solution of rare gases(51), or some perfluorocarbon gases(136), in water at various temperatures with data calculated by means of the SPT. In a tentative made to extend the theory at aqueous solutions where solute-solute interactions have to be considered and hydrophobic interactions are operative, it has been shown that the dependence of the partial molar volumes and enthalpies of hard-sphere solutes in water on concentration and temperature are due to the anomalous trends the $\delta\beta/\delta P$ and $\delta\beta/\delta T$ (β coefficient of isothermal compressibility) of pure water present, rather than to the solute structural effects(137).

The results obtained by applying the SPT to non polar solutes show that, although the structure of the water does not appear explicitely in the theory, it is implicitly considered when density, expansivity and compressibility of bulk water are introduced in the equation used for computing the thermodynamic properties of aqueous solutions, and, as a consequence, the particular properties the latter present with respect to the non aqueous ones are due to the particular properties the water has with respect to other solvents. For this reason an improvement of the theory is implicit if other experimental data for water are introduced. This has been suggested and made by Stillinger(138) who incorporated in the theory also the experimental surface tension and the radial distribution function of pure water so to take into account also the highly directional interactions operating in liquid water. The improved Stillinger version and the Pierotti's original form(125) have been both used(130) to compare the trends the ΔX_h^o functions calculated for hard sphere molecules exihibit as the σ_2 values are increased. Both the Stillinger and Pierotti versions of SPT agree satisfactorily when the solute hard sphere diameter is <4 Å, but for higher diameters the ΔG_h^o values, i.e. the solubility, in the Stillinger's model is dominated by the enthalpy rather than the entropy contribution and its value is, for the same σ_2 value, greater that in the Pierotti model.

Besides this uncertainty about the way the thermodynamic functions for the formation of the cavity depend on the size of the solute, the way the SPT was applied to evaluate the hydrophobic interactions(139) is not in agreement with the very interesting results recently obtained by the theory developed by Pratt and Chandler(122). These results were confirmed by the Monte Carlo simulation technique applied by Pangali-Rao-Berne(121) for studying the hydrophobic interactions between two apolar (Lennard-Jones) spheres dissolved in the Rahman-Stillinger(108) liquid water, by the Monte Carlo simulation technique applied by Swaninathan-Beveridge(120) for studying the CH_4-CH_4 interactions between two molecules of methane dissolved in 214 water molecules, the system being described by potential functions derived from ab initio quantum mechanical calculations(109c, 115) and finally by the molecular dynamics simulation applied by Geiger-Rahman-Stillinger(119) to a system constituted by two Lennard-Jones solute particles and 214 water molecules interacting according to the "ST2" pair potential model(108d).

In any case it resulted that there are two relatively stable configurations: in the first the two non polar solutes are in contact inside the same water cage, in the other one molecule of water is interposed between the two solute molecules. This picture of hydrophobic interactions is coherent with that formulated to interprete the results of a statistical mechanical study of the hydrophobic interaction(140) in which is used a potential of average force obtained as a sum of the vacuum potential and a Gurney energy function(141).

Various theoretical approaches to the study of the pure water and aqueous infinitely dilute solutions or dilute solutions of non-polar solutes have been analyzed in the well known book by Ben-Naim (4b) and in some more recent papers of the same author(142). Chan et alii(143) have given a summary of the distinctive characteristic of the thermodynamic properties of aqueous solutions of non polar molecules with respect to the aqueous solutions of polar molecules and some suggestions about the way the properties of the aqueous solutions could be justified with an "ad hoc" molecular distribution function.

The limiting solute partial molar enthalpies and entropies for a model of aqueous solution in which an interstitial lattice model (144a) or lattice model(144b) for fluid water is modified when a non-polar solute is introduced, have been calculated and then compared with those calculated when the model solvent is not hydrogen bonded(144a) or with those experimentally determined for non-polar gases in water(144b).

Hermann(145) used a method based on the first order perturbation theory of liquid mixture in order to calculate the solubilities of hydrocarbons in water and the hydrophobic interaction.

A mean field theory of solvent structure has been employed by Marcelja(146) to describe the effect of solvent correlation on solute-solute interactions of both hydrophobic and hydrophilic solutes. The interactions between hydrophilic solutes in water has also been considered in a group of papers(141,147-150) where the heats of dilution and of the mixing at constant molality for various non electrolytes (alcohols, amides, sugars, urea, aminoacids and peptides) are interpreted in the framework of the McMillan-Mayer theory(151) and the enthalpy effects arising from interactions between each functional group on one molecule and every functional group on the other molecule are evaluated.

A review article on the interaction in aqueous solution has been edited by Scheraga(152), also the specific interactions between water and polar groups are there mentioned, but the chief point of the theoretical approaches to the aqueous solutions has been devoted to the study of the hydrophobic hydration and hydrophobic interactions. Despite the lack of theoretical studies about the way simple organic compounds having one or more than one hydrophilic centres interact with water, molecular dynamics studies on an alanine dipeptide(153) dissolved in 195 water molecules, described by a modification of the St2 model of Stillinger and Rahman(108d) and Monte Carlo simulations of glycine(154) dissolved in 200 water molecules interacting according to the water-water Matsuoka(109c) interaction potential have been reported. Hodes, Némethy and Scheraga have presented(155) a hydration model designed to account for the interactions of both polar and non-polar groups of the solute with surrounding water molecules with the purpose to study the effect of the hydration on the conformations of peptides in aqueous solution.

6. SUMMARY AND CONCLUSION

To summarize, the recent studies on the thermodynamics of dilute aqueous solutions of organic compounds have noticeably improved our knowledge about the behaviour of organic molecules in water, in particular in the case of the non-aromatic molecules. Much can be today said about the incidence the interactions of the water with hydrocarbon groups or hydrophilic centers have on the thermodynamic hydration functions and on the partial molar properties of simple monofunctional compounds or of compounds having functional groups at least three or four bonds apart, that is when the interactions with the water of the group taken into consideration occur in local scale. Also the entities of the solvent mediated interactions between two hydrophilic centres are known. Such interactions produce large effects not easy to rationalize, consequently the hope to frame the behaviour of organic molecules in water in some relatively simple scheme has unfortunately to be given up. It seems in fact now very unlikely that the knowledge of localized interactions between water and single hydrophilic centres allows to predict the behaviour in water of macromolecules rich in hydrophilic groups. Nevertheless the experimental studies of the thermodynamic behaviour of organic molecules in water deserve to be continued and more attention had to be devoted to aromatic and heteroaromatic compounds and to the polyfunctional compounds with functional groups put at given distances and inserted in various molecular frameworks. Moreover it would be desirable that the measurements were always extended to a large range of temperature; in fact while many data are now at disposal for the temperature of 25°C, the trends of thermodynamic properties of organic molecules in the whole range of existence of the liquid aqueous solutions are almost non-existent.

As far as the studies on the hydrophobic hydration and hydrophobic interactions are concerned, the assumption recently made(93) that the rare gases behave in water in a very different way with respect to the hydrocarbons is stimulating. Such hypothesis has been made on the sole basis of the comparison of the different trends the ΔG_h^o of rare gases present with respect to the ΔG_h^o of n-alkanes when both are plotted against a parameter related to the solute volume. It would be important to verify if such differences in the trends remain when other thermodynamic properties are taken into consideration. If so, the question arises whether there is a specific effect of interaction between hydrogen atom bonded to the carbon atom and water or the observed behaviour is due only to the greater sizes the hydrocarbon molecules, except the methane, have with respect to the rare gases molecules.

The simulation through the computer of the aqueous systems is very interesting, and has given many promising results, in particular as far as the structures of the water and aqueous solutions are regarded. A good reproduction of thermodynamic data is however still far away. The systems considered are constituted by a non polar solute molecule in a definite number of water molecules, in general less than three hundred. In spite of the differences on the choice of the number of water molecules, of the pairwise water-water and water-solute intermolecular potentials, of the ensemble, of the use of Monte Carlo or dynamic molecular methods and finally of the type of non-polar molecule, the resulting structures of the solution around the solute is in any case in substantial agreement with classical models proposed some time ago(156) for aqueous solutions of non polar solutes: i.e. a fluctuating quasi-clathrate cage of water molecules is formed around the solute molecule as a consequence of the strengthening of the hydrogen bonding of water induced by the non polar molecule.

Until now simulation methods have not been applied to the study of molecules containing a hydrophilic centre, it will be very interesting to see what structural information these methods will give on the state of water around the hydrophilic centres.

REFERENCES

1. H. L. Friedman and C. V. Krishnan, in "Water. A Comprehensive Treatise", F. Franks, ed., Vol. 3, Chap. 1, Plenum Press, New York, N. Y. (1973).
2. E. Arnett and D. R. McKelvey, in "Solute-Solvent Interactions", D. F. Coetze and C. D. Ritche, ed., Chap. 6, Marcel Dekker, New York, N. Y. (1969).
3. (a) S. Cabani, in "Calorimetry", R. Barbucci ed., Scuola Universitaria, Firenze, Italy (1979); (b) Conference de Thermodynamique Chimique - Thermodinamique des solutions electrolitiques et non electrolitiques - Clermont-Ferrand, France, 20-21 Octobre 1977 (Societé Chimique de France - Division de Chimie Analytique - Groupe de Thermodynamique Experimentale).
4. (a) A. Ben Naim, in "Water and Aqueous Solutions" R. A. Horne, ed., Chap. 11, Wiley Interscience, New York, N. Y. (1972); (b) in "Water and Aqueous Solutions, Introduction to a Molecular Theory", Chap. 4, Plenum Press, New York, N. Y. (1974); (c) in "Solution and Solubilities", M. R. J. Dack, ed., Vol. VIII of Technique of Chemistry, Chap. 1, Wiley Interscience, New York, N. Y. (1975); (d) J. Phys. Chem. 82:792 (1978).
5. (a) J. A. V. Butler, C. N. Ramchandani, D. W. Thomson,

J. Chem. Soc. 280 (1935); (b) J. A. V. Butler and C. N. Ramchandani, J. Chem. Soc. 952 (1935); (c) J. A. V. Butler and W. S. Reid, J. Chem. Soc. 1171 (1936); (d) J. A. V. Butler, Trans Faraday Soc. 33:229 (1937).

6. (a) F. Franks, in "Physico Chemical Processes in Mixed Aqueous Solvents", F. Frankds, ed., Heineman, London, p. 50 (1967); (b) in "Hydrogen Bonded Solvent System", A. K. Covington and P. Jones, ed., Taylor and Francis, London, p. 31 (1968); (c) in "Water. A Comprehensive Treatise", F. Franks, ed., Vol. 4, Chap. 1, Plenum Press, New York, N. Y. (1975); (d) F. Franks and D. S. Reid, in "Water. A Comprehensive Treatise", F. Franks, ed., Vol. 2, Chap. 5, Plenum Press, New York, N. Y. (1973).
7. S. Cabani, G. Conti, and E. Matteoli, Zeit. für Physik. Chem. Neue Folge Bd. 115, S, 121 (1979).
8. P. A. Rice, R. P. Gale, and A. J. Barduhn, J. Chem. Eng. Data, 21:204 (1976).
9. S. Yamamoto, J. B. Alsonskas, and T.E. Crozier, J. Chem. Eng. Data, 21:78 (1976).
10. A. Hvidt, R. Moss, and G. Nielsen, Acta Chem. Scand. B32: 274 (1978).
11. (a) G. Roux, G. Perron, and J. E. Desnovers, J. Solution Chem. 7:639 (1978); (b) Can. J. Chem. 56:2808 (1978); (c) G. Perron and J. E. Desnoyers, Fluid Phase Equilibria, 2:239 (1979).
12. S. Cabani, G. Conti, and E. Matteoli, J. Solution Chem. 5: 751 (1976).
13. C. DeVisser, W. J. M. Heuvessland, L. A. Dunn, and G. Somsen, J. Chem. Soc. Faraday Tans. I, 74:1159 (1978).
14. M. Aquam-Yuan, D. McKay, and W. Y. Shiu, J. Chem. Eng. Data, 24:30 (1979).
15. D. McKay and W. Y. Shiu, J. Chem. Eng. Data, 22:399 (1977).
16. R. L. Brown and S. P. Wasijk, J. Res. Natl. Bur. Stand. 78A:453 (1974).
17. W. G. Green and W. S. Frank, J. Solution Chem. 8:187 (1979).
18. (a) S. Cabani, G. Conti, V. Mollica, and L. Lepori, J. Chem. Soc. Faraday Trans. I, 71:1943 (1975); (b) S. Cabani, V. Mollica, and L. Lepori, J. Chem. Soc. Faraday Trans. I, 71:1154 (1975).
19. E. M. Arnett and B. Chawla, in press.
20. J. A. Abe, K. Nakanishi, and H. Tauhara, J. Chem. Thermodynamics, 10:483 (1978).
21. R. Wolfenden, J. Am. Chem. Soc. 98:1987 (1976).
22. (a) N. Nichols and I. Wadsö, J. Chem. Thermodynamics, 7:329, (1975); (b) S. J. Gill, N. F. Nichols, and I. Wadsö, J. Chem. Thermodynamics, 7:175 (1975).
23. R. Bury, M. Lucas, and P. Barberi, J. Chim. Phys. 75:575 (1978).

24. D. Richon and A. Viallard, Can. J. Chem. 54:2584 (1976).
25. E. M. Arnett, B. Chawla, L. Bell, M. Taagepera, W. J. Hehre, and R. W. Taft, J. Am. Chem. Soc. 99:5729 (1977).
26. (a) C. L. Liotta, H. P. Hopkins, P. T. Kassudia, J. Am. Chem Soc. 96:7153 (1974); (b) C. L. Liotta, E. M. Perdue, and H. P. Hopkins, J. Am. Chem. Soc. 96:7981 (1974); (c) C. L. Liotta, A. Abidaud, and H. P. Hopkins, J. Am. Chem. Soc. 94:8624 (1972).
27. C. H. Rochester and J. R. Symonds, J. Chem. Soc. Faraday Trans. I, 69:1577 (1973).
28. K. Kusano, J. Suurkuusk, and I. Wadsö, J. Chem. Thermodynamics, 5:757 (1973).
29. S. J. Gill, N. F. Nichols, and I. Wadsö, J. Chem. Thermodynamics, 8:445 (1976).
30. S. Cabani and P. Gianni, J. Chem. Soc. Faraday Trans. I, 75:1184 (1979).
31. C. Jolicoeur and G. Lacroix, Can. J. Chem. 54:624 (1976).
32. R. Sköld, J. Suurkuusk, and I. Wadsö, J. Chem. Thermodynamics, 8:1075 (1976).
33. D. J. T. Hill and L. R. White, Aust. J. Chem. 27:1905 (1974).
34. O. Kiyohara, G. Perron, and J. E. Desnoyers, Can. J. Chem. 53:2591 (1975).
35. (a) S. Cabani, G. Conti, A. Martinelli, and E. Matteoli, J. Chem. Soc. Faraday Trans. I, 69:2112 (1973); (b) S. Cabani, G. Conti, and E. Matteoli, J. Solution Chem. 5:125 (1976).
36. O. Kiyohara, G. Perron, and J. E. Desnoyers, Can. J. Chem. 53:3263 (1975).
37. (a) C. De Visser, G. Perron, J. E. Desnoyers, and G. Somsen, J. Chem. Eng. Data, 22:74 (1977); (b) C. De Visser, P. Pel, and G. Somsen, J. Solution Chem. 6:571 (1977).
38. (a) O. Enea, P. P. Singh, and L. G. Hepler, J. Solution Chem. 6:719 (1977); (b) O. Enea, C. Jolicoeur, and L. G. Hepler, Can. J. Chem. in press.
39. H. P. Hopkins, W. C. Duer, and F. J. Millero, J. Solution Chem. 5:263 (1976).
40. C. Jolicoeur, P. Picker, and G. Perron, Can. J. Chem. 53:3634 (1975).
41. S. Cabani, S. T. Lobo, and E. Matteoli, J. Solution Chem. 8:5 (1979).
42. C. Di Paola and B. Belleau, Can. J. Chem. 55:3825 (1977).
43. N. Nichols, R. Sköld, C. Spink, and I. Wadsö, J. Chem. Thermodynamics, 8:993 (1976).
44. O. D. Bonner and P. J. Cerutti, J. Chem. Thermodynamics, 8:105 (1976).
45. J. H. Stern and L. R. Beeninga, J. Phys. Chem. 79:582 (1975).
46. P. R. Philip, G. Perron, and J. E. Desnoyers, Can. J. Chem. 52:1709 (1974).
47. (a) E. Wilhelm, R. Battino, and R. J. Wilcock, Chem. Rev.

77:219 (1977); (b) E. Wilhelm and R. Battino, Chem. Rev. 73:1 (1973).
48. D. M. Alexander, D. J. T. Hill, and L. R. White, Aust. J. Chem. 24:1143 (1971).
49. R. D. Wauchope and R. Haque, Can. J. Chem. 50:433 (1972).
50. C. E. Klotz and B. B. Benson, J. Phys. Chem. 67:933 (1963).
51. B. Benson and D. Krause, J. Chem. Phys. 64:689 (1976).
52. R. N. Potter and M. A. Clynne, J. Solution Chem. 7:837 (1978).
53. (a) . Ben Naim, J. Wilf, and M. Yaacobi, J. Phys. Chem. 77:95 (1979). (b) A. Ben Naim and M. Yaacobi, ibid. 78:170 (1974).
54. C. O. Zerpa, P. Bennett, H. Pharmawardham, V. R. Parrish, and J. Sloan, J. Chem. Eng. Data, 24:26 (1979).
55. H. Høiland and E. Vikingstad, Acta Chem. Scand. A-30, 182 (1976).
56. T. Nakajima, T. Komatsu, and T. Nakagawa, Bull. Chem. Soc. Jap. 48:783 (1975).
57. S. Cabani, G. Conti, and L. Lepori, J. Phys. Chem. 78:1030 (1974).
58. J. T. Edward, P. G. Farrell, and F. Shahidi, J. Chem. Soc. Faraday Trans. I, 73:705 (1977).
59. (a) M. Manabe and M. Koda, Bull. Chem. Soc. Jap. 48:2367 (1975); (b) M. Manabe and M. Koda, Mem. Niihama, Tech. Coll. 12:68 (1976).
60. L. Lepori and V. Mollica, J. Chem. Eng. Data, 23:65 (1978).
61. S. Terasawa, H. Itsuki, and S. Arakawa, J. Phys. Chem. 79:2345 (1975).
62. M. V. Kaukgud and K. J. Patil, J. Phys. Chem. 78:714 (1974).
63. F. Shahidi, P. G. Farrell, and J. T. Edward, J. Chem. Soc. Faraday Trans. I, 73:715 (1977).
64. V. Mollica and L. Lepori, Chim. Ind. (Milan), 59:877 (1977).
65. P. De Luca and T. V. Rebagay, J. Phys. Chem. 79:2493 (1975).
66. (a) H. Høiland, Acta Chem. Scand. A-28, 6:699 (1974); (b) H. Høiland, Acta Chem. Scand. 27:2687 (1973); (c) H. Høiland and E. Vikingstad, J. Chem. Soc. Faraday Trans. I, 71:2007 (1975); (d) H. Høiland, J. Chem. Soc. Faraday Trans. I, 71:797 (1975).
67. M. Palma and J. P. Morel, J. Chem. Phys. 73:645 (1976).
68. S. Harada, T. Nakajima, T. Komatsu, J. Solution Chem. 7:463 (1978).
69. L. Lepori and V. Mollica, J. Polym. Sci. A-2, 16:1123 (1978).

70. A. Ray and G. Nemethy, J. Chem. Eng. Data, 18:309 (1973).
71. S. Harada and T. Nakagawa, J. Solution Chem. 8:267 (1979).
72. H. Høiland and H. Holvik, J. Solution Chem. 7:587 (1978).
73. F. Shahidi, P. G. Farrell, and J. T. Edward, J. Solution Chem. 5:807 (1976).
74. A. Lo Surdo, C. Shin, and F. J. Millero, J. Chem. Eng. Data, 23:197 (1978).
75. T. T. Herskovits and T. M. Kelly, J. Phys. Chem. 77:381 (1973).
76. W. Y. Wen, N. Takeguchi, and D. P. Wilson, J. Solution Chem. 3:103 (1974).
77. (a) S. Cabani, V. Mollica, L. Lepori, and S. T. Lobo, J. Phys. Chem. 81:982 (1977); (b) ibid. 81:987 (1977); (c) S. Cabani, L. Lepori, and E. Matteoli, Chim. Ind. (Milan), 58:221 (1976).
78. C. De Visser, G. Perron, and J. E. Desnoyers, Can. J. Chem. 55:856 (1977).
79. S. Cabani, G. Conti, and E. Matteoli, J. Solution Chem. 8:11 (1979).
80. T. Nakajima, T. Komatsu, and T. Nakagawa, Bull. Chem. Soc. Japan, 48:788 (1975).
81. H. Høiland and E. Vikingstad, Acta Chem. Scand. A-30, 692 (1976).
82. (a) M. V. Kaulgud and K. S. Rao, J. Chem. Soc. Faraday Trans. I, 75:2237 (1979); (b) M. V. Kaulgud, ibid, 75:2246 (1979).
83. (a) D. D. MacDonald, M. E. Estep, M. D. Smith, and J. B. Hyne, J. Solution Chem. 3:713 (1974); (b) D. D. MacDonald, B. Dolan, and J. B. Hyne, J. Solution Chem. 5:405 (1976); (c) D. D. MacDonald, A. MacLean, and J. B. Hyne,, J. Solution Chem. 7:63 (1978); (d) D. D. MacDonald, A. MacLean, and J. B. Hyne, J. Solution Chem. 8:97 (1979).
84. (a) S. V. Subrahamanyam and N. M. Moorthy, Zeit. Für Phisik. Chem. Neue Folge, 88, S, 116 (1974); (b) J. Solution Chem. 4:347 (1975).
85. (a) S. W. Benson, "Thermochemical Kinetics", John Wiley, New York (1976); (b) R. C. Reid, J. M. Prausnitz, and T. K. Sherwood, "The Properties of Gases and Liquids", McGra-Hill Book Co., New York (1978).
86. I. M. Barcaly and J. A. V. Butler, Trans. Faraday Soc. 34:1445 (1938).

87. H. S. Frank and W. M. Evans, J. Chem. Phys. 13:507 (1945).
88. R. Lumry and S. Rajender, Biopolymers, 9:1125 (1970).
89. J. Konicek and I. Wadsö, Acta Chem. Scand. 25:1541 (1971).
90. N. Nichols, R. Sköld, C. Spink, J. Suurkuusk, and I. Wadsö, J. Chem. Thermodynamics, 8:1081 (1976).
91. (a) S. Cabani, G. Conti, and L. Lepori, J. Phys. Chem. 76:1338 (1972); (b) S. Cabani, G. Conti, L. Lepori, and L. Leva, J. Phys. Chem. 76:1343 (1972); (c) S. Cabani, G. Conti, and L. Lepori, Trans. Faraday Soc. 67:1943 (1971).
92. R. Wolfenden and C. A. Lewis, J. Theor. Biol. 59:231 (1976).
93. M. H. Abraham, J. Am. Chem. Soc. 101:5477 (1979).
94. C. Jolicoeur, J. B. Boileau, S. Bazinte, and P. Picker, Can. J. Chem. 53:716 (1975).
95. C. Jolicoeur and J. Boileau, Can. J. Chem. 56:2707 (1978).
96. J. Hine and P. K. Mookerjee, J. Org. Chem. 40:292 (1975).
97. S. Cabani, G. Conti, and E. Matteoli, J. Chem. Soc. Faraday Trans. I, 74:2408 (1978).
98. F. Franks, J. E. Ravenhill, and D. S. Reid, J. Solution Chem. 1:3 (1972).
99. J. P. Guthrie, Can. J. Chem. 55:3700 (1977).
100. S. Cabani, P. Gianni, V. Mollica, and L. Lepori, Primo Convegno Nazionale di Calorimetria ed Analisi Termica AICAT, p. 115, Firenze (Italy), 17-19 Dicembre (1979).
101. A. Bondi, J. Phys. Chem. 68:441 (1964).
102. M. R. Tiné, "Volumetric Properties of Aminoacids and Polypeptides in Aqueous Solution" Thesis. Institute of Physical Chemistry, University of Pisa (Italy), (1977).
103. J. T. Edward and P. G. Farrel, Can. J. Chem. 53:2965 (1975).
104. S. J. Gill and I. Wadsö, Proc. Natl. Acad. Sci. U.S.A., 73:2955 (1976).
105. R. D. Cramer III, J. Am. Chem. Soc. 99:5408 (1977).
106. J. A. Barker and R. O. Watts, Chem. Phys. Letters, 3:144 (1969).
107. V. G. Dashevsky and G. M. Sarkisov, Mol. Phys. 27:1271 (1974).
108. (a) A. Rahman and F. H. Stillinger, J. Chem. Phys. 55:3336 (1971); (b) A. Ben Naim and F. H. Stillinger, in "Structure and Transport Processes in Water and Aqueous Solutions", A. H. Horne, ed., Chap. 8, p. 295, Wiley Inter., N. Y. (1972); (c) A. Rahman and F. H. Stillinger, J. Am. Chem. Soc. 95:7943 (1973); (d) F. H. Stillinger and

A. Rahman, J. Chem. Phys. 60:1545 (1974); (e) G. N. Sarkisov, V. G. Dashevsky, and G. G. Malenkov, J. Mol. Phys. 27:1249 (1974); (f) R. O. Watts, Mol. Phys. 28:1069 (1974); (g) L. L. Shipman and H. A. Scheraga, J. Phys. Chem. 79:909 (1974); (h) J. C. Owicki, L. L. Shipman and H. A. Scheraga, J. Phys. Chem. 79:1794 (1975).

109. (a) H. Popkie, K. Kistanmacher, and E. Clementi, J. Chem. Phys. 59:1325 (1973); (b) H. Kistanmacher, H. Popkie, E. Clementi, and R. O. Watts, J. Chem. Phys. 60:4455 (1974); (c) G. C. Lie and E. Clementi, J. Chem. Phys. 62:2135 (1975); (d) O. Matsouka, E. Clementi, and M. Yoshimine, J. Chem. Phys. 64:2314 (1976); (f) S. Swaminathan and D. L. Beveridge, J. Am. Chem. Soc. 99:8392 (1977); (g) M. Mezei, S. Swaminathan, and D. L. Beveridge, J. Am. Chem. Soc. 100:3255 (1978); (h) W. L. Jorgensen, J. Am. Chem. Soc. 101:2011 (1979); ibid. 101:2016 (1979).

110. F. Stillinger and H. L. Lemberg, J. Chem. Phys. 62:1340 (1975).

111. A. J. Ladd, Mol. Phys. 33:1038 (1977).

112. (a) J. C. Owicki and H. A. Scheraga, J. Am. Chem. Soc. 99:7403 (1977); (b) ibid. 99:7413 (1977).

113. F. H. Stillinger, in "Adv. Chem. Phys." I. Prigogine and S. A. Rice, ed., Vol. XXXI, p. 1-101, Wiley Inter., New York (1975).

114. D. H. Wood, in "Water. A Comprehensive Treatise", F. Franks, ed., Vol. 6, Chap. 6, Plenum Press, New York (197).

115. S. Swaminathan, S. W. Harrison, and D. L. Beveridge, J. Am. Chem. Soc. 78:5705 (1978).

116. S. W. Harrison, S. Swaminthan, D. L. Beveridge, and R. Ditchfield, Int. J. Quantum Chem. XIV:319 (1978).

117. S. Okazaki, K. Nakanishi, and H. Touhara, J. Chem. Phys. 71:2421 (1979).

118. G. Alagona and A. Tani, J. Chem. Phys. 72:580 (1980).

119. A. Geiger, A. Rahman, and F. H. Stillinger, J. Chem. Phys. 70:263 (1979).

120. S. Swaminathan and D. L. Beveridge, J. Am. Chem. Soc. 101:5832 (1979).

121. (a) C. Pangali, M. Rao, and A. J. Berne, J. Chem. Phys. 71:2975 (1979); (b) ibid. 71:2982 (1979).

122. L. R. Pratt and D. Chandler, J. Chem. Phys. 67:3683 (1977).

123. (a) H. Reiss, H. Frisch, and J. L. Lebowitz, J. Chem. Phys. 31:369 (1959); (b) H. Reiss, H. Frisch, H. L. Helfman, and J. L. Lebowitz, ibid. 32:119 (1960); (c) H. Reiss, Adv. Chem. Phys. IX:1 (1966).

124. J. L. Lebowitz and J. S. Rowlinson, J. Chem. Phys. 41:133 (1964).

125. (a) R. A. Pierotti, J. Phys. Chem. 69:281 (1965); (b) ibid. 71:2366 (1967).

126. M. Lucas, in "L'eau et les systèmes biologiques" ed., A. Alfsen ed A. J. Berteaud, Centre National de la Recher-

che Scientifique, Paris (1976).
127. R. A. Pierotti, Chem. Rev. 76:717 (1976).
128. H. De Voe, J. Am. Chem. Soc. 98:1724 (1976).
129. (a) N. F. Carnahan and K. E. Starlin, J. Chem. Phys. 51:635 (1969); (b) T. Boublik, J. Chem. Phys. 53:471 (1970); (c) L. V. Woodcock, J. Chem. Soc. Faraday Trans. II, 72:731 (1976).
130. (a) M. Lucas, J. Phys. Chem. 80:359 (1976); (b) M. Lucas and R. Bury, J. Phys. Chem. 80:999 (1976).
131. E. Wilhelm and R. Battino, J. Chem. Phys. 56:563 (1972).
132. K. Arakawa, K. Tokiwano, N. Ohtomo, and H. Uedaira, Bull. Chem. Soc. Jpn. 52:2483 (1979).
133. P. R. Philip and C. Jolicoeur, J. Solution Chem. 4:105 (1975).
134. C. Jolicoeur and G. Lacroix, Can. J. Chem. 51:3051 (1973).
135. L. D. de Ligny and N. G. van der Veen, J. Solution Chem. 4:841 (1975).
136. W. Wen and J. A. Muccitelli, J. Solution Chem. 8:225 (1979);
137. M. Lucas, J. Phys. Chem. 77:2479 (1973).
138. F. H. Stillinger, J. Solution Chem. 2:141 (1973).
139. A. Ben Naim and R. Tenne, J. Chem. Phys. 67:627 (1977).
140. A. H. Clark, F. Franks, M. D. Pedley, and D. S. Reid, J. Chem. Soc. Faraday Trans. I, 73:290 (1977).
141. H. L. Friedman and C. V. Krishnan, J. Solution Chem. 2:119 (1973).
142. (a) A. Ben Naim, Biopolymers, 14:1337 (1975); (b) J. Phys. Chem. 79:1268 (1975); (c) ibid. 82:874 (1978).
143. (a) D. Y. Chan, D. J. Mitchell, B. W. Ninham and B. A. Pailthorpe, in "Water. A Comprehensive Treatise", F. Franks, ed., Vol. 6, Chap. 5, p. 239, Plenum Press, New York (1979); (b) J. Chem. Soc. Faraday Trans. II, 74:2050 (1978).
144. (a) H. A. Sallouta and G. M. Bell, Mol. Phys. 32:839 (1976); (b) B. G. L. Wilson and G. M. Bell, J. Chem. Soc. Faraday Trans. II, 74:1702 (1978).
145. (a) R. B. Hermann, J. Phys. Chem. 79:163 (1975); (b) in "Molecular and Quantum Pharmacology", E. Bergmann and B. Pullman, eds., p. 44, Reidel Publ. Co., Dordrecht, Holland (1974).
146. S. Marcelja, D. J. Mitchell, B. W. Ninham, and M. J. Sculley, J. Chem. Soc. Faraday Trans. II, 73:630 (1977).
147. (a) R. B. Cassel and R. H. Wood, J. Phys. Chem. 78:2465 (1974); (b) J. J. Savage and R. H. Wood, J. Solution Chem. 5:733 (1976); (c) B. Y. Okamoto and R. H. Wood, J. Chem. Soc. Faraday Trans. I, 74:1990 (1978).
148. F. Franks, M. Pedley, and D. S. Reid, J. Chem. Soc. Faraday Trans. I, 72:359 (1976).
149. T. H. Lilley and R. P. Scott, J. Chem. Soc. Faraday Trans. I, 72:184 (1976).

150. G. Barone, G. Castronuovo, V. Crescenzi, V. Elis, and E. Rizzo, J. Solution Chem. 7:179 (1978).
151. W. G. McMillan and J. E. Mayer, J. Chem. Phys. 13:276 (1945).
152. H. A. Scheraga, Acc. Chem. Res. 12:7 (1979).
153. (a) P. J. Rossky, M. Karplus, and A. Rahman, Biopolymers, 18:825 (1979); (b) P. J. Rossky and M. Karplus, J. Am. Chem. Soc. 101:1913 (1979).
154. S. Romano and E. Clementi, Intern. J. Quantum Chem. XIV: 839 (1978).
155. Z. I. Hodes, G. Némethy, and H. A. Scheraga, Biopolymers, 18:1565 (1979); ibid. 18:1611 (1979).
156. (a) H. S. Frank and M. W. Evans, J. Chem. Phys. 13:507 (1945); (b) W. Kauzmann, Adv. Protein Chem. 14:1 (1959); (c) G. Némethy and H. A. Scheraga, J. Chem. Phys. 36: 3401 (1962); (d) D. N. Glew, J. Phys. Chem. 66:605 (1962); (e) D. N. Glew, H. D. Mak, and N. S. Rath, in "Hydrogen-Bonded Systems" A. K. Covington and P. Jones, eds., p. 195, Taylor and Francis, London (1968).

CLASSICAL IONIC FLUIDS IN THE MEAN SPHERICAL APPROXIMATION[†]

Roberto Triolo[§]

Oak Ridge National Laboratory, Oak Ridge, TN 37830, USA

and Antonio M. Floriano

Istituto di Chimica Fisica, Università di Palermo, Italy, and Department of Chemistry, University of Alberta, Canada

Abstract — The recently obtained analytical solution of the mean spherical approximation has been used to calculate thermodynamic and structural properties of aqueous solutions of asymmetric electrolytes. The same approximation has also been used to calculate structure functions of pure and mixed molten salts. The agreement between experimental or "quasi-experimental" structure functions and those obtained within the framework of the MSA is quite good especially when the ionic radii are obtained by fitting the long wavelength limit of the structure functions to the isothermal compressibility of the system, under the condition that the diameter ratio is the same as in the crystal.

1. INTRODUCTION

Electrolyte solutions and molten salts are typical ionic fluids of great interest both from a practical and from a theoretical point of view. Despite many efforts and decades of both experimental and theoretical work in this area, our ability to describe "real" systems is still rudimentary. Nevertheless, in the last decade or so, the theoretical work has received new incentive with the acquisition of new mathematical techniques, proven to be very useful in dealing with long range potentials like those involved in the treatment of ionic fluids.

[†]Research sponsored by the Division of Chemical Sciences/Office of Basic Energy Sciences, U. S. Department of Energy under contract W-7405-eng-26 with the Union Carbide Corporation.
[§]On leave of absence from University of Palermo (Italy).

What follows will concern electrolyte solutions as well as molten salts. In fact, as we will see later, within the framework of the McMillan-Mayer theory(1), there is no difference in the mathematical treatment of a dilute aqueous solution of a given electrolyte and the corresponding molten salt. Of course, the density, temperature and potential energy will be different, but in both cases, the model to be used will be the same. It should then not be surprising that the next section starts with a discussion of the McMillan-Mayer and Debye-Hückel theories(2) for dilute systems of charged particles. The Debye-Hückel theory (DH) has been the most successful theory of electrolyte solutions and some of the modern approximations are simple extensions of DH theory, which are statistically consistent.

2. McMILLAN-MAYER AND DEBYE-HÜCKEL THEORIES

The concept behind the DH theory was not new, in that Milner (3a), almost a decade before, formulated a theory of ionic solutions based on the concept of "ionic atmosphere". He, however, was unable to solve the proposed equations. Double layer theories(3b,3c), which used the same concept, also preceded the DH theory. The merit of Debye and Hückel was to introduce several approximations that made an analytical solution for the theory possible. The starting point of the DH theory is the assumption that the excess of thermodynamic properties of electrolyte solutions (when compared with non-electrolyte solutions) is due only to the Coulombic interactions between the ions. It is then necessary to calculate the average electrostatic potential at the surface of a given ion (taken as reference) due to all the other ions. These other ions constitute the "ionic atmosphere". Once this potential is known, it is evidently possible to calculate all the thermodynamic properties of the system. Indicating with $z_i e$ and $z_j e$ the charge of the reference ion (i) and of an arbitrary ion (j) in the "ionic atmosphere", respectively, the effective interaction energy between the two ions will be

$$U_{ij}(r_{ij}) = z_i z_j e^2/\varepsilon \, r_{ij} \tag{1}$$

where r_{ij} is the distance between the two ions and ε is the macroscopic dielectric constant of the solvent considered to be a continuum. Strictly speaking, Debye and Hückel have already introduced approximation at this stage, in that they neglected the molecular structure of the solvent; moreover they ruled out any other electrostatic interactions (ion-dipole, dipole-dipole), and more important, they neglected the non-Coulombic interactions, including, evidently, the short range non-Coulombic interactions, i.e., hard core repulsions(4). At this point another approximation was introduced, in assuming that all the ions in the "ionic atmosphere" were point charges; consequently it was possible to define an average local density $\rho_j(r)$ which was a continuous function of r, and an average electrostatic potential $\psi_j(r)$, in such a way they are related to each other through the (nonlinearized) Poisson-Boltzman equation (PBE)

$$\nabla^2 \psi_j(r) = \kappa^2 \psi_j(r) \qquad (2)$$

Finally in order to linearize the PBE, Debye and Hückel assumed that $z_j e \psi_i << \beta^{-1}$, where $\beta^{-1} = k_B T$, with k_B = Boltzman constant and T is the temperature in degrees Kelvin. After some rather simple algebra, Debye and Hückel could prove that the logarithm of the activity coefficients ($\gamma_\pm$) should vary linearly with the square root of the ionic strength (I) in the very dilute regime, and in general, the relationship between $\ln \gamma_\pm$ and ionic strength would be

$$\ln \gamma_\pm = \frac{S I^{\frac{1}{2}}}{1 + \kappa a} \qquad (3)$$

where κ^{-1}, having the dimension of a distance, is interpreted as the distance (from the reference ion) where the charge density of the ionic atmosphere has its maximum value and a is the minimum distance of approach, and S is a constant. At this point it is clear that for extremely high dilutions, the DH theory must hold, because none of the approximations made is important in the high dilution regime. Increasing the concentration causes the ions to come closer together and hence electrostatic interactions may become even greater than the thermal energy ($k_B T$); as a consequence the linearized PBE becomes nothing else but a crude approximation at high ionic strength; furthermore the closer the ions are the stronger the "hard core interactions", neglected in the model become.

To conclude this section on the DH theory, we would like to point out that these last two criticisms (neglecting short range repulsive interactions and linearizing the PBE) are the only valid criticisms. In fact the McMillan-Mayer theory (MMM) showed that, provided a correct definition of the "effective interaction potential" is given, the molecular structure of the solvent needs not to be considered explicitly(1) in calculating the thermodynamic properties of ionic solutions. This conclusion has very important consequences: the first one is that, as the number density of ions in a typical electrolyte solutions is of the order of 10^{-3} ions/Å^3,[†] then the solution can be considered as a dilute ionic gas; as a consequence the theories available for gases can be used for ionic fluids, provided the "effective potential" (more often called potential of the mean force at infinite dilution) takes the place ot the gas-gas interaction potential. Strictly this is true only in the limit of infinite dilution, but will hold also at finite concentrations, provided the chemical potential of the solvent in the given solution is the same as in the infinitely dilute solutions. This actually

---

[†]Along this paper 1 Å = 100 pm.

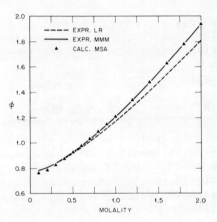

Fig. 1. Comparison between osmotic coefficients in the LR systems (---), and in the MMM system (——). Triangles are points calculated within the MSA.

means that all the thermodynamic properties calculated within the framework of the MMM theory are given not at 1 atm (as the commonly used Lewis-Randall (LR) thermodynamic properties), but rather under the osmotic pressure of the solution itself(6), the excess of pressure being necessary to raise the activity of the solvent to its value in an infinitely dilute solution at the same temperature and 1 atm of pressure. Values of factors for the LR to MMM system conversion have been recently calculated(7) for a variety of 1-1, 2-1, and 3-1 aqueous electrolytes, for osmotic and activity coefficients. Usually these corrections are small (less than 10% in a typical 2.0 M solution), but nevertheless greater than the standard deviation of the fit between calculated and experimental result, as shown in Figure 1. We will conclude this section by giving two references which should be consulted by readers interested in a lucid and rigorous treatment of the DH and the MMM theories; they are a recent contribution by L. Blum(8) and another contribution by H. L. Friedman and W. D. T. Dale(9).

3. THE MEAN SPHERICAL APPROXIMATION (MSA)

The starting point of many modern theories of fluids is the Ornstein-Zernicke equation (OZ)(10), which relates the direct correlation function $c_{ij}(r)$ and the total correlation function $h_{ij}(r)$

$$h_{ij}(r) = c_{ij}(r) + \Sigma_k \rho_k \int d\bar{r}' h_{ij}(r') c_{ki}(|\bar{r}'-\bar{r}|) \qquad (4)$$

where $h_{ij}(r) = g_{ij}(r) - 1$, with $g_{ij}(r)$ indicating the pair correlation function; the summation in (4) is extended over all the different ions. The interested reader is urged to refer to the papers by L. Blum, et al.(8,11), for details concerning the primitive model electrolytes in the MSA; in this section we will primarily be concerned with the practical solution of the MSA for an arbitrary mixture of charged hard spheres in a continuum of dielectric constant ε. The only restrictions will be that the principle of electroneutrality will not be violated and that the particle diameters will be additive. For such a system, the effective potential energy for a pair i, j of ions will be of the form

$$[U_{ij}(r)]_e = [U_{ij}(r)]_{nesr} + [U_{ij}(r)]_{eC} \tag{5}$$

where the first term of the right hand side of equation (5) corresponds to the non electrostatic short range interactions (hard core repulsions), while the second term corresponds to the effective Coulombic interactions. In the MSA, this last term is a purely Coulombic term ($z_i z_j e^2/\varepsilon r_{ij}$). We call it here "effective", because in part of the following discussion, the distance parameter r_{ij} will be treated as adjustable and will also be considered a function of the concentration and temperature, in order to include all the effects left out from the theory. For more details we refer to other papers of the present authors(5,12).

For the non-electrostatic short range potential we will use the hard sphere potential

$$U_{ij}(r) = 0 \quad \text{for} \quad r > \sigma_{ij}$$
$$U_{ij}(r) = \infty \quad \text{for} \quad r < \sigma_{ij} \tag{6}$$

where σ_{ij} is the minimum distance of approach between particles i and particles j. Equations (6) lead to the MSA boundary condition

$$h_{ij}(r) = g_{ij}(r) - 1 = -1 \quad \text{for} \quad r < \sigma_{ij}$$

while the other closure for the OZ equation concerns the direct correlation function

$$c_{ij}(r) = -\beta U_{ij}(r) \quad \text{for} \quad r > \sigma_{ij}$$

Given these closures, the OZ equation can be solved in terms of the so-called Baxter factor correlation function matrix Q(r)(11), whose elements can be easily found once the Γ parameter (in a sense the equivalent of the parameter κ of the DH theory) is found. Γ can be obtained by solving (for example by means of the Newton-Raphson method) the equation

$$\Gamma = (\alpha/2) \left[\Sigma_k [\rho_k/(1 + \Gamma\sigma_k)^2] \left[z_k - \pi\sigma_k^2 P_n/2\Delta \right]^2 \right]^{1/2} \tag{7}$$

where α is similar to the plasma parameter and P_n is an explicit function of the diameter σ_i, the charge z_i and the number density ρ_i of the different species in solution; finally Δ is a function of the packing fraction.

Thermodynamic Properties. The knowledge of Γ allows the evaluation of the excess (E^{ex} of A^{ex}) of the thermodynamic functions E (internal energy) and A (free energy). These excess functions are related to each other through the equation(11,18):

$$\beta A^{ex} = \beta E^{ex} + \Gamma^3/3 \qquad (8)$$

with

$$\beta E^{ex} = -(\beta e^2/\varepsilon)\left[\Gamma \Sigma_k \rho_k z_k^2 /(1 + \Gamma \sigma_k) + \pi \Omega P_n^2 /2\Delta\right] \qquad (9)$$

where Ω depends on the same variables as Δ and P_n (5,11,12). The knowledge of A^{ex} is important to calculate the excess MSA osmotic coefficient ϕ^{ex} and the excess MSA activity coefficient $\ln\gamma^{ex}$ that, when added to the corresponding hard spheres contributions (ϕ^{HS} and $\ln\gamma^{HS}$) will give the total MSA osmotic coefficient and activity coefficient, respectively,

$$\phi^{MSA} = \phi^{HS} + \phi^{ex} \qquad (10)$$

$$\ln\gamma^{MSA} = \ln\gamma^{HS} + \ln\gamma^{ex} \qquad (11)$$

Results for 1-1 Electrolytes. Here we present some of the most recent data obtained for electrolyte solutions within the framework of the MSA. We have shown(5,12) that using the MSA as an empirical theory, osmotic coefficients and activity coefficients of 1-1 electrolytes can be calculated within the experimental accuracy if a density dependent cation radius is used in connection with Pauling crystallographic radii for the anions. Figures 2 and 3 show the kind of

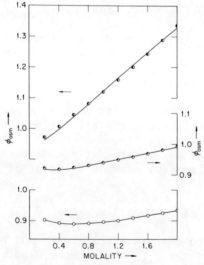

Fig. 2. MSA osmotic coefficients for three typical electrolytes: ($\bigcirc$)CsCl, ($\lefthalfcircle$)KBr, ($\newmoon$)LiI.

IONIC FLUIDS IN THE MEAN SPHERICAL APPROXIMATION

agreement obtained for three typical electrolytes. The agreement can be noticeably improved if ε is treated as an adjustable parameter(12a,12d). Excellent agreement with experimental osmotic and activity coefficients has been found(12c) for aqueous solutions of HCl in the concentration range 0-2 M and in the temperature range 0 - 60°C. Values of the partial molal heat capacity and of the partial molal enthalpy (Table I) have been found in very good agreement with the corresponding experimental quantities(13). A temperature as well as concentration dependent radius was used, in

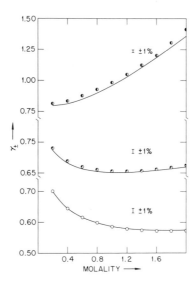

Fig. 3. MSA activity coefficients for the same electrolytes of Figure 2.

Table I. Values of Partial Molal Enthalpy (L_2) and Partial Molal Termal Capacity (J_2) for Aqueous Solutions of HCl at Various Temperatures[a]

	Temperature, °C						S,%	a	10^4 xb	10^6 xc	Concentration (m)
	0	15	25	40	50	60					
L_2[b]	138	173	200	244	276	311	0.5	-0.30	6	-2.0	0.2
J_2[c]	2.2	2.5	2.8	3.1	3.4	3.6					
L_2[b]	417	509	576	688	770	850	0.5	-0.13	1.0	-4.5	1.0
J_2[c]	5.7	6.5	7.0	7.9	8.5	0.1					

[a] The equation relating to the temperature T (K) is: $\ln \gamma = a + bT + cT^2$. [b] Cal mol^{-1}. [c] Cal mol^{-1} K^{-1}. Where 1 cal = 4.187 J.

this case, for the cation. Recently we obtained similar results for aqueous solutions of NaCl in the same concentration range and in the temperature range 0-100°C. The results of this study will be soon published. Here, in Figure 4, we give the temperature dependence of the zero concentration hard sphere diameter of Na^+ and H_3O^+ ions, respectively. As can be noted, the two cations show a different trend; this fact is not easy to explain, and we believe that more calculations in similar systems must be performed before an explanation can be found. Finally we would like to show how it is possible to recover the "cube-root law" within the framework of the MSA. Figure 5 shows the logarithm of the activity coefficient calculated for a set of parameters corresponding to aqueous solutions of NaF(5) in the concentration range 0.001 - 0.3 M. As can be noted, there is an excellent linear correlation between $\ln\gamma_\pm$ and the cube root of the concentration; all this without any a priori assumption concerning the distribution of ions in solution. Evidently at higher concentrations the linear correlations between $\ln\gamma$ and $M^{1/3}$ will be lost. Also in the case of bigger ions the linear correlation will be lost, or at least, the concentration range in which the correlation is noted, will be smaller. We conclude that the existence of a linear relationship between $\ln\gamma_\pm$ and $M^{1/3}$ does not support the so-called "lattice model" of electrolyte solutions.

<u>Asymmetric Electrolytes</u>. We have performed extensive calculations on asymmetric 2-1 and 1-2 electrolytes and we have found (Table II) that the agreement with the experiments is not as good

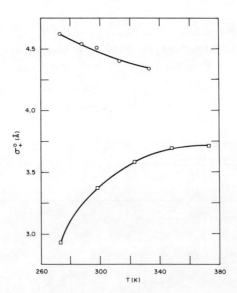

Fig. 4. Temperature dependence of the infinite dilution hard core diameters used to fit osmotic coefficients of aqueous HCl (○) and aqueous NaCl (□), respectively.

Table II. Best Hard Core Diameters Found by Least-Squares Refinement Against Experimental McMillan-Mayer Osmotic Coefficients. All data refer to aqueous solutions unless otherwise stated. SD% is the standard deviation percent. See the text for the meaning of σ_+^o and α.

Salt	σ_+^o	α	SD%	Conc.Range (M)	Temperature	Reference
LiCl	4.44 ± 0.02	77 ± 4	0.18	0.2 - 2.0	298.16	(12[b])
LiBr	4.36 ± 0.02	38 ± 6	0.21	0.2 - 2.0	"	"
LiI	4.98 ± 0.07	126 ± 15	0.73	0.2 - 2.0	"	"
NaF	3.29 ± 0.04	477 ± 21	0.11	0.1 - 1.0	"	"
NaCl	3.33 ± 0.05	132 ± 15	0.30	0.2 - 2.0	"	"
NaBr	3.42 ± 0.05	101 ± 14	0.30	0.2 - 2.0	"	"
NaI	3.62 ± 0.04	101 ± 10	0.25	0.2 - 2.0	"	"
KF	3.91 ± 0.04	71 ± 10	0.23	0.2 - 2.0	"	"
KCl	2.84 ± 0.04	176 ± 14	0.21	0.2 - 2.0	"	"
KBr	2.82 ± 0.05	184 ± 15	0.24	0.2 - 2.0	"	"
KI	2.96 ± 0.04	216 ± 11	0.20	0.2 - 2.0	"	"
RbF	4.73 ± 0.02	148 ± 38	0.14	0.2 - 2.0	"	"
RbCl	2.57 ± 0.02	121 ± 98	0.12	0.2 - 2.0	"	"
RbBr	2.34 ± 0.02	180 ± 10	0.11	0.2 - 2.0	"	"
RbI	1.99 ± 0.02	194 ± 12	0.12	0.2 - 2.0	"	"
CsF	4.82 ± 0.05	78 ± 10	0.40	0.2 - 2.0	"	"
CsCl	2.02 ± 0.02	32 ± 8	0.07	0.2 - 2.0	"	"
CsBr	1.73 ± 0.03	71 ± 20	0.15	0.2 - 2.0	"	"
CsI	1.51 ± 0.01	276 ± 4	0.04	0.2 - 2.0	"	"
HCl	4.62 ± 0.02	80 ± 5	0.19	0.0 - 2.0	273.16	(12[c])
HCl	4.54 ± 0.02	83 ± 5	0.18	0.0 - 2.0	288.16	"
HCl	4.51 ± 0.02	88 ± 5	0.17	0.0 - 2.0	298.16	"
HCl	4.43 ± 0.02	90 ± 4	0.16	0.0 - 2.0	313.16	"
HCl	4.40 ± 0.02	95 ± 5	0.16	0.0 - 2.0	323.16	"
HCl	4.34 ± 0.02	98 ± 4	0.14	0.0 - 2.0	333.16	"
HCl[†]	4.87 ± 0.02	94 ± 4	0.20	0.0 - 2.0	298.16	"
HCl[††]	5.73 ± 0.06	124 ± 9	0.75	0.0 - 2.0	298.16	"
NaCl	2.93 ± 0.05	150 ± 16	0.30	0.0 - 2.0	273.16	this work
NaCl	3.37 ± 0.04	143 ± 12	0.30	0.0 - 2.0	298.16	"
NaCl	3.58 ± 0.03	137 ± 9	0.30	0.0 - 2.0	323.16	"
NaCl	3.69 ± 0.03	130 ± 8	0.20	0.0 - 2.0	348.16	"
NaCl	3.71 ± 0.03	134 ± 8	0.20	0.0 - 2.0	373.16	"
$MgCl_2$	5.86 ± 0.04	75 ± 7	0.9	0.2 - 2.0	298.16	this work
$MgBr_2$	6.39 ± 0.03	107 ± 4	0.6	0.2 - 2.0	"	"
MgI_2	6.51 ± 0.02	120 ± 3	0.5	0.2 - 2.0	"	"

[†]Mixed solvent: 20% dioxane and 80% water (by weight)
[††]Mixed solvent: 45% Dioxane and 55% water (by weight)

Table II. continued

Salt	σ_+^o	α	SD%	Conc.Range (M)	Temperature	Reference
$CaCl_2$	5.46 ± 0.04	82 ± 7	0.70	0.2 - 2.0	298.16	this work
$CaBr_2$	5.82 ± 0.04	91 ± 7	0.80	0.2 - 2.0	"	"
CaI_2	6.07 ± 0.04	122 ± 6	0.90	0.2 - 2.0	"	"
$SrCl_2$	5.22 ± 0.05	81 ± 10	0.90	0.2 - 2.0	"	"
$SrBr_2$	5.46 ± 0.05	86 ± 9	0.90	0.2 - 2.0	"	"
SrI_2	5.75 ± 0.05	83 ± 9	1.00	0.2 - 2.0	"	"
$BaCl_2$	4.93 ± 0.06	178 ± 13	0.80	0.2 - 2.0	"	"
$BaBr_2$	5.16 ± 0.05	126 ± 9	0.76	0.2 - 2.0	"	"
BaI_2	5.72 ± 0.13	85 ± 25	2.70	0.2 - 2.0	"	"

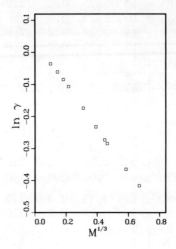

Fig. 5. The "cube-root law" in the MSA.

as in the case of 1-1 electrolytes, even when using density dependent radii. MSA might be expected to become less accurate with increasing asymmetry of the electrolytes. We have found that this is not always correct; in fact we have calculated(12f) osmotic and activity coefficients of aqueous solutions of rare earth chlorides and compared the calculated values with the experimental results of Spedding and co-workers (Table III). Figure 1 shows the kind of agreement obtained with a two parameter fit. The interesting point is that the MSA cation radii are very close to those obtained for the hydrated rare earth cations. The success so far obtained with the 3-1 electrolytes examined, prompted us to try more asymmetric salts, as both thermodynamic and structural information were available for a model 6-1 electrolyte in the HNC approximation(15). The HNC is probably(16) the best approximation available today for ionic fluids. The only problem with it is the high degree of complexity of the calculation

Table III. Best Hard Core Fitting Parameters for Several 3-1 Electrolytes.
Data refer to aqueous solutions at 298.16 °K in the concentration range 0.1 - 2.0 M

Salt	σ_+^o	α	SD%
$LaCl_3$	6.39 ± 0.03	91 ± 5	0.08
$PrCl_3$	6.39 ± 0.04	94 ± 5	0.12
$NdCl_3$	6.37 ± 0.04	87 ± 6	0.12
$SmCl_3$	6.46 ± 0.03	93 ± 5	0.11
$EuCl_3$	6.55 ± 0.03	98 ± 5	0.10
$GdCl_3$	6.60 ± 0.03	95 ± 4	0.09
$TbCl_3$	6.64 ± 0.03	91 ± 4	0.09
$DyCl_3$	6.67 ± 0.03	90 ± 4	0.09
$HoCl_3$	6.70 ± 0.03	94 ± 5	0.11
$ErCl_3$	6.66 ± 0.03	88 ± 4	0.10
$TmCl_3$	6.65 ± 0.02	87 ± 3	0.05
$YbCl_3$	6.64 ± 0.03	86 ± 4	0.08
$LuCl_3$	6.64 ± 0.03	87 ± 5	0.11
YCl_3	6.72 ± 0.02	92 ± 3	0.05

involved. Comparison with the HNC approximation is a means to assess the accuracy of our calculations. Figure 6 shows the comparison betwenn the two approximations for two different models of a 6-1 electrolyte. In the first model ("equal diameter" case), cations of charge +1e and anions of charge -6e have the same diameter ($\sigma_- = \sigma_+ = 4.2$ Å). In the second model ("unequal diameter" case), the anions are twice as big as the cations ($\sigma_- = 2\sigma_+ = 8.4$ Å). While the agreement between MSA and HNC is disappointing in the equal diameter case, it is indeed good for the unequal diameter case, which seems to be a more realistic one. We intend to use the MSA to calculate thermodynamic and structural properties of micellar systems (of which the present 6-1 model can be considered an embryo), by systematically expanding the anion and increasing also its charge. We also hope to be able to use the HNC approximation, for comparison. The purpose of the planned work will also be to assess the effect of the radius-ratio and of the charge on the thermodynamics and on the structural properties of these solutions.

From the preliminary work done so far, we can conclude that, as far as the thermodynamic properties are concerned, the MSA approximation gives excellent results for low surface charge density systems both for symmetric and asymmetric electrolytes. In high surface charge density systems, the MSA can lead to serious discrepancies. Finally we want to summarize the results of calculations done using the MSA as an empirical theory, and fitting the experimental results (converted to the MMM systems) to a model of charged hard spheres, with concentration dependent diameters. The diameters

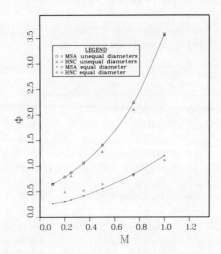

Fig. 6. Comparison between HCN and MSA osmotic coefficients for aqueous solutions of two different models of a 6 - 1 electrolyte.

of the anions have been kept constant at the corresponding Pauling values, while the cationic diameters (in Å) were supposed to vary with the number density, according to the formula(12c)

$$\sigma_+ = \sigma_+^0 (1 - \alpha\rho), \quad \text{where} \quad \rho = 6.0225 \; 10^{-4} \; M.$$

The results of such calculations are summarized in Tables II and III.

Structural Properties. A good way to define the concept of "Structure" of fluids is through the pair correlation function $g_{ij}(r)$(17). Recently Blum and Høye(18), found the analytical solution for the Laplace transform of the correlation function $g_{ij}(r)$

$$L_{ij}(s) = \int_0^\infty dr \; e^{-sr} \; r \; g_{ij}(r) \qquad (12)$$

It can be shown(18) that, in the MSA, the complex quantity $L_{ij}(s)$ is given by the sum of three terms

$$L_{ij}(s) = \mu L_{ij}^{PY}(s) + L_{ij}^e(s) + L_{ij}^c(s) \qquad (13)$$

where the superscripts PY, e and c refer to Percus-Yevick, electrostatic and cross term, respectively; μ is a proportionality term which is a function of Γ, ξ and α; s is a pure imaginary number ($s = ik$, where $i = (-1)^{1/2}$ and k is the modulus of the momentum transfer vector, or wave vector). For a system of uncharged particles, the electrostatic and the cross term are equal to zero and the Laplace transform of $g_{ij}(r)$ is equal to the Laplace transform of $g_{ij}^{PY}(r)$, where $g_{ij}^{PY}(r)$ is the Percus-Yevick solution(19) for the pair correlation function. We want to stress the fact that, in the limit of zero charge, the MSA and the PY theory coincide, and that at low concentration the MSA results and the HNC results are generally coincident at least for systems with low values of $\alpha(20)$. These features seem to make the MSA appealing in the treatment of complicated systems dominated by hard core interactions, but in which the charge might play some role. This is, for example, the case of micellar fluids, a field that we plan to explore in the near future. Some preliminary results have been shown in the section concerning the thermodynamic properties of electrolyte solutions. A complete study will be published later.

The two terms $L_{ij}^e(s)$ and $L_{ij}^c(s)$ are complicated but explicit functions of the same parameters as $L_{ij}^{PY}(s)$ and of the charges of the ions, and can be easily calculated with a computer capable of handling complex numbers. At this point we would like to introduce some functions of great importance in the field of the structure of liquids and amorphous materials: one of these is the "partial structure function" $h_{ij}(k)$ defined as the Fourier inverse of total pair correlation function $h_{ij}(r) = g_{ij}(r) - 1$. These functions are usually obtained, whenever possible(16,21), by means of scattering experiments. For a system composed of n different particles, there are $n(n+1)/2$ such functions, of which $n(n-1)/2$ are "cross" functions ($i \neq j$), and the rest are "self" functions ($i = j$); the total structure function, easily obtained by measuring the intensity of the beam (X-rays, neutrons, electrons, etc) scattered by the sample at different angles (or better at different values of the wave vector k), turns out to be a linear combination of such partial structure functions. In turn, these last functions and the pair correlation functions $g_{ij}(r)$ are related by a Fourier transform, so that when one of them is known, the other can be calculated. Unfortunately, up to now, there is no way to find (in the MSA), an analytical solution for the partial structure functions; as we have seen before, it is indeed possible to get the analytical solution of the Laplace transform of $g_{ij}(r)$, and this, in turn, is related to the partial structure functions $h_{ij}(k)$, so once $L_{ij}(s)$ is calculated, also $h_{ij}(k)$ is known and hence $g_{ij}(r)$ can be obtained. Ideally one has to calculate $L_{ij}(ik)$ and $L_{ij}(-ik)$; then h_{ij} will be given by

$$h_{ij}(k) = 2\pi(\rho_i\rho_j)^{1/2}\left[L_{ij}(-ik) - L_{ij}(ik)\right]/ik \tag{14}$$

In practice, as

$$L_{ij}(\pm ik) = \text{Re} \pm i\text{Im}$$

then

$$h_{ij}(k) = 4\pi(\rho_i\rho_j)^{1/2}\text{Im}/k \tag{15}$$

where Im is the imaginary part of the Laplace transform of the correlation function $g_{ij}(r)$, ρ_i and ρ_j are the number densities of particles i and j, respectively.

Another important function is the modified partial structure function $S_{ij}(k) = h_{ij}(k) + \delta_{ij}$, where δ_{ij} is the Kronecker delta function, equal to 1 for the "self" terms, and zero for the "cross" terms. In terms of the pair correlation functions $g_{ij}(r)$, the $S_{ij}(k)$'s are given by

$$S_{ij}(k) = \delta_{ij} + 4\pi(\rho_i\rho_j)^{1/2} \int_0^\infty [g_{ij}(r) - 1] j_o(kr) r^2 dr \qquad (16)$$

with $j_o(kr)$ indicating the spherical Bessel function of 0th order; the $S_{ij}(k)$'s are also related to the direct correlation functions $c_{ij}(k)$; for example, in the case of a two component system, we have

$$S_{ij}(k) = [\delta_{ij} + (-1)^{\delta_{ij}} c_{ij}(k)]/B(k) \qquad (17)$$

where

$$B(k) = [1 - c_{11}(k)][1 - c_{22}(k)] - c_{12}^2(k) \qquad (18)$$

The $S_{ij}(k)$'s are important functions whose values are related to basic thermodynamic properties. In particular, the limiting value of $S_{ij}(k)$ for $k \to 0$, is proportional to the fluctuation of the particles densities(21),

$$S_{ij}(0) = \lim_{k \to o} S_{ij}(k) = \frac{<\Delta Ni \; \Delta Nj>}{(<\Delta Ni \; \Delta Nj>)^{1/2}} \qquad (19)$$

Bathia and Thornton(22) introduced three new partial structure functions which are linear combinations of the above $S_{ij}(k)$'s. These new functions are the density-density partial structure function $[S_{nn}(k)]$, the concentration-concentration partial structure function $[S_{cc}(k)]$ and their cross term $S_{nc}(k)$. The $S_{cc}(k)$'s are also called (in the case of charged systems), charge-charge partial structure functions. In the case of a symmetric two-component system (for example molten NaCl), the three new functions will be given by

$$\rho_o S_{nn}(k) = \rho_1 S_{11}(k) + \rho_2 S_{22}(k) + 2(\rho_1\rho_2)^{1/2} S_{12}(k) \qquad (20)$$

$$\rho_o S_{cc}(k) = S_{nn}(k) - 4(\rho_1\rho_2)^{1/2} S_{12}(k) \qquad (21)$$

$$\rho_o S_{nc}(k) = \rho_o [S_{11}(k) - S_{22}(k)]/? \qquad (22)$$

where ρ_o is the total number density.

These functions are very interesting in that they are related, by simple mathematical expressions, to important thermodynamic properties, especially $S_{nn}(0)$ and $S_{cc}(0)$.

IONIC FLUIDS IN THE MEAN SPHERICAL APPROXIMATION

Results. This section on results will be divided in two parts. The first part concerns the high density regime (molten salts) and the second concerns the low density regime (electrolyte solutions).

Results for 1-1 Molten Salts. The two most important parameters influencing the partial structure functions and the correlation functions of molten salts are the radius-ratio (rr) and the plasma parameter. Abramo, et al.(23) using a numerical solution of the MSA for an MX-type fluid of charged hard spheres of different diamters, have carefully examined these aspects, for molten 1-1 salts, concentrating their attention mainly on the effect of the ion size and the radius-ratio. They also reported, in some cases, comparison with experimental results. The general agreement is fair, in some cases excellent and some generalizations are possible.

Whenever the radius-ratio is changed, the structure function $S_{ii}(k)$ of the smaller ion is seriously affected both from the quantitative and the qualitative point of view. This is shown in Figure 7 where we present our own calculations of the three partial structure functions for molten NaCl at 1148°K, obtained for two sets of radii: in case 1 we used Pauling diameters, in case 2 we used σ_+ = 2.06 Å and σ_- = 2.80 Å

With their extensive analysis of molten alkali halides, the above mentioned authors have shown that for molten salts whose ionic radii are close, the structure function is dominated by

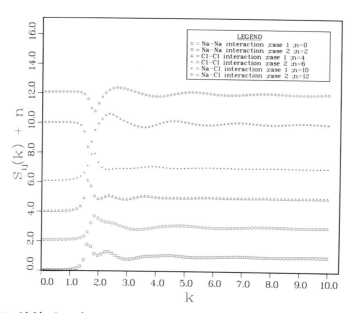

Fig. 7. Modified pair structure functions for two models of molten NaCl at 1148°K.

charge-charge correlation, while by increasing the difference in ion size, the ordering in the subsystem of smaller ions becomes progressively dominated by density-density correlations. One important parameter in the calculations is the plasma parameter. In actual calculations at a certain radius-ratio, the plasma parameter can be varied either by varying the charge on the ions, or by changing the temperature. Although the former way is less realistic, it has the main advantage that knowledge of the variation of the density with the temperature is not necessary. Besides, as the results are a function only of Γ, the ρ's and the hard sphere diameters, the scheme by which Γ is varied is immaterial. We have calculated the partial structure functions S_{Na-Na}, S_{Na-Cl}, S_{Cl-Cl} together with S_{cc} and S_{nn} for a model of NaCl at 1148°K. Results are shown in Figures 8-11. For these calculations we have used Pauling radii for both Na^+ and Cl^-; the charge of the ions has been increased gradually form 0% to 100% of the real charge; correspondingly the plasma parameter varies from 0 to approximately 50. As can be noted the charges hardly change the shape of the density-density structure function (Figure 8), while there is a strong effect on the charge-charge structure (Figure 9). A very interesting fact can be noted if the three $S_{ij}(k)$'s for charge = 0 (Figure 10) and for 1% of the full charge (Figure 11) are compared. (It should be remembered that in the limit of zero charge the solution of the MSA is coincident with the PY equation for a mixture of hard spheres with additive diameters)

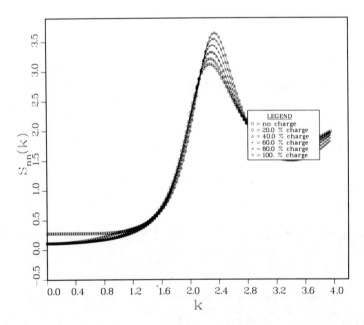

Fig. 8. Density-density structure functions for molten NaCl at 1148°K for different values of the plasma parameter (see text for details).

IONIC FLUIDS IN THE MEAN SPHERICAL APPROXIMATION 57

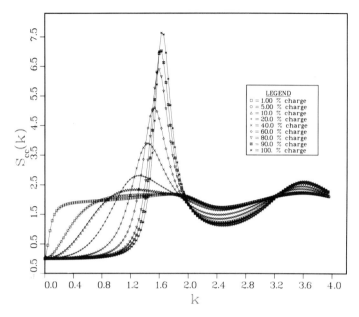

Fig. 9. Charge-charge structure functions for molten NaCl at 1148°K, for different values of the plasma parameter.

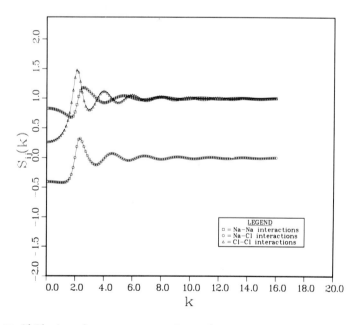

Fig. 10. Modified pair structure functions for molten NaCl at 1148°K for the value of plasma parameter equal to zero.

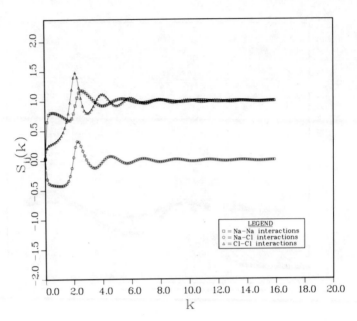

Fig. 11. Modified pair structure functions for molten NaCl at 1148°K for a small value of the plasma parameter ($\cong 0.005$).

The difference between a given $S_{ij}(k)$ in the charged system (MSA) and in the uncharged system (MSA = PY) are negligible in all the k range examined, except in the long wavelength limit ($k \to 0$). The effect of switching on the charges (even if only 1% of one electron unit), is to drastically reduce the particle fluctuations relative to the uncharged system. Hence, according to equation (19) all the $S_{ij}(0)$'s of the system will tend to assume values close to zero. This is due to the fact that to induce a fluctuation in a charged system it is necessary to work against the strong Coulombic potential, while it is relatively easy to induce concentration fluctuations in a system of uncharged hard spheres.

Results of Mixtures of 1-1 Molten Salts. We are not aware of any MSA calculation for mixtures of molten salts. One of us has recently completed(24) calculations of the partial structure functions $S_{ij}(k)$ for various LiBr-KBr mixtures, for which molecular dynamics simulations were available(25).

Some results are shown in Figure 12 and Figure 13; the various $S_{ij}(k)$'s are shifted along the Y axis for reasons of clarity. We present here only one "self" term [S_{Br-Br} -Figure 12] and one "cross" term [S_{K-Br} - Figure 13]. As far as the Br-Br interactions are concerned, the S_{ij}'s in pure molten salts are completely different from the corresponding S_{ij}'s in mixtures of LiBr and KBr.

IONIC FLUIDS IN THE MEAN SPHERICAL APPROXIMATION

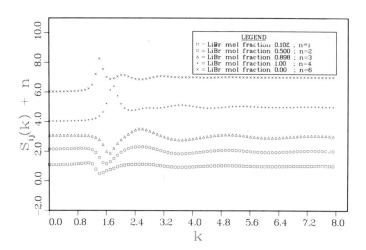

Fig. 12. Br-Br modified structure functions for mixtures of molten LiBr and KBr at 1020°K.

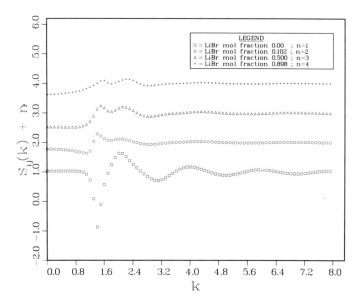

Fig. 13. K-Br modified structure functions for mixtures of molten LiBr and KBr at 1020°K.

While this is generally true for all the cross terms[†], it does not hold for the other "self" terms [namely S_{Li-Li} and S_{K-K}]. We can then conclude that the behavior just shown, is related to the variation of the cation-anion interactions caused by the composition variation. This point of view agrees also with the calculated(25) and experimental(29) dynamic properties of these mixtures. We do not present here any correlation functions for the systems under examination. In fact, we have found that by Fourier inversion of the $S_{ij}(k)$'s so far calculated, we obtain $g_{ij}(r)$'s which show periodic spurious peaks affecting the main peak and the long range tail of the correlation functions. This seems to be due(30) to the discontinuities of the direct correlation functions. We are currently writing a code to perform the necessary corrections. Here we present only the contact correlation functions that can be calculated analytically(8). The results of such calculations are shown in Table IV. It can be noted that MSA has a tendency to understimate the contact terms for the smallest ions in the systems (Li^+ and K^+ in the mixtures, Li^+ in pure LiBr and K^+ in pure KBr); this fact is a weakness of the MSA and has been pointed out in other similar cases(8,23,26,27).

Results for Asymmetric Molten Salts. To the best of our knowledge, only two papers have been published concerning asymmetric molten salts in the MSA(26,27). In one case(27), it has been possible to compare the MSA results with previous MonteCarlo (MC) calculations(28) of De Leeuw on CaF_2 and $SrCl_2$ at 1773°K. For each salt, two sets of radii have been used in the calculations, following the suggestion of Abramo et al.(23). The overall agreement between MSA $g_{ij}(r)$ (and the corresponding coordination numbers) and MC $g_{ij}(r)$ is fair; evidently the agreement improves when the radius-ratio used in the MC calculations is the same as that used for the MSA calculations.

Figure 14 shows the three structure functions for CaF_2 at 1773°K calculated using one set of ionic radii ($\sigma_+ = 1.94$ Å, $\sigma_- = 2.18$ Å) taken from reference(27). As can be noted, the first minimum in the Ca-F interaction function falls at a value of k equal to the one where the first maximum of the Ca-Ca and F-F interactions are located (~ 2.2 Å^{-1}). This condition is typical of fluids with strong coulombic interaction; in fact according to equation (21) the charge-charge structure function $S_{cc}(k)$ will show a sharp peak at ~ 2.2 Å^{-1} due to the reinforcement of the three structure functions.

[†] Of the three "cross" terms [S_{Li-Br}, S_{K-Br} and S_{Li-K}], only the first two can be calculated for both the pure components and their mixtures, for obvious reasons.

IONIC FLUIDS IN THE MEAN SPHERICAL APPROXIMATION

Results for Electrolyte Solutions. Aqueous $NiCl_2$ solutions are of particular interest because elastic neutron scattering data (31), as well as spectroscopic information(32) have been interpreted assuming that the ions, in concentrated solutions, are charge oriented(31). Other models assume formation of inner sphere chloride complexes, ruled out by a recent EXAFS investigation(33). Quirke and Soper(34), on the other hand, argued that a neutral hard spheres model might describe the Ni-Ni structure function, and indeed have been able to reproduce, qualitatively, the general features of $S_{NiNi}(k)$, using the PY equation. We have redone these calculations,

Table IV. List of $g_{ij}(\sigma_{ij})$ for Molten LiBr - KBr Mixtures at $1020°K$†

i	j	Mole Fraction of LiBr				
		1.0	0.898	0.500	0.102	0.00
1	1	-8.81	-9.18	-10.8	-12.7	
1	2	7.08	-3.16	-3.98	-4.94	
1	3		7.29	8.19	9.21	
2	2	1.21	-0.27	-0.74	-1.30	-1.47
2	3		5.34	5.80	6.36	6.51
3	3		1.12	0.69	0.17	0.02

† 1, 2 and 3 refers to Li, K and Br, respectively.

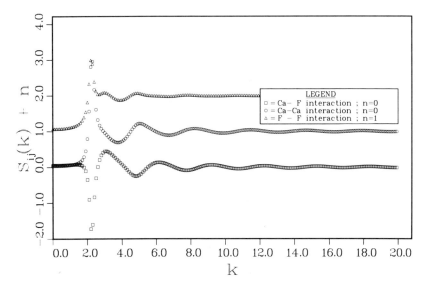

Fig. 14. Modified pair structure functions for molten CaF_2 at $1773°K$.

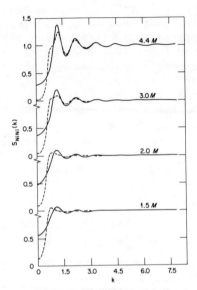

Fig. 15. PY (———) vs MSA (-- --) modified Ni-Ni structure function for aqueous $NiCl_2$ solutions at 298.16°K.

but we also included the case of charged hard spheres. Results of these calculations are shown in Figure 15. As can be noted, for values of k greater than $\sim 3\text{Å}^{-1}$ the PY results (uncharged particles) and the MSA results (charged particles) are, for all practical purposes, indistinguishable. For k smaller than $\sim 3\text{Å}^{-1}$ there are small but significant differences between the charged and the uncharged hard sphere structure functions: first, due to the fact that the strong coulombic interaction reduces the particles fluctuations in the charged system (when compared with the uncharged), the MSA $S_{NiNi}(0)$'s are always smaller than the corresponding PY $S_{NiNi}(0)$'s; in addition to this, while the PY $S_{NiNi}(k)$'s show only one peak whose position is essentially concentration independent ($k_o \simeq 1 \text{Å}^{-1}$), in the case of the charged system, for concentrations below ~ 2 M, there is one peak (0.7 - 0.8 Å^{-1}) which, for concentrations above ~ 2 M develops into a pre-peak followed by the main peak centered at $\sim 1 \text{Å}^{-1}$. For low concentrations the position of the pre-peak coincides with the experimental one(31), while for M greater than ~ 2, its position becomes essentially concentration independent. In other words it seems that, for concentrations up to ~ 2M, the charges play a fundamental role in the structural properties of aqueous $NiCl_2$ solutions, while for concentrations greater than 2M also the hard core repulsions become important, in determining the structural properties. [Another explanation of the "peculiar" concentration behavior of the pre-peak might be the fact that at these high concentrations, the MSA does not describe well the structural properties of the above mentioned solutions]. Whatever the explanation, we think we can agree with Quirke and Soper(34) that in order to explain the general features of $S_{NiNi}(k)$ there is no need to introduce the concept of quasi-lattice, which, on the other hand, seems necessary to explain both spectroscopic and inelastic neutron scattering experiments.

4. CONCLUSION

The MSA, despite the fact that it is the lowest order approximation in the optimized cluster series(8,35), can be a very useful tool for modern chemists. Its key feature is the fact that it can be solved analytically, and improved by including corrections and the use of more realistic Hamiltonian models(8,9); in addition both the thermodynamic and the structural properties can be expressed in terms of a scaling parameter Γ (in a sense analogous to the DH parameter κ). When used as an empirical theory, the MSA has been shown to be able to describe surprisingly well the thermodynamic properties of solutions of symmetric and asymmetric electrolytes at least up to concentrations $\sim$2 M. When high density ionic fluids (molten salts) are considered, the agreement between experimental or "quasi-experimental" (MC and/or MD) structure functions and those obtained within the framework of the MSA, is quite good, especially when the ionic radii are selected according to Abramo et al.(23).

Acknowledgments. Professor L. Blum (University of Puerto Rico), Dr. E. Johnson (ORNL), and Dr. R. E. Meyer (ORNL) critically read the manuscript; interaction with them has been fruitful and very rewarding. We want to thank the Italian CRRNSM for a grant used for some of the calculations presented here and done at the Computer Centre of the University of Palermo (Italy). Finally, we want to thank the organization committee of the Fifth I.U.P.A.C International Symposium on Solute-Solute-Solvent Interactions for inviting this paper.

REFERENCES

1. W. G. McMillan and J. E. Mayer, J. Chem. Phys. 13:276 (1945).
2. P. Debye and E. Hückel, Phys. Z. 24:185 (1923).
3. (a) S. R. Milner, Phil Mag. 23:551 (1912); Phil. Mag. 25:742 (1913). (b) L. Gouy, Ann. Chim. Phys. 8:291 (1906); J. Physique, 9:457 (1910); Ann. Phys. 7:129 (1917). (c) D. L. Chapman, Phil Mag. 25:457 (1913).
4. One of the versions of DH theory used a model of charged points. It has been proven that such a system is not thermodynamically stable unless a short range repulsion is provided (H. A. Kramers, Proc. Royal Acad. Amsterdam, 30:145 (1927); E. Lieb and J. A. Lebowitz, Adv. Math. 9:316 (1972)). Any textbook of physical chemistry or electrochemistry is a good reference for the different versions of the DH theory.

5. (a) R. Triolo, R. Grigera, and L. Blum, J. Phys. Chem. 80:1858 (1976). (b) H. C. Andersen, in "Modern Aspects of Electrochemistry", 11:1 (1975). (c) J. C. Rasaiah, J. Sol. Chem. 2:301 (1973).
6. H. L. Friedman, J. Sol. Chem. 1:387 (1972).
7. (a) G. Accardi, Thesis, Univ. of Palermo (Italy). (b) R. Triolo, unpublished data.
8. L. Blum, in "Theoretical Chemistry; Advances and Perspective", H. Eyring and D. Henderson ed. vol. 5, Academic Press, in press.
9. H. L. Friedman and W. D. T. Dale, in "Modern Theoretical Chemistry", B. J. Berne ed. vol. VI, Plenum, New York (1977).
10. For one component systems, a derivation of (4) using physical intuition was given by Ornstein and Zernicke, Proc. Akad. Sci. Amsterdam, 17:703 (1914), and a rigorous mathematical derivation was given by Percus, Bull. Am. Phys. Soc. 7:407 (1962); the generalization of the one-component system equation to mixtures can be easily done by using an interpretation of Hiroike and Morita, J. Chem. Phys. 52:5489 (1970). An excellent reference on the equilibrium theory of classical fluids is a lecture note and reprint series edited by H. L. Frish and J. L. Lebowitz, "The Equilibrium Theory of Classical Fluids", W. A. Benjamin, Inc., Publ. New York (1964).
11. (a) L. Blum and H. Tibavisco, unpublished results.
    (b) L. Blum, J. Chem. Phys. 61:2129 (1974); Mol. Phys. 30:1529 (1975); J. Stat. Phys. 19:201 (1978); J. Stat. Phys. 18:451 (1978).
12. (a) R. Triolo, M. A. Floriano, and L. Blum, J. Chem. Phys. 67:5956 (1977). (b) R. Triolo, L. Blum, and M. A. Floriano, J. Phys. Chem. 82:1368 (1978). (c) R. Triolo, M. A. Floriano, I. Buffo, and L. Blum, Ann. Chim. 67:433 (1977).
13. R. A. Greeley, W. T. Smith, Jr., M. H. Lietzke, and R. W. Stoughton, J. Phys. Chem. 64:1445 (1960).
14. F. H. Spedding, H. O. Weber, V. W. Saeger, H. H. Petheram, J. A. Rard, and A. Habenschuss, J. Chem. Eng. Data, 21:341 (1976).
15. D. Elkoubi, P. Turq, and J. P. Hansen, Chem. Phys. Lett. 52:493 (1977).
16. While this paper was being written, E. Johnson using the three particles O.Z. equation, has derived a new method, J. Chem. Phys. 72:3010 (1980), that will allow the calculations of pair and triplet correlation functions of classical systems, including ionic solutions and molten salts. This method, being rigorous, does not contain any approximation and should give "better"

    results than HNC. However, so far, the method has not yet been applied to charged systems.
17. R. Triolo and A. D'Aprano, Gazz. Chim. Ital. 102:0000 (1979).
18. L. Blum and J. A. Høye, J. Phys. Chem. 81:1311 (1977).
19. J. K. Percus and G. Yevick, Phys. Rev. 110:1 (1958); Phys. Rev. 118:290 (1944).
20. J. C. Rasaiah, D. V. Card, and J. P. Vallean, J. Chem. Phys. 56:248 (1972).
21. (a) J. L. Beeby, J. Phys. C. Solid State Phys. 6:2262 (1973). (b) N. H. March and M. P. Tosi, "Atomic Dynamics in Liquids", McMillan, New York (1976).
22. (a) A. B. Bathia and D. E. Thornton, Phys. Rev. B2:3004 (1970). (b) A. B. Bathia and J. K. Ratti, Phys. Chem. Lig. 6:201 (1976).
23. (a) M. C. Abramo, C. Caccamo, G. Pizzimenti, M. Parrinello, and M. P. Tosi, J. Phys. C. Solid State Phys. 9:L593 (1976); J. Chem. Phys. 68:2889 (1978).(b) M. C. Abramo, C. Caccamo and G. Pizzimenti, Phys. Chem. Liq. 6:167 (1977).
24. R. Triolo, in preparation.
25. F. Lantelme and P. Turq, Mol. Phys. 38:1003 (1979).
26. L. B. Bhuiyan, N. H. March, M. C. Abramo, C. Caccamo, and G. Pizzimenti, Mol. Phys. 34:755 (1977).
27. L. B. Bhuiyan, Mol. Phys. 38:1737 (1979).
28. S. W. DeLeeuw, Mol. Phys. 36:103 (1978).
29. (a) M. Chelma, F. Lantelme, and O. P. Mehta, J. Chim. Phys. Suppl. 136 (1969). (b) J. Perie, M. Chelma, and M. Gignoux, Bull. Soc. Chem. Fr. 1249 (1961).
30. E. Johnson, private communication.
31. J. E. Enderby, Proc. Royal Soc. Lond. 345:107 (1975).
32. M. P. Fontana, G. Maisano, P. Migliardo, and F. Wanderlingh, Sol. State Comm. 23:489 (1977).
33. P. Lagarde, A. Fontaine, D. Raoux, A. Sadoc, and P. Migliardo, J. Chem. Phys. 72:3061 (1980).
34. N. Quirke and A. K. Soper, J. Phys. C. 10:1802 (1977).
35. H. C. Andersen and D. Chandler, J. Chem. Phys. 57:1818 (1972).

A COMPARISON BETWEEN STRUCTURES OF AQUA AND AMMINE COMPLEXES IN SOLUTIONS AS STUDIED BY AN X-RAY DIFFRACTION METHOD

Hitoshi Ohtaki

Department of Electronic Chemistry
Tokyo Institute of Technology at Nagatsuta
Nagatsuta-cho, Midori-ku, Yokohama, 227 Japan

Abstract - The structures of aqua and ammine complexes of transition metals in aqueous solutions were determined by the X-ray diffraction method at 25°C. The structures of some ethylenediamine complexes were also studied under the same experimental conditions. All of the aqua complexes studied, except the silver(I) complex, have six water molecules in their first hydration sphere. Silver(I) ion has only two water molecules in the linear form. In ammoniacal aqueous solutions, zinc(II) ion combines with at most four ammonia molecules to form the tetrahedral complex. The tetraamminecopper(II) complex has two more water molecules at the apices of a distorted octahedron. A higher-ammine complex of copper(II) ion having more than four ammonia molecules within the complex is formed in an almost saturated aqueous ammonia solution. Distortion of the octahedral structure of the higher-ammine complex is pronounced compared with that of the tetraamminediaqua complex. The diamminesilver(I) complex has a much shorter Ag-NH$_3$ bond than the Ag-OH$_2$ bond within the aqua complex. Structures of bis- and tris(ethylenediamine) complexes of Ni^{2+}, Cu^{2+}, Zn^{2+} and Cd^{2+} ions were determined and the bond lengths within these complexes were compared with those found for the aqua and ammine complexes. A brief discussion was made for the chelate effect of ethylenediamine molecules from the structural point of view.

1. INTRODUCTION

In these decades huge accumulations for thermodynamic data have been achieved for complex formation reactions of metal ions in solution, and structures of a large number of metal complexes have been investigated by various methods. Nevertheless, interpretations of thermodynamic and kinetic data in connection with structural information of the complexes are usually qualitative, because of insufficiency of a quantitative knowledge for structures of the complexes, e.g., atomic distances and charge distribution within the complexes, potential functions of atom pairs in the systems, and molecular arrangements in the solution phase. Structural data found in the solid phase are usually used, if necessary, by assuming that structures of chemical species in the liquid phase are the same as those in the solid state. It is obvious, however, that the assumption is too simple, and sometimes even erroneous.

X-Ray diffraction methods, as well as X-ray absorption and neutron diffraction techniques, which have been applied to crystallographic investigations, are very useful for exploring molecular structures and molecular arrangements in solution.

In this paper, we describe structures of the aqua and ammine complexes of some metal ions determined by the X-ray diffraction method, and variations of bond lengths within the complexes are discussed. Some results for structures of ethylenediamine complexes of metal ions are included, and a brief discussion is made for the chelate effect of ethylenediamine molecules.

2. EXPERIMENTAL

Reagents. Aqueous solutions of silver(I) and divalent transition metal perchlorate were used for investigating the structures of the hydrated metal ions(1-4). Copper sulfate(1,2), zinc sulfate(1) and silver(I) nitrate(3) were also used in the structural studies of the aqua metal complexes. A small amount of acid was added to the solutions to prevent hydrolysis of the metal ions. Indium(III) perchlorate was used for the study of the structural determination of the aqua indium(III) complex(5), but no further investigation has been carried out for other complexes except the hexachloroindate complex(6), for which we will not describe here.

Metal chlorides or nitrates were dissolved in ammoniacal aqueous solutions. Structures of ammine complexes thus formed of copper(II)(7), zinc(II)(8), cadmium(II)(9) and silver(I)(3) ions were determined.

Ethylenediamine solutions of nickel(II)(10), copper(II)(10), zinc(II)(11) and cadmium(II)(12) nitrates were used for determination of structures of the bis- and tris(ethylenediamine) complexes of the relevant metal ions.

All the chemicals were of reagent grade and the metal salts were recrystallized at least twice from water. Concentration of the metal ions in the samples were analyzed by two different methods. The composition of the test solutions is listed in Table I.

Table I. Composition of sample solutions (mol dm^{-3}).

a) Aqueous solutions

	Mn(ClO$_4$)$_2$	Fe(ClO$_4$)$_2$	Co(ClO$_4$)$_2$	Ni(ClO$_4$)$_2$	Cu(ClO$_4$)$_2$	Zn(ClO$_4$)$_2$
M	2.612	2.173	2.672	2.491	3.550	2.886
Cl	5.818	4.499	5.730	5.430	7.537	6.170
O	65.93	64.10	66.66	66.61	69.86	67.65
H	85.48	92.36	88.00	90.23	81.12	86.27

	CuSO$_4$	ZnSO$_4$	Cd(ClO$_4$)$_2$	AgClO$_4$	AgNO$_3$	In(ClO$_4$)$_3$
M	1.368	2.797	2.920	4.235	3.453	3.023
S	1.371	2.809	-	-	-	-
Cl	-	-	6.303	4.336	-	9.388
O	60.51	64.87	66.02	61.50	59.44	71.61
N	-	-	-	-	3.453	-
H	110.1	107.1	81.15	88.42	98.16	68.44

b) Ammonia solutions

	CuCl$_2$ A	CuCl$_2$ B	CuCl$_2$ C	ZnCl$_2$ A	ZnCl$_2$ B	ZnCl$_2$ C	CdCl$_2$	AgNO$_3$
M	2.108	2.180	1.541	2.679	2.364	2.217	1.336	3.967
Cl	4.217	4.360	3.083	5.359	4.727	4.434	2.671	-
O	39.60	39.91	30.70	37.99	32.30	32.61	36.31	45.58
N	10.21	10.87	17.30	10.09	11.79	14.66	12.53	14.08
H	109.9	112.4	113.3	106.3	99.96	109.2	110.2	97.69
$\frac{NH_3}{M}$	4.84	4.98	11.23	3.77	4.99	6.61	9.38	2.55

c) Ethylenediamine solutions

	Ni(NO$_3$)$_2$ A	Ni(NO$_3$)$_2$ B	Ni(NO$_3$)$_2$ C	Ni(NO$_3$)$_2$ D	Cu(NO$_3$)$_2$ A	Cu(NO$_3$)$_2$ B	Zn(NO$_3$)$_2$ A	Zn(NO$_3$)$_2$ B
M	1.476	1.356	1.717	1.787	1.460	1.230	1.846	1.295
O	42.69	42.34	50.13	50.09	49.73	37.82	47.11	49.28
N	13.79	13.46	10.54	10.34	9.011	15.10	15.52	8.422
C	10.84	10.75	7.113	6.762	6.091	12.64	11.83	5.832
H	111.0	111.4	106.1	92.27	106.3	111.5	119.4	106.4
$\frac{en}{M}$	3.67	3.96	2.07	1.89	2.09	5.14	3.20	2.25

Table I. Composition of sample solutions (mol dm^{-3}), (continued).

	Cd(NO$_3$)$_2$	
	A	B
M	1.916	1.760
O	38.03	46.95
N	16.81	11.22
C	12.98	7.702
H	105.0	103.6
$\frac{en}{M}$	3.39	2.19

<u>Method of Measurements</u>. All the measurements were carried out in a thermostated room at 25°C by the use of a θ-θ type X-ray diffractometer (JEOL Co., Tokyo). Monochromatized MoKα (λ = 0.7107 Å) X-ray beam was used for the diffraction measurements. The apparatus was equipped with an LiF crystal with a curve surface and a scintillation counter with a pulse-height analyzer. Time required to accumulate 40000 counts at each angle was recorded. The whole angle range was scanned twice. The observed range of scattering angle (2θ) was 2° to 140°, corresponding to the range 0.31 Å^{-1} < s < 16.6 Å^{-1} [†] (s = $4\pi\lambda^{-1}\sin\theta$).

<u>Treatment of Intensity Data</u>. After corrections for back-ground, polarization and absorption in the sample, the measured intensities were scaled to absolute intensities by the conventional methods. The coherent and incoherent scattering factors of the atoms in the sample were quoted from the literature (see each reference).

The radial distribution function D(r) was calculated by means of the Fourier transform according to Equation (1).

$$D(r) = 4\pi r^2 \rho_0 + 2r\pi^{-1} \int_0^{s_{max}} s \cdot i(s) M(s) \sin(rs) ds \qquad (1)$$

where the reduced intensities i(s) are given as

$$i(s) = I(s) - \sum_i n_i \{(f_i(s) + \Delta f_i')^2 + (\Delta f_i'')^2 + \Phi(s) I_i^{inco}(s)\} \qquad (2)$$

The theoretical scattering intensities due to atom-pairs of all possible combinations in the system were given as follows:

$$i_{calc}(s) = \sum_i \sum_{\substack{j \\ i \neq j}} n_{ij} \{(f_i(s) + \Delta f_i')(f_j(s) + \Delta f_j') + (\Delta f_i'')(\Delta f_j'')\}$$

$$\cdot \frac{\sin(r_{ij}s)}{(r_{ij}s)} \cdot \exp(-b_{ij}s^2) \qquad (3)$$

[†] 1 Å = 100 pm.

STRUCTURES OF AQUA AND AMMINE COMPLEXES IS SOLUTION 71

From shapes and positions of peaks in the radial distribution curve obtained, approximate values of distances between atoms i and j (atomic distances r_{ij}), numbers of atom-pairs (frequency factors n_{ij}) and mean amplitudes of the bonds (temperature factors $b_{ij} = <\Delta r_{ij}^2>/2$) were determined, and the values were finally optimized by the least-squares method so as to minimize the sum $U = \Sigma s^2 \{i(s) - i_{calc}(s)\}^2$.

Symbols. Symbols used in the above descriptions are as follows: ρ_0; the bulk electron density in the stoichiometrically unit volume, s_{max}; the maximum value of s accessible in the experiments, i(s); the reduced intensity, M(s); $\{\Sigma n_i f_i^2(0)/\Sigma n_i f_i^2(s)\}\exp(-ks^2)$, n_i; the number of atom i in the stoichiometric volume containing one metal atom, k; damping factor, f_i; the scattering factor of atom i, $\Delta f_i'$; the real part of the anomalous dispersion, $\Delta f_i''$; the imaginary part of the anomalous dispersion, I(s); the scaled intensity after correction for back-ground, polarization and absorption in the sample, $\Phi(s)$; the fraction of the total incoherent scattering reaching the counter, $I_i^{inco}(s)$; the incoherent scattering of atom i.

3. RESULTS AND DISCUSSION

Radial distribution curves and structural intensities multiplied by s (i.e., s·i(s)) were analyzed by the method described in each reference. Details of the analytical procedures were described in the previous papers(1-8,11,12). Bond distances within the complexes studied are summarized in Tables II - IV.

Aqua complexes. It is obvious from the results in Table II that the metal-water bond lenghts within the octahedral complexes of aqua divalent-metal ions in solution change in a manner as expected from

Table II. The structures and the bond distances within aqua complexes.

Metal	$M(H_2O)_n^{z+}$	$r_{M-OH_2}/Å$	Metal	$M(H_2O)_n^{z+}$	$r_{M-OH_2}/Å$
Mn^{2+}	$Mn(H_2O)_6^{2+}$	2.20[1]	Zn^{2+}	$Zn(H_2O)_6^{2+}$	2.08[1]
Fe^{2+}	$Fe(H_2O)_6^{2+}$	2.12[1]	Cd^{2+}	$Cd(H_2O)_6^{2+}$	2.31[4]
Co^{2+}	$Co(H_2O)_6^{2+}$	2.08[1]	Ag^+	$Ag(H_2O)_2^+$	2.41[3]c
Ni^{2+}	$Ni(H_2O)_6^{2+}$	2.04[1]			2.45[3]d
Cu^{2+}	$Cu(H_2O)_6^{2+}$	$\{1.94^{1,2)a}, 2.43\}$ $\{1.94^{1,2)b}, 2.38\}$	In^{3+}	$In(H_2O)_6^{3+}$	2.15[5]

[a]Copper perchlorate, [b]Copper sulfate, [c]Silver perchlorate, [d]Silver nitrate.

Table III. The structures and the bond distances within ammine complexes.

Metal	Structure	$r_{M-L}/\text{Å}$
Cu^{2+}	$Cu(NH_3)_4(H_2O)_2^{2+}$	$\begin{cases} 2.03 \text{ (Cu-N)}[7] \\ 2.33 \text{ (Cu-O)} \end{cases}$
	$Cu(NH_3)_5(H_2O)^{2+}$ or $Cu(NH_3)_6^{2+}$	$\begin{cases} 1.93 \text{ (Cu-N(eq))}[7] \\ 2.30 \text{ (Cu-N(ax) and/} \\ \text{or Cu-O(ax))} \end{cases}$
Zn^{2+}	$Zn(NH_3)_4^{2+}$	2.03 (Zn-N)[8]
	$ZnCl(NH_3)_3^+$	$\begin{cases} 2.00 \text{ (Zn-N)}[8] \\ 2.30 \text{ (Zn-Cl)} \end{cases}$
Cd^{2+}	$Cd(NH_3)_6^{2+}$	2.37 (Cd-N)[9]
Ag^+	$Ag(NH_3)_2^+$	2.22 (Ag-N)[3]

Table IV. The structures and the bond distances within ethylenediamine complexes.

Metal	Structure	$r_{M-L}/\text{Å}$
Ni^{2+}	$Ni(H_2O)_2(en)_2^{2+}$	$\begin{cases} 2.10 \text{ (Ni-N, Ni-O)}[10] \\ 2.82 \text{ (Ni}\cdots\text{C)} \end{cases}$
	$Ni(en)_3^{2+}$	$\begin{cases} 2.20 \text{ (Ni-N)}[10] \\ 2.88 \text{ (Ni}\cdots\text{C)} \end{cases}$
Cu^{2+}	$Cu(H_2O)_2(en)_2^{2+}$	$\begin{cases} 1.93 \text{ (Cu-N(eq))}[10] \\ 2.92 \text{ (Cu-O(ax))} \\ 2.89 \text{ (Cu}\cdots\text{C)} \end{cases}$
	$Cu(en)_3^{2+}$	$\begin{cases} 1.92 \text{ (Cu-N(eq))}[10] \\ 2.22 \text{ (Cu-N(ax))} \\ 2.83 \text{ (Cu}\cdots\text{C)} \end{cases}$
Zn^{2+}	$Zn(en)_2^{2+}$	$\begin{cases} 2.13 \text{ (Zn-N)}[11] \\ 2.89 \text{ (Zn}\cdots\text{C)} \end{cases}$
	$Zn(en)_3^{2+}$	$\begin{cases} 2.28 \text{ (Zn-N)}[11] \\ 3.00 \text{ (Zn}\cdots\text{C)} \end{cases}$
Cd^{2+}	$Cd(en)_2^{2+}$	$\begin{cases} 2.34 \text{ (Cd-N)}[12] \\ 3.04 \text{ (Cd}\cdots\text{C)} \end{cases}$
	$Cd(en)_3^{2+}$	$\begin{cases} 2.37 \text{ (Cd-N)}[12] \\ 3.18 \text{ (Cd}\cdots\text{C)} \end{cases}$

the ligand field theory, that is, the M^{2+}-OH_2 bond length decreases with an increase in the atomic number from Mn^{2+} to Ni^{2+} and then increases to Zn^{2+} after passing through a minimum which appears at Cu^{2+} which has a distorted structure due to the Jahn-Teller effect.

Thus it is found that the observed hydration energies of these ions (13) are closely related to the change in the bond lengths.

An application of a simple electrostatic theory examined by Garrick(14), followed by Basolo and Pearson(15), to these ions leads to the following relation:

$$\Delta H_b^\circ = -6z(\mu_o + \mu_i)/r^2 + 6A(\mu_o + \mu_i)^2/(2r)^3 + 6\mu_i^2/(2\alpha) + 6B/r^9 \quad (4)$$

where z denotes the charge of the metal ion, μ_o and μ_i the permanent and induced dipole moments of water molecules, respectively, α the polarizability of water molecules (1.43 Å^3), A a geometrical factor which is 2.37 for the regular octahedron, and B is a constant. μ_i can be evaluated by the following equation:

$$\mu_i = (\mu_o + \alpha z r^{-2})/\{1 + \sum_{p=1}^{p=N-1} \alpha(1 + \cos^2\Theta_p)R_{pq}^{-3}\} - \mu_o \quad (5)$$

where Θ_p represents an angle metal-H$_2$O(p)-H$_2$O(q) and R_{pq} is the distance between the water molecules p and q in the first coordination sphere of the aqua complex.

For an octahedral complex deformed along the axis with the ratio of the two different metal-water bond distances, J = r(ax)/r(eq), equation (5) may be modified by introducing ω for the angle metal-H$_2$O(eq)-H$_2$O(ax) as follows:

$$\mu_i(ax) = (\mu_o + \alpha z J^{-2} R_{pq}^{-2})/[1 + \alpha\{4(\sin\omega\cdot\sin(90°-\omega))$$
$$+ 2\cos\omega\cdot\cos(90°-\omega))(1 + J^2)^{-3/2}R_{pq}^{-3} + (4J^3 R_{pq}^3)^{-1}\}] - \mu_o \quad (6)$$

for the induced dipole moment of a water molecule on the vertical (deformed) axis, and

$$\mu_i(eq) = (\mu_o + \alpha z R_{pq}^{-2})/[1 + \alpha\{(1 + \cos^2(45°))(\sqrt{2}R_{pq})^{-1} + (4R_{pq}^3)^{-1}$$
$$+ 2(\sin\omega\cdot\sin(90°-\omega) + 2\cos\omega\cdot\cos(90°-\omega))(1 + J^2)^{-3/2}R_{pq}^{-3}\}]$$
$$- \mu_o \quad (7)$$

for the induced dipole moment of a water molecule on the equatorial position.

The calculated bond energies ΔH_b° of the hydrated divalent transition metal ion change in parallel with the hydration energies of the ions, ΔH_h°(13) (see Figure 1). The difference between the two curves is ca. 500 kJ mol^{-1}, independent of the ions, which corresponds to the energy of transfer of a doubly charged ion having an ionic radius of about 3.5 Å ($\simeq r_{metal} + 2r_{water}$, r: radius) from vacuum

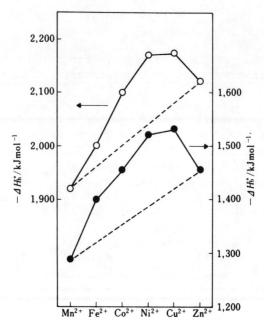

Fig. 1. Comparison between the hydration energies, $-\Delta H_h^\circ$, and calculated bond energies, $-\Delta H_b^\circ$, for Mn^{2+}, Fe^{2+}, Co^{2+}, Ni^{2+}, Cu^{2+} and Zn^{2+}. Open symbols represent $-\Delta H_h^\circ$ and filled ones $-\Delta H_b^\circ$.

to a continuous medium with a relative dielectric constant of 80 at 25°C(1).

In these calculations covalent interactions such as charge transfer from the ligand to a central metal ion are not sufficiently taken into account. Nevertheless, the parallelism found between ΔH_h° observed and ΔH_b° thus calculated may show that the crystal field stabilization energies of the complexes are substantially reflected to the changes in the bond lengths between the central metal ions and the ligand water molecules.

<u>Ammine complexes</u>. For ammine complexes, situations are more complicate than those in the case of aqua complexes, because the coordination numbers of metal ions sometimes change depending on the number of ammonia molecules coordinated.

Zinc(II) ion forms the tetraammine complex in an aqueous solution containing a large amount of ammonia, but the hexaamminezinc(II) complex is never formed in an aqueous medium at room temperature. The $Zn-NH_3$ bond length within the tetrahedral complex is 2.03 Å(8) (see Table III), which is slightly shorter than the $Zn-OH_2$ bond within the aqua zinc(II) ion, the former ligand having a similar size to that of the latter. The reason why the $Zn-NH_3$ bond is shorter than the Zn-OH$_2$ bond may partly be attributed to a stronger covalent bonding of ammonia molecules than water toward zinc(II) ion, but a structural change from an octahedral in the aqua zinc(II) complex to a tetrahedron in the tetraamminezinc(II) complex may also contribute to the shortening(16). The triamminemonochlorozinc(II) complex is formed in a zinc(II) chloride solution of an ammoniacal medium, which has a tetrahedral structure with the Zn-NH$_3$ and Zn-Cl bond distances of 2.00 Å and 2.30 Å, respectively(8).

The hexaamminecadmium(II) complex (the Cd-NH$_3$ distance is 2.37 Å(9)) has even a slightly longer bond distances than the hexaaquacadmium(II) complex (the Cd-OH$_2$ distance is 2.31 Å), although the enthalpy change of the complex formation is largely negative ($\Delta H° = -88$ kJ mol^{-1}).

The bond length between Ag$^+$ and NH$_3$ is remarkably shorter than that between Ag$^+$ and H$_2$O. The diamminesilver(I) complex, which is linear, has the Ag-NH$_3$ bond of 2.22 Å, while the Ag-OH$_2$ bond length within the diaquasilver(I) ion is 2.41 - 2.45 Å(3). The soft metal ion may react with ammonia molecules to form stronger and thus, shorter covalent bonds than with water molecules.

In copper(II) ammine complexes(7), the Cu-NH$_3$ bond length changes in a complicate manner with changing number of ammonia molecules coordinated. The tetraamminecopper(II) complex has a distorted octahedral structure in which the bond length between copper(II) ion and an ammonia molecule situated at the equatorial position is 2.03 Å, the bond length being longer by 0.09 Å than the Cu-OH$_2$ bond within the aqua copper(II) complex. On the other hand, the Cu-OH$_2$(ax) bond within the tetraamminecopper(II) complex shortens to 2.33 Å from 2.43 Å of the Cu-OH$_2$(ax) bond within the aqua complex. The ratio $J = r(ax)/r(eq)$ of the tetraammine complex is 1.15, while the value of the aqua complex is 1.25. Since it is generally accepted that coordination of ammonia to copper(II) ion is stronger than that of water in aqueous solutions, lengthening of the Cu-L(eq) bond by changing L from H$_2$O to NH$_3$ might not be due to weakening of the Cu-NH$_3$(eq) bond compared with the Cu-OH$_2$(eq) bond within the aqua complex.

Although no general theory has been proposed to interpret the bond length variation of copper(II) complexes by changing ligands, an interpretation can be given as follows: the Jahn-Teller effect on coordination of ligands to copper(II) ion may be most pronounced when only one kind of ligands are bound, if interactions between the ligands in the first coordination shell and solvent molecules can be neglected and those between the central metal ion and ligand molecules are essentially the same, independent of molecules when a mixed ligand complex is formed. A less distorted structure of the tetraamminediaquacopper(II) complex may be due to a less symmetrical structure of the complex.

Table V. Comparison between the M-N bond lengths and thermodynamic data of ammine and ethylenediamine complexes[a].

Complex	$\nu_{M-N}(cm^{-1})$	$r_{M-N}(Å)$	$\Delta H°(kJ\ mol^{-1})$	$T\Delta S°(KJ\ mol^{-1}\ K^{-1})$
$Zn(NH_3)_4^{2+}$	429	2.03	-61.9	-7.48
$Zn(en)_2^{2+}$	447	2.13	-49.8	13.1
$Zn(en)_3^{2+}$	427	2.28	-71.6	2.49
$Cd(NH_3)_6^{2+}$	341	2.37	-88	-60
$Cd(en)_2^{2+}$	414	2.34	-54	-2.1
$Cd(en)_3^{2+}$	404	2.37	-84	-19
$Cu(NH_3)_4(H_2O)_2^{2+}$	-	2.03	-83.7	-11.7
$Cu(NH_3)_5(H_2O)^{2+}$ or $Cu(NH_3)_6^{2+}$	-	{1.93, 2.30}	-97.1	-28.3
$Cu(en)_2(H_2O)_2^{2+}$	-	1.93	-119	-0.873
$Cu(en)_3^{2+}$	-	{1.92, 2.22}	-	-

[a] S. J. Ashcroft and C. T. Mortimer, "Thermochemistry of Transition Metal Complexes", Academic Press Inc. London (1970).

When we add more ammonia to a solution containing the tetraamminecopper(II) complex, a higher-ammine complex having much intensive blue color forms. Bjerrum(17) concluded that this species is the pentaamminecopper(II) complex from potentiometric titrations. The X-ray diffraction method can not distinguish between nitrogen and oxygen atoms, and it is found that, based on equilibrium data, the complex has either five nitrogen and one oxygen atoms or six nitrogen atoms coordinated to the copper(II) ion. The bond distances within the higher-ammine complex are given in Table III, and thus the J-value is 1.19. The $Cu-NH_3(eq)$ bond shortens when the axial water molecule(s) within the tetraammine complex are replaced with ammonia molecule(s). The increased J-value from 1.15 in the tetraammine complex to 1.19 shows an enhanced Jahn-Teller distortion of the higher-complex. A shorter $Cu-NH_3(ax)$ bond within the higher-ammine complex may be due to a stronger interaction between the copper(II) ion and the ammonia molecules.

Ethylenediamine complexes. Structures of chelate complexes in solution have never been determined by means of X-ray and neutron diffraction. The first examination has been done for ethylenediamine complexes of zinc(II)(11), followed by cadmium(II)(12), nickel(II) and copper(II) ions(10). Since details of the method of measurements and treatment of data for the ethylenediamine complexes are described elsewhere(11,12), the only results found are shown in the paper (see Table IV), from which we will discuss structural changes of the complexes by formation of chelate rings.

In Table V are shown overall changes of enthalpies and entropies of formation of the ammine and ethylenediamine complexes of zinc(II), cadmium(II) and copper(II) ions, together with the lengths and frequencies of the totally symmetric vibration of the M-N bonds.

Stabilities of ethylenediamine complexes, compared with those of ammine complexes, are usually attributed to an increase in entropies by chelate formation, and enthalpy changes do not play an essential role in the stabilization, as far as the ligand nitrogen atoms of the chelating agent are regarded as the same in their coordinating power to the central ion as ammonia molecules. A higher frequencies of M-N bonds within the ethylenediamine complexes than those of the corresponding ammine complexes having the same number of the coordinated nitrogen atoms may be due to contributions of N-C vibrations coupled with the M-N vibrations. Discussion for the changes in frequencies of the M-N bonds, therefore, should be done for complexes with the same ligand.

For zinc(II)-complexes having four nitrogen atoms as the ligands, the length of the Zn-N bond within the $Zn(NH_3)_4^{2+}$ complex is shorter than that within the $Zn(en)_2^{2+}$ complex, the fact suggesting that the former complex has a stronger bond than the latter. The enthalpy change of the formation of the $Zn(NH_3)_4^{2+}$ complex is more negative than of the $Zn(en)_2^{2+}$ complex, and this result supports the above consideration. Stabilization of the $Zn(en)_2^{2+}$ complex more than the $Zn(NH_3)_4^{2+}$ complex in water is obviously due to a large entropy change in the former complex than in the latter. The Zn-N bond within the bis-complex is longer than that of the tris-complex. This result agrees with a general trend of bond-length variations of complexes (18). A comparison between the frequencies of the Zn-N bonds within $Zn(en)_2^{2+}$ and $Zn(en)_3^{2+}$ complexes shows that the former complex with a shorter M-N bond may have a stronger bond than the latter. The enthalpy change of the bis-complex divided by the number of the ligand atoms is -24.9 kJ mol^{-1}, which is more negative than the corresponding quantity of the tris-complex, -21.8 kJ mol^{-1}. The bonds in the former complex might thus be stronger than those of the latter.

For the cadmium-ethylenediamine complexes, the lengths of the M-N bonds are practically independent of the coordination number of the cadmium ion. The values of enthalpy changes of each step of the reactions are also invariable (-14 kJ mol^{-1} for each Cd-N bond), which are the same as that of the $Cd(NH_3)_6^{2+}$ complex. Therefore, the Cd-N bond strength of these complexes are almost the same, irrespective of their structures.

In the copper(II)-ethylenediamine complexes, interactions between Cu^{2+} and NH_2-groups within ethylenediamine molecules are so strong that $\Delta H°$ of the $Cu(en)_2(H_2O)^{2+}$ complex is more negative than that of the $Cu(NH_3)_4(H_2O)_2^{2+}$ complex. Since the Cu-N bonds within the ethylenediamine complex become strong, the bond length of the axial $Cu-OH_2$ bonds within the bis-complex increases to 2.9 Å (J = 1.5), which is much longer than the corresponding $Cu-OH_2$ bonds within the $Cu(H_2O)_6^{2+}$ and $Cu(NH_3)_4(H_2O)_2^{2+}$ complex. Since the $Cu-OH_2$ bond is very long in the bis(ethylenediamine)copper(II) complex, the complex can be regarded as a square-planar complex, rather than distorted octahedral one. The length of 1.92 - 1.93 Å may be a limiting value of the Cu-N contact.

Acknowledgment. The author express his sincere thanks to Dr. Georg Johansson of the Royal Institute of Technology for his kind cooperation at the first stage of the investigation of our work. The author also thanks Dr. Masunobu Maeda, Dr. Toshio Yamaguchi, Mr. Takeshi Akaishi and Dr. Tadao Fujita for their help. The work has been partially supported by a Grant-in-Aid for Scientific Research from the Ministry of Education, Science and Culture (No. 343011). The Yamada Kagaku Shinko Zaidan also gives the author a support.

REFERENCES

1. H. Ohtaki, T. Yamaguchi, and M. Maeda, Bull. Chem. Soc. Jpn. 49:701 (1976); "Analytical Chemistry - Essays in Memory of Anders Ringbom", ed. E. Wänninen, p. 163, Pergamon (1977).
2. H. Ohtaki and M. Maeda, Bull. Chem. Soc. Jpn. 47:2197 (1974).
3. M. Maeda, M. Maegawa, T. Yamaguchi, and H. Ohtaki, Bull. Chem. Soc. Jpn. 52:2545 (1979).
4. H. Ohtaki and M. Maeda, Bull. Chem. Soc. Jpn. 47:2217 (1974).
5. M. Maeda and H. Ohtaki, Bull. Chem. Soc. Jpn. 50:1893 (1977).
6. H. Ohtaki and M. Maeda, Asahi Glass Kogyo Gijutsu Shoreikai Hokoku, 28:221 (1976).
7. T. Yamaguchi and H. Ohtaki, Bull. Chem. Soc. Jpn. 52:415 (1979).
8. T. Yamaguchi and H. Ohtaki, Bull. Chem. Soc. Jpn. 51:3227 (1978).
9. T. Yamaguchi and H. Ohtaki, Bull. Chem. Soc. Jpn. 52:1223 (1979).
10. T. Fujita and H. Ohtaki, to be published.
11. T. Fujita, T. Yamaguchi, and H. Ohtaki, Bull. Chem. Soc. Jpn. 52:3539 (1979).
12. T. Fujita and H. Ohtaki, Bull. Chem. Soc. Jpn. 53:930 (1980).

13. D. R. Rosseinsky, Chem. Rev. 65:467 (1967).
14. F. J. Garrick, Phil.Mag. 9:131 (1930); 10:71 (1930).
15. F. Basolo and R. G. Pearson, "Mechanisms of Inorganic Reactions", 2nd Ed., John Wiley & Sons (1967).
16. R. D. Shannon, Acta Crystallogr. Sect. A, 751 (1976).
17. J. Bjerrum, Kgl. Danske Videnskab Selskab, Mat.-Fys. Medd. 11:No.5 (1931); "Metal Ammine Formation in Aqueous Solution", P. Haase and Son, Copenhagen(1941).
18. V. Gutmann, "The Donor-Acceptor Approach to Molecular Interactions" Plenum, New York (1978).

COPPER(II) CHELATE COMPLEXES-SOLUTE AND/OR SOLVENT INTERACTIONS

N. D. Yordanov

Institute of Organic Chemistry
Bulgarian Academy of Sciences
1113 Sofia, Bulgaria

Abstract - Some problems of the specific interactions between copper(II) complexes and Lewis bases and/or acids added to the "inert" solvents, or present as a neat solvent are discussed. Emphasis is placed on changes in the structure of the complexes as a result of the above interactions. The main conclusions are transferred to some chemical reactions in which the title complexes are used as catalysts or inhibitors.

It is well know that copper(II) labile complexes are most widely spread in nature as catalysts of oxidation-reduction reactions and in this sense probably they take second place after iron complexes. On the other hand, however, the copper chelate complexes do not act as catalysts of oxidation, but very often are among the most effective inhibitors of the oxidation reactions. The differences in their catalytic properties may be explained by the fact that all four coordination positions of the central ion are occupied by the chelate ligands. This was demonstrated by Siegel(1) by studying the catalase and peroxidase action of some copper chelate complexes (see scheme 1 in the next page). It was established that the above rate constants sharply decrease in the order I>>II>III of the following scheme and almost reach zero on coordination of the four ligand atoms.

For the above reason the neutral copper chelate complexes should not show catalytic activity. Let us examine the following examples. Skibida et al.(2) have studied cumene oxidation in presence of copper(II) acetylacetonate complex at about 100°C and have established that the complex does not show catalytic activity. The addition of small ammounts of o-phenanthroline, however, results in conside-

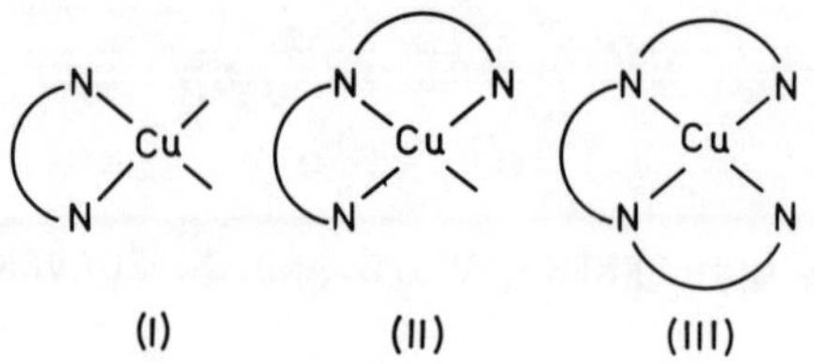

Scheme 1

rable increase of the catalytic properties. It should be noted that the catalytically active complex so obtained and the copper o-phenanthroline complex are different at least in the intial stage of the reaction and, therefore, a simple ligand exchange does not account for the observed properties. The following principal question arises: what is the role of o-phenanthroline? What is the structure of the catalytically active complex? In this case o-phenanthroline is one of the solutes, and we turn our attention to the solute(copper complex)- solute(o-phenanthroline)interaction. This is one of the main questions to be discussed further on in this paper.

The second problem, which we wish to discuss is the copper chelate complexes-solvent interactions. Our studies have shown(3) that the oxidation of methanol to formaldehyde in presence of copper(II) acetylacetonate complexes proceeds irrespective of the fact that copper chelate complexes do not act as catalysts of oxidation. The display of catalytic properties, according to us, is attributed to the reaction of copper acetylacetonate complex with alcohols(in presence of KOH). This results in a quantitative dimer formation(4) (Scheme 2).

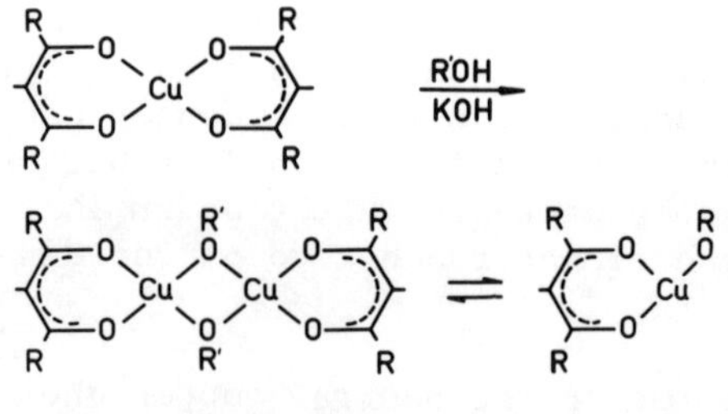

Scheme 2

which may be at equilibrium with monomers in solution; therefore free coordination positions in the first coordination metal ion sphere are obtained.

The knowledge of the solvent or solute effect on the catalytic activity of a metal complex provides broad possibilities for its regulation. Thus the study(5) of the catalytic activity "in vitro" of the tyrosine enzyme (its structure is not yet well known) of DOPA oxidation shows that the rate of reaction increases on addition of the stable free radical 2,2,6,6-tetramethyl-4-oxopiperidine-1-oxyl(TMPO) and viceversa it is inhibited if some complexing agents are added. The results obtained "in vivo" have shown an increased melanin accumulation(TMPO)- also in the skin of test animals and a reverse decrease of its amount with the use of different complexing agents. The question of the reaction mechanism is not yet clear, but is probably connected with some forms of skin cancer.

In all cases mentioned above, the catalytically active complexes are obtained "in situ" and often their existence is not taken into account or it is a comparatively little studied question. In some cases the question may not at all be unequivocally solved, taking into consideration that it is possible, in the course of a reaction for a very small part of the initial complex, to turn into a catalitically active form, which cannot be established due to the considerably greater amount of the initial complex. In view of this fact, the metal complex-solute and/or solvent interactions will be discussed further in this work as direct non - catalytic reactions.

With the aim of achieving greater simplicity we will differentiate between the solute and the solvent, depending on their donor or acceptor properties.

When the solute or the solvent show donor properties, several types of products of the copper chelate complex reaction are known (6).

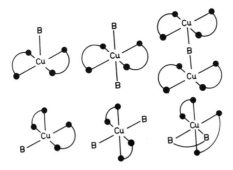

Scheme 3

As it is seen in scheme 3, the solute or solvent donor molecules are always coordinated to the metal ion, which usually is the strongest acceptor centre of the complex molecule.

The first three cases of axial base coordination (scheme 3) do not lead to significant changes in the structure of the initial complex, i.e. they do not directly set free coordination positions, but they are very important in cases which will be studied further. Particular attention should be paid to the last three cases depicted in scheme 3 where the chelate ligands are shifted by the plane of the complex and one or two "free" coordination positions are obtained. Therefore such "adducts" may show catalytic activity. In particular, the last case is closely similar to the example with o-phenanthroline, which "promote" the catalytic activity of copper acetylacetonate complex during cumene oxidation. Similar structures were proved by the X-ray analysis(7,8) and also by EPR investigations(9).

Very often it is possible that oxidation-reduction reactions occur between the copper chelate complex and the base; as a result the chelate ligands are removed from the first coordination sphere and new complexes showing catalytic activity are obtained. Similar oxidation-reduction reactions were established in the reaction of acetylacetonate, oxine, thioxine, dithiocarbamate, dithiophosphate, etc., complexes with strong Lewis bases, such as triphenylphosphine (10), piperidine, pirrolidine(11), ammonia(6,12) etc.

Without going into the details of the mechanism of the above described redox reactions, it should be noted that their study shows(6) an adduct formation with axial base coordination, resulting into a change in the electronic structure of the initial complex which facilitates the formation of an adduct with equatorial base coordination. Furthermore, the oxidation-reduction reaction may occur depending on the donor and acceptor properties of the reactants (6).

Reactions in which the solute or the solvent show acceptor properties are related to the second type. They are widely known as elementary stages of the "electrophilic catalysis"; a typical example is the bromination of the acetacetic ester in presence of copper ions(13) (see scheme 4 in the next page). As is known, for these reaction it is assumed that, owing to the complex formation between the substrate and the copper, the donor properties of acetacetic ester oxygen atoms are strongly reduced. This is the reason why other secondary important donor ligand centres, such as the γ-carbon atom, attain primary importance, and the bromine atom is coordinated to the latter. However this rule is not always followed. Depending on the donor atoms in the chromophore two cases may be distinguished(14): if oxygen or nitrogen atoms are involved in the chromophore, the coordination of the acceptor molecule is always on the ligand and the metal ion is reduced. Thus, for example, it was established that free ligand radicals (see scheme 5 in the next page) were formed in the reaction of NO_2 with acetylacetonate

COPPER(II) CHELATE COMPLEXES

(15,16) or oxime(16) copper(II) complexes. The same result was obtained in the reaction of the complexes with tetranitromethane(17).

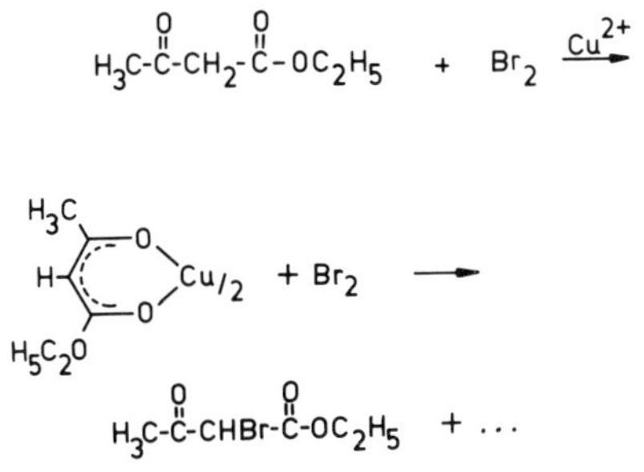

$$H_3C-\overset{O}{\underset{\|}{C}}-CH_2-\overset{O}{\underset{\|}{C}}-OC_2H_5 \;+\; Br_2 \;\xrightarrow{Cu^{2+}}$$

[structure of Cu(acac)-type chelate] $+ Br_2 \longrightarrow$

$$H_3C-\overset{O}{\underset{\|}{C}}-CHBr-\overset{O}{\underset{\|}{C}}-OC_2H_5 \;+\; \ldots$$

Scheme 4

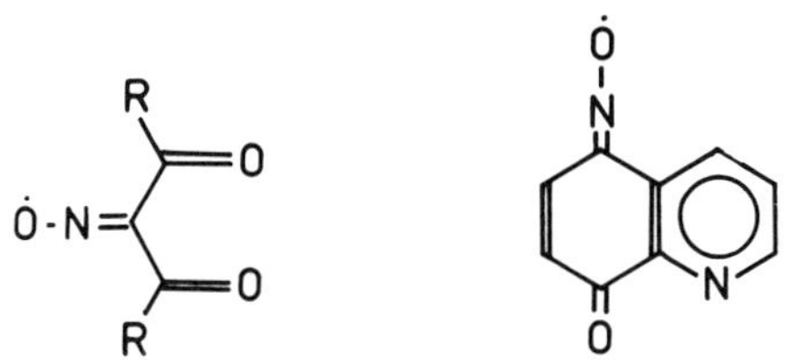

Scheme 5

Similar results are also obtained when phenylisocyanate or chloral (18,19) is the acceptor molecule. The following reaction mechanism was suggested(19).(Scheme 6).

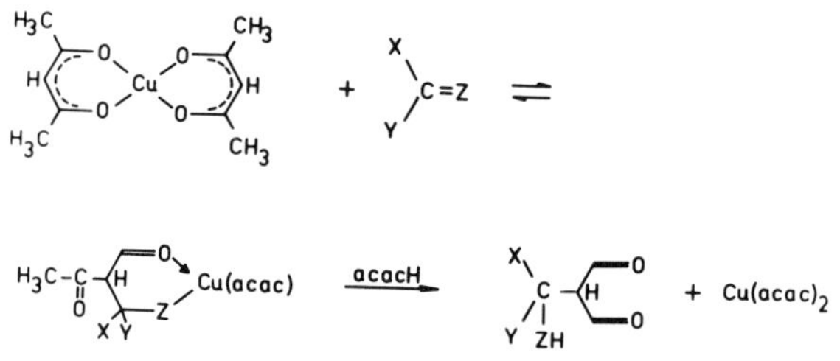

Scheme 6

We consider these results very important as it is known that copper(II) acetylacetonate complexes act as catalysts in chloral polymerization(20) and polyurethane formation(21) (in presence of alcohols). In connection with these results the following question arises again: which is actually the catalyst of chloral polymerization, the copper bis-acetylacetonate complex or its reaction product with the substrate?

It is seen from these studies that when the donor ligand atoms are involved in the chromophore and they are more strong donors, such as sulphur or selenium, the acceptor molecule attacks the sulphur atom(14). In the initial stage of the reaction, mixed-ligand complexes and esters or ligand disulphides(14) are obtained.

In a number of cases, it was established that copper chelate complexes react with some inorganic salts affording associates or new unusual complexes. (Scheme 7).

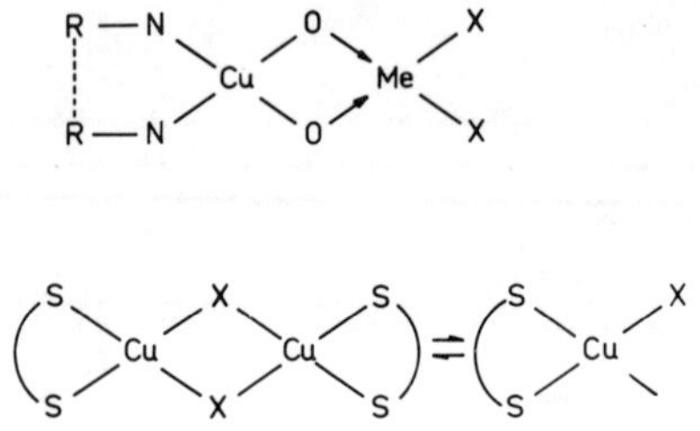

(X = Cl, Br)

Scheme 7

Thus the results from the studies of copper bis-dithiocarbamate complex reactions with tin(22) antimony(22) and copper(23) halides show formation of copper(II) mixed-ligand complexes.

New unusual complexes or associates may also be obtained in the reaction of copper chelate complexes with some other metal chelate complexes. In this connection, the following examples are of interest: Drago et al.(24) have established that Cu(hfacac)$_2$ and Co(salen) interact exchanging their ligands and thus forming the associate Cu(salen).Co(hfacac)$_2$. The analogous reaction with Cu(acac)$_2$ could be not established, because the Cu(acac)$_2$ complex has weaker acceptor properties. The reaction between bis(dithiocarbamato)copper(II) and bis(acetylacetonato)cobalt(II) is accompanied also by ligand exchange, which leads to the Cu(acac)$_2$.Co^{II}(dtc)$_2$(25). A characteristic of Co(dtc)$_2$ is that it cannot be obtained by the standard methods since the extraction from water solutions of CoCl$_2$ and Nadtc always yields Co^{III}(dtc)$_3$. An interesting reaction was established between copper(II) thiosemicarbazone complexes and nickel(II) analogues(26), in which the unpaired copper electron is shifted from the $d_{x^2-y^2}$ orbital to the d_{z^2}.

As it was already pointed out, the copper labile complexes are highly effective catalysts of oxidation. In some cases, however, this reaction is not desirable, therefore the deactivation of the latter is an essential problem(27). The non-transition metal chelate complexes are generally used as "metal deactivators". The deactivation of the catalytically active metal ions represents a ligand exchange, which leads to chelate complexes. Copper chelate complexes do not act as catalysts of oxidation but prove to be highly effective inhibitors of the reaction. As inhibitors of oxidation they possess the property of decomposing hydroperoxides obtained in the course of oxidation. Studies of the effect of various solutes on the rate of organic hydroperoxide decomposition have shown(28,29) that it is dependent on them. As it is seen in Figure 1, in all cases the Lewis bases lead to significant increase of the decomposition rate of copper dithiophosphate complex and hydroperoxide, respectively, as compared to the case when they are not present. It is noteworthy that in presence of piridine the above reaction is increased more than 10^4 times(28).

Recently analogous results were obtained in the study of copper chelate complexes containing hydrocarbons(30). It is well known that the latter are weak Lewis acids and direct reaction with copper chelate complexes is not established. However, the addition of small amounts of bases causes the decomposition of copper chelate complexes if they contain sulphur in their chromophore. In complexes containing oxygen and/or nitrogen atoms in the chromophore, similar reaction did not occur. The formation of copper(II) halide and a resin-like organic residue was established as a final reaction product.

The following mechanism is suggested for the last two solute-solute-solvent reactions (Scheme 8):

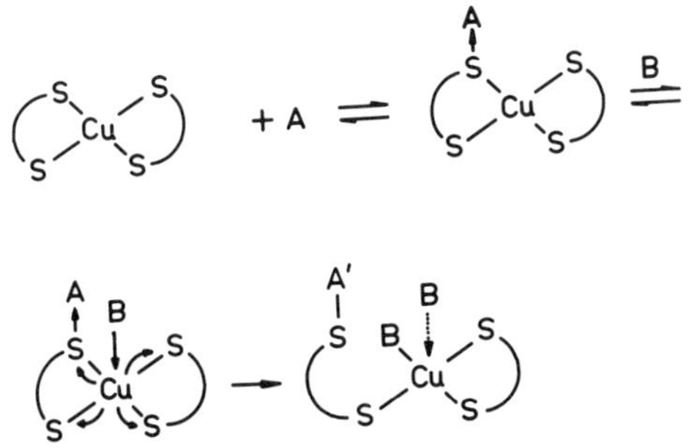

Scheme 8

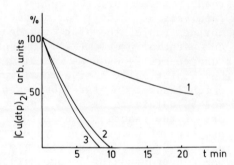

Fig. 1. Decomposition rate of bis(dithiophosphato)copper(II).

the acceptor molecules of the solvent(solute) are coordinate to the donor atoms of the chromophore(sulphur atoms here). The obtained complex is rather weak, however, and so the reaction proceeds comparatively at a low rate (it is zero for halogen containing hydrocarbons). The presence of Lewis bases in the reaction mixture leads to adduct formation with axial base coordination (first 3 cases in scheme 3). The electronic density on the sulphur atoms increases, the possibility of an electron transfer from the sulphur atom to the acceptor molecule is increased and the complex is destroyed.

In conclusion it is worth noting that the question of the role of the solute-solute-solvent interactions with participation of metal complexes is of particular importance and it is not yet solved. In view of this the focusing of the efforts of many workers on it is justified.

REFERENCES

1. H. Siegel, Angew. Chem. Intl. Edn. 8:167 (1969).
2. O. N. Emanuel, A. M. Sakharov, and I. P. Skibida, Izv. Akad. Nauk USSR ser. Chem. 2692 (1975).
3. N. D. Yordanov, unpublished results.
4. J. A. Bertrand and R. I. Kaplan, Inorg. Chem. 4:1657 (1965); Z. Olejnic, R. Grobelny, and B. Jezowska-Trzebiatowska, Bull. Polish Acad. Sci. Chem. 24:883 (1976).
5. Z. D. Raikov, P. M. Blagoeva, and N. D. Yordanov, Compt. rend. Acad. Sci. Bulg. 29:709, 881 (1976).
6. N. D. Yordanov and D. Shopov, in "Theory, Structure, Properties of Complex Compound", B. Jezowska-Trzebiatowska, L. Legendziewicz, and M. F. Rudolf, eds., Polish Scientific Publishers, p. 313 (1979).
7. M. A. Bush and D. F. Fenton, J. Chem. Soc.(A), 2446 (1971);

M. A. Busch, D. E. Fenton, R. S. Nyholm, and M. R. Truter, Chem. Comm. 1335 (1970).
8. M. V. Veidis, G. H. Schreiber, T. E. Gough, and G. J. Palenik, J. Am. Chem. Soc. 91:1859 (1969).
9. N. D. Yordanov and J. B. Raynor, submitted for publication.
10. N. D. Yordanov and D. Shopov, Inorg. Nucl. Chem. Lett. 9: 19 (1973).
11. D. Rehorek and Ph. Thomas, Z. Chem. 13:477 (1973).
12. N. D. Yordanov and D. Shopov, Proc. II Intl. Symp. Solute-Solute-Solvent Interactions, Leuven, p. 225 (1974).
13. F. Basolo and R. G. Pearson, "Mechanism of Inorganic Reactions. A Study of Metal Complexes in Solution", J. Wiley, New York (1967); J. Candlin, K. Taylor, and D. Thompson, "Reactions of Transition Metal Complexes" (Russian Edn.), Mir, Moscow (1970).
14. N. D. Yordanov, to be published.
15. W. Wolodarski, J. Faniran, and J. K. S. Wan, Canad. J. Chem. 51:4872 (1973).
16. N. D. Yordanov, V. Terziev, and D. Shopov, Proc. XIX I.C.C.C. Prague, Rep. N. 133a (1978).
17. N. D. Yordanov, unpublished results.
18. K. Uehara, Y. Ohashi, and M. Tanaka, Bull. Chem. Soc. Japan, 49:1447 (1976).
19. Yu. N. Nizelskii and T. E. Lipatova, Theoret. Exsp. Khim. 6:838 (1970); T. E. Lipatova and Yu. N. Nizelskii, Doklady Akad. Nauk USSR, 175:128 (1967).
20. T. Otsu, Y. Nishikawa, and S. Watanuma, Macromol. Chem. 115:278 (1968).
21. J. Fisher, Tetrahedron, 19 Suppl. 97 (1963); L. Weisfeld, J. Appl. Polym. Sci. 5:421 (1961); A. I. Volozin, D. P. Kozmina, and S. N. Danilov, Zh. Prikl. Khim. 37:2077 (1964); T. E. Lipatova and Yu. N. Nizelskii, in "Metal Complex Catalysis", Kiev, Naukova Dumka, p. 23 (1977).
22. N. D. Yordanov, V. Terziev, V. Iliev, and D. Shopov, Compt. rend. Acad. Sci. Bulg. 3°:675 (1977).
23. N. D. Yordanov, Theoret. Exsp.Khim. 9:70 (1973); N. D. Yordanov and D. Shopov, J. Inorg. Nucl. Chem. 38:137 (1976); N. D. Yordanov, J. Mol. Struct. 47:107 (1978).
24. N. B. O'Brian, T. O. Maier, I. C. Paul, and R. S. Drago, J. Am. Chem. Soc. 95:6640 (1973).
25. N. D. Yordanov and V. Alexiev, to be published.
26. J. A. De Bolfo, T. D. Smith, J. F. Boas, and J. R. Pilbrow, Austr. J. Chem. 29:2583 (1976).
27. K. U. Ingold, in "Advances in Chemistry", Series 75, "Oxidation of Organic Compounds - I", p. 320 (1968).
28. D. Shopov and N. D. Yordanov, Proc. XIV I.C.C.C., Canada, p. 236 (1972).
29. N. D. Yordanov, unpublished data.
30. N. D. Yordanov, V. Iliev, V. Terziev, and D. Shopov, Inorg. Chem. submitted for publication.

PHOTO-INDUCED LIGAND-SOLVENT INTERCHANGE IN TRANSITION METAL

COMPLEXES

L. G. Vanquickenborne and A. Ceulemans

Department of Chemistry
Celestijnenlaan 200 F
3030 Heverlee, Belgium

Abstract - A large number of stereochemical data on d^3- and d^6-photosubstitutions can be rationalized by assuming a dissociative reaction mechanism. The relevant orbital and state correlation diagrams reveal the reason for the difference in behavior between d^3- and d^6-systems. Both for the axially and the equatorially labilized Cr^{III} complexes, it can be shown that the stereomobile reaction is an allowed process, whereas the stereoretent reaction is a forbidden process. In d^6-systems, stereomobility as well as stereoretention is observed; the specific reaction path is related to the numerical value of certain ligand field parameters. So far, it has been possible to rationalize all the experimental data within the framework of the present methodology.

1. INTRODUCTION

Essentially, the photochemistry of transition metal complexes (1,2) can be subdivided into two main categories. On the one hand, one has the photoredox or electron reactions, which are generally induced by charge transfer excitations. On the other hand, one has the photosubstitution or the ligand transfer reactions, which can often be induced by ligand field excitations(3,4).

In the domain of photosubstitution reactions, the systematization of the experimental data began in 1967 by the formulation of Adamson's empirical rules(5). Consider an arbitrary complex, characterized by an octahedral skeleton, and where the six ligands are not all identical, as for instance trans-ML_4AB (Figure 1). In a number of cases, excitation in one of the ligand field bands leads to the replacement of one of the ligands by a solvent molecule.

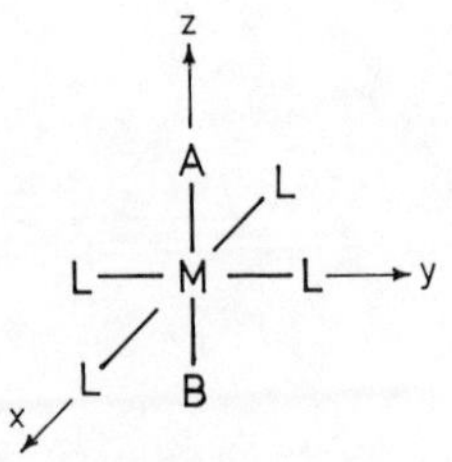

$(Dq)_x = (Dq)_y = (Dq)_L$

$(Dq)_z = \frac{1}{2}[(Dq)_A + (Dq)_B]$

Fig. 1. A hexa-coordinated ML_4AB complex.

The first question to be asked is which one of the ligands will be expelled: L, A or B. In an attempt to answer this question, Adamson first associates an average spectrochemical strenght to each bond axis; this is simply the arithmetic mean of the two relevant 10 Dq-values. Subsequently, he predicts the leaving ligand by means of a two-step procedure. In his first rule, he claims that the leaving ligand is situated on the axis, characterized by the weakest ligand field. This formulation introduces the notion of a "labilized axis": in the C_{4v}-complex of Figure 1, the first rule leads to the prediction of either axial or equatorial labilization. Adamson's second rule states that, of the two ligands. situated on the labilized axis, the leaving ligand is the one exhibiting the strongest ligand field.

These rules are very simple, and - surprisingly - much more general than originally thought. The fact that they can be expressed by using nothing more than the spectrochemical parameters Dq, suggests that they can be tackled by using simple ligand field theory(6,7).

The systematization of the experimental data has been extended so as to include also the stereochemistry of the reactions(9). In 1978, we suggested(10) the validity of Table I. At the time, this Table seemed to accomodate nearly all the available experimental facts - that was approximately 35 different photosubstitution reactions, both of d^3- and d^6-systems. For instance, in the case of axial labilization, a trans-complex of Cr^{III} generally gives rise to a cis-product upon photosubstitution (stereomobility) while a trans-complex of Co^{III} yields a trans-product (stereoretention). Here again, the rules look very simple, and, at first sight, there appears to be a connection between the stereochemistry and Adamson's labilized axis terminology. Therefore, it is very natural to attempt a ligand field analysis of the proposed rules.

Table I. Empirical Symmary of the Stereochemistry (1978)

	d^3: Cr^{III}	d^6: Co^{III}, Rh^{III}, Ir^{III}
Axial labilization of ML_5X or trans-ML_4XY	stereomobility[a]	stereoretention
Equatorial labilization of cis-ML_4X_2	?	stereomobility

[a] ref. 8

2. METHODS AND HYPOTHESES

In studying chemical reactions, ligand field theory should preferably be used in its Angular Overlap version. Indeed, if one wants to simulate a chemical reaction, one will have to modify a given structure by displacing individual ligands. In this case, it will be useful to have a perturbation Hamiltonian of the form

$$V = \sum_i \sum_L V^L(i)$$

where i numbers the electrons and L the ligands. V is written explicitly as a sum of individual ligand perturbations, and changes in a very transparent way as a function of the ligand positions. Table II shows the ligand field parameters for a number of Cr^{III}-L-interactions, taken from various sources in the literature.

By using numbers such as the ones given in Table II, it is possible to make semi-empirical ligand field predictions, both on the leaving ligand, and on the stereochemistry(11). In the present paper, we will not consider the leaving ligand problem any further. Instead we will confine our attention exclusively to the discussion of the stereochemistry.

In order to discuss the photochemistry, one has to make a hypothesis on the reaction mechanism. Both a dissociative mechanism - characterized by a five-coordinated intermediate - and an associative mechanism - characterized by a seven-coordinated intermediate - are quite well conceivable. For several reasons, which will not be elaborated upon here, we have adopted a dissociative reaction mechanism(12): formally, the reaction will be dissected into three separate processes. In a first step, we consider the dissociation of the leaving ligand, generating a square pyramidal five-coordinated intermediate. Next, if we want to account for the stereomobility, which is observed in many cases, we will have to consider the possibility of isomerization of this five-coordinated fragment. The most obvious, and in fact, the only reasonable alternative structure is a trigonal bipyramid. Finally we have the association of a solvent molecule S to one of these two polyhedron types, in order to obtain the final hexa-coordinated substitution product. It is obvious that these three processes may actually proceed in a more or less concerted way, and not simply consecutively.

Table II. Spectrochemical parameters for Cr^{III}-L-interactions (μm^{-1}) See ref. 11 and references therein.

Ligand	$10Dq=3\sigma-4\pi$	σ	π	π/σ
I^-	1.115	0.458	0.065	0.14
Br^-	1.185	0.508	0.085	0.17
Cl^-	1.315	0.558	0.090	0.16
H_2O	1.583	0.594	0.050	0.08
F^-	1.610	0.763	0.170	0.22
NCS^-	1.772	0.641	0.038	0.06
OH^-	1.881	0.812	0.139	0.17
NH_3;en	2.155	0.718	0.	0.
CN^-	2.661	0.848	-0.029	-0.03

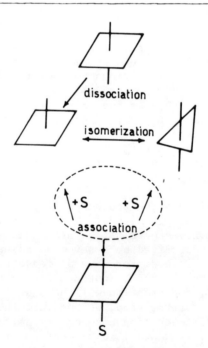

Fig. 2. A proposal for a three-step substitution reaction.

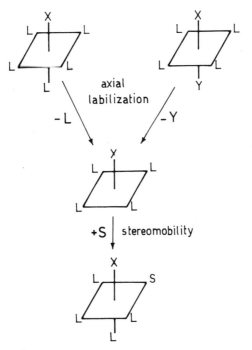

Fig. 3. Summary of the experimental facts on the axial labilization of Cr^{III}-d^3-complexes.

3. DISCUSSION OF d^3-SYSTEMS

The axial labilization of Cr^{III} complexes can be summarized as in Figure 3. The mono-acido and the trans-diacido-complexes consitute the bulk of the experimental data. In the former case, L is the leaving ligand; in the latter case, Y is the leaving ligand. But the same penta-coordinated structure is supposed to be the reaction intermediate; since the photosubstitution is characterized by a clearly pronounced stereomobility, S comes in cis from X.

The central square pyramid (SP) has two distinct distortion possibilities, leading to two different trigonal bipyramids, one with X in equatorial position ($TBPX_{eq}$) and one with X in apical position ($TBPX_{ap}$). In principle $TBPX_{ap}$ appears to be the most promising intermediate. Indeed, the subsequent addition of a solvent molecule S to either one of the three sides of the equatorial trian-

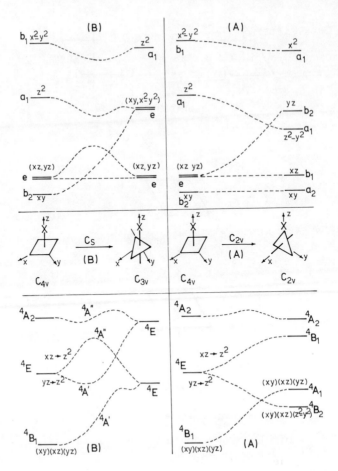

Fig. 4. The two distortion possibilities of the square pyramid in d^3-systems; orbital and state correlation diagrams.

gle gives a cis-product, which is precisely what is found experimentally. The TBPX$_{eq}$ on the other hand has two access ways leading to a cis-product, but one access way leading to a trans-product. On purely statistical grounds, the formation of TBPX$_{eq}$ would be expected to lead to a cis/trans ratio of 2/1 instead of 100% cis-product. Figure 4 shows an orbital and state correlation diagram for the two relevant distortions. Reaction path (A) leads to TBPX$_{eq}$, and reaction path (B) leads to TBPX$_{ap}$.

In both cases, the point of departure is identical, and of C_{4v}-symmetry. The orbital energy order is quite simply understood from a comparison of this structure with an octahedral molecule: the two σ-orbitals are now energetically very different because one ligand has been removed along the z-axis, thereby stabilizing z^2

with respect to x^2-y^2. Similarly, the π-orbitals are split because in general X is a π-donor, while L is an amine ligand. The abscissas represent an appropriate distortion coordinate, corresponding to a concerted angular displacement of the ligands. In (B) the result has trigonal symmetry (C_{3v}) while in (A) the result has only C_{2v} symmetry. Notice that, when X = L, the trigonal skeleton also imposes a C_{3v}-symmetry in case (A). The energy splittings between the (a_2, b_2) and (b_1, a_2) set result entirely from the parameter differences between X and L.

The detailed correlation lines are different in both cases. Indeed, in the Woodward-Hoffmann sense, the symmetry conserved during the reaction is only C_s in (B), but it is C_{2v} in (A). One of the most obvious consequences is that the xy-orbital remains the lowest orbital along the entire reaction path in case (A), but moves up to higher energy in case (B), thereby crossing two other orbitals. Another difference shows up in the interactions between the σ-orbitals. Along reaction path (B), the z^2- and x^2-y^2-levels interchange their orbital composition. But along reaction path (A), z^2 transforms into an orbital which is approximately described as z^2-y^2, while x^2-y^2 transforms into an orbital which is best designated as x^2. This is of course related to the fact that the approximate trigonal axis in TBPX$_{eq}$ is the x-axis in case (A).

When the orbitals are to be populated by three electrons as in the Cr^{III} case, the state diagrams of Figure 4 are readily obtained.

The ground state is a quartet in all cases, as can be expected from Hund's rules. The first excited state in the square pyramid is a degenerate ^4E state, with a configuration corresponding to the one-electron excitation (xz, yz) $\to z^2$. This ^4E is directly correlated to the lowest excited quartet state of the hexacoordinated complex, which has been identified as the photo-active state; therefore the properties of this ^4E state are essential in understanding the substitution behavior.

In both cases, (A) and (B), the degeneracy of ^4E is lifted in the course of the distortion; but obviously, the formation of TBPX$_{eq}$ will be favored with respect to TBPX$_{ap}$. In case (A), the first excited ^4E will behave as a typical stereomobile state: its $^4B_2$ component will spontaneously transform into the ground state of the appropriate trigonal bipyramid. This can readily be understood from the orbital composition of $^4B_2$: z^2 is populated, while yz is not. The orbital correlation diagram shows that this is the optimal way to stabilize the system along this particular bending coordinate. Therefore, to the extent that these correlation diagrams have any predictive value, one has to conclude that the resulting trigonal bipyramid will have X in the equatorial position. Obviously, in order to explain the observed stereomobility, some other factor has to be invoked.

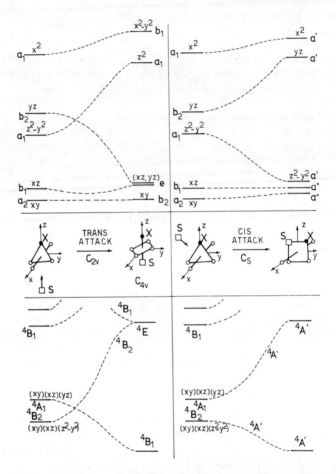

Fig. 5. Correlation diagram for cis- and trans-attack of a solvent molecule S on the five-coordinated fragment.

This factor can only be found from an analysis of the details of the association reaction. Figure 5 shows the correlation diagrams for both cis- and trans-attack, where the solvent molecule S approaches in cis or trans of the heteroligand X. In both cases, the reaction path consists of an appropriate angular displacement of two equatorial ligands, accompanied by a concerted approach of S.

The most striking difference between the two modes of attack lies in the evolution of the orbitals yz and z^2-y^2. Indeed, yz decreases in energy when a trans-product is formed, while it increases in energy upon the formation of a cis-product. The z^2-y^2 orbital is characterized by precisely the opposite behavior. This can be understood from the directional properties of these two orbitals (z^2-y^2 points towards the heteroligand X). A solvent trans-attack will induce a very strong σ^*-interaction with z^2-y^2, while leaving yz unaffected. This effect will be reinforced by the concerted rearrangement of the other two equatorial ligands, thereby destabilizing z^2-y^2 even more, and at the same time decreasing the π^*-interactions with yz. The same line of reasoning shows that cis-attack has the opposite effect: it destabilizes yz, it stabilizes z^2-y^2.

The corresponding state correlation diagrams illustrate this feature very nicely. The ground state of the trigonal bipyramid is $^4B_2$, characterized by the configuration $(xy)(xz)(z^2-y^2)$ — that is by the occupation of the three lowest orbitals by one electron. The first excited state, $^4A_1$ is characterized by the occupation $(xy)(xz)(yz)$. The difference between the two states is thus precisely the occupation or the non-occupation of the two crucial orbitals yz and (z^2-y^2). Therefore, it is hardly surprising that the state correlation diagrams reflect the orbital diagrams quite faithfully.

As a consequence, cis-attack on the $^4B_2$ ground state of the trigonal bipyramid leads to the ground state of the hexacoordinated complex without energy barrier; a trans-attack on the same state correlates with an excited state of the product. This simple rationalization of the experimentally observed facts can be formulated as a selection rule in the Woodward-Hoffman sense. The principle of conservation of orbital symmetry leads to the conclusion that the cis-attack is an allowed process, while the trans-attack is a forbidden process.

Alternatively, looking back at the spatial arrangements of orbitals and ligands, one might interpret the association reactions in terms of Fukui's frontier orbitals(13). A ligand approaching a molecular fragment, will preferably do so along an access path containing vacant d-orbitals. The donor levels of the ligand act as HOMO's, while the metal d-orbitals act as LUMO's.

4. DISCUSSION OF d^6-SYSTEMS

For Co^{III}, Rh^{III} and Ir^{III} complexes, both axial and equatorial labilization are observed and quite well documented(10). Table II suggests that axial labilization gives rise to stereoretention, and equatorial labilization to stereomobility; this implies a clearly pronounced trans-preference(14) for the photosubstitution products. Within the framework of a dissociative mechanism a number of five-coordinated intermediates have to be considered. On the basis of Figure 6, it is not obvious why a trans-preference should be observed: in d^3-systems, all TBP-structures gave rise to cis-addition products. Are the d^6-trigonal bipyramids drastically different? Or do the square pyramids (SP) play a decisive role?

In order to answer these questions, correlation diagrams should be constructed for the transformation of the different five-coordinated structures into each other. Orbital correlation diagrams for $Rh(NH_3)_4Cl^{2+}$ are shown in Figure 7, as a typical example of the behavior of ML_4X-fragments. The transition $SPX_{ap} \rightarrow TBPX_{ap}$ has not been included because it can easily be shown to be a forbidden process (see also Figure 4B). The symmetry conserved along the three distortion paths under consideration are C_{2v}, C_s and C_s.

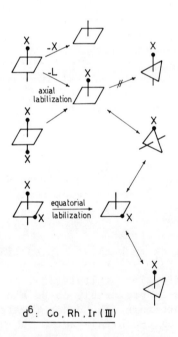

Fig. 6. The penta-coordinated fragments relevant to the study of d^6-systems.

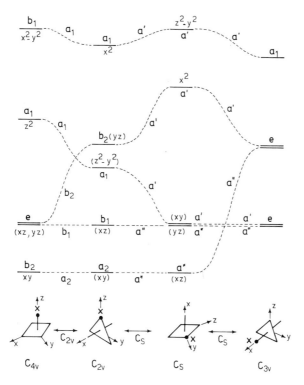

Fig. 7. Orbital correlation diagram for the distortions in d^6-systems.

There is one important difference between the square pyramids and the trigonal bipyramids. The ground configuration of the square pyramids is simply given by the double occupation of the three lowest orbitals, thus giving rise to a singlet state. But in the trigonal bipyramids, the two middle orbitals are so close (in TBPX$_{ap}$ they are even degenerate) that the ground configuration has one electron in each one of these orbitals, thus leading to a triplet ground state. This is also shown in the state correlation diagram of Figure 8: the singlet SP-ground state corresponds to the first excited state of the TBP-structures. However, the most important correlation line is not the lowest singlet curve, but the lowest triplet curve. Indeed, whether we start from a trans- or from a cis-complex, the photoactive state corresponds to the first excited triplet state, that is ^{3}E, or ^{3}A'. So the main question is how this triplet state will associate with a solvent molecule in order to result in a stable hexa-coordinated molecule, and to complete the photosubstitution reaction. It turns out that none of the four considered fragments has favorable association possibilities. Indeed, for the square pyramids, the solvent approach along the negative z-axis is unfavorable because of the occupation of z^2. For the trigonal bipyramids, both yz and z^2-y^2 are occupied (in marked contrast with the d^3-systems!) As a matter of fact, the only structures that are able to associate readily with a solvent molecule, are the square pyramidal ground states. The reason is that they have vacant

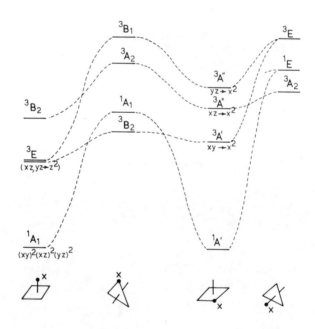

Fig. 8. State correlation diagram for the distortions in d^6-systems.

z^2-orbitals, and therefore, they are characterized by an access direction in trans of the axial ligand. Therefore, even if the transformation from one structure to another is possible on the triplet potential surface, eventually, the five coordinated fragment will be carried into one of the SP ground states by intersystem crossing. And from there hexa-coordination will be restored by association of a solvent molecule in trans of the axial ligand.

The specific evolution of the triplet potential surface is characterized by two remarkable features. First, one notices again the presence of an electronic selection rule: the transition $SPX_{eq} \to TBPX_{ap}$ is a forbidden process, while $SPX_{eq} \to TBPX_{eq}$ is an allowed process. Second, the two square pyramids are separated from each other by a potential barrier in $TBPX_{eq}$. The height of this barrier determines whether or not stereomobility will be possible. Moreover, the relative position of the triplet levels in the two SP's (^3E and $^3A'$) determines which one of those will act as an energy trap, from where the ground state is populated. In the case of Figure 8, SPX_{ap} has the lowest triplet energy. In our opinion, this fact is at the basis of the stereoretention of the trans-complexes, and of the stereomobility of the cis-complexes.

Basically, the energy difference between the two states $^3E(SPX_{ap})$ and $^3A'(SPX_{eq})$ can be traced back to the energy difference between two orbitals, $z^2(SPX_{ap})$ and $x^2(SPX_{eq})$. This energy difference is in first order simply proportional to $(\sigma_X - \sigma_L)$. In 1978, when we first proposed the present scheme(10), only complexes with $\sigma_X < \sigma_L$ were known, and the observed trans-preference was readily accounted for.

In the last two years, several papers appeared where the alternative situation was realised(16,17) : $\sigma_X > \sigma_L$. For instance $Rh(NH_3)_4(OH)Cl^+$, $Ir(en)_2(OH)Cl^+$, where the halide Cl^- is the leaving ligand, give rise to a penta-coordinated fragment, where the heteroligand X = OH^- is the strongest σ-donor (see also Table II). In those cases, whether one starts from a cis- or from a trans-complex, the photoproduct is cis. Within the framework of Figure 8, the explanation is very simple: now $^3A'$ drops below ^3E, and $^3A'$ serves as a trap, whose association with a solvent molecule leads to a cis-product.

Another way to test the here proposed scheme is to prepare complexes with five strong ligands in the intermediate, like for instance cis- or trans-$Rh(en)_2NH_3Cl^{2+}$, giving rise to a RhN_5 fragment. This structure is characterized by a comparatively high barrier in the TBP-structure; therefore, one expects stereoretention in any case: cis → cis and trans → trans. This is indeed, what has been found experimentally(15).

5. CONCLUSION

So far, the stereochemistry of some 75 reactions on d^3- and d^6-systems has been unraveled. The here present framework appears to be able to rationalise most, if not all the stereochemical data. The suggestions of Table I are obviously too limited and especially for d^6-systems, both cis- and trans-complexes can exhibit either stereoretention or stereomobility, depending on the relative value of certain ligand field parameters.

The fact that orbital selection rules appear to be operative in these reactions may be an indication that the Woodward-Hoffmann approach can be quite fruitful, not only in the theory of organic reactions, but also in the study of transition metal complexes.

REFERENCES

1. V. Balzani and V. Carassiti, "Photochemistry of Coordination Compounds", Academic Press, New York (1970).
2. A. W. Adamson and P. D. Fleischauer, eds.,"Concepts of Inorganic Photochemistry", Wiley Interscience, New York (1975).
3. E. Zinato, ref. 2, chapter 4.
4. A. W. Adamson, Pure and Applied Chem. 51:313 (1979).
5. A. W. Adamson, J. Phys. Chem. 71:798 (1967).
6. M. Wrighton, H. B. Gray, and G. S. Hammond, Mol. Photochem. 5:164 (1973).
7. J. I. Zink, J. Am. Chem. Soc. 94:8039 (1972).
8. A. D. Kirk, Mol. Photochem. 5:127 (1973).
9. F. Scandola, in "Rearrangements in Ground and Excited States", P. De Mayo ed., Academic Press, New York, (1980).
10. L. G. Vanquickenborne and A. Ceulemans, Inorg. Chem. 17: 2730 (1978).
11. L. G. Vanquickenborne and A. Ceulemans, J. Am. Chem. Soc. 99:2208 (1977).
12. L. G. Vanquickenborne and A. Ceulemans, J. Am. Chem. Soc. 100:475 (1978).
13. K. Fukui, Top. Curr. Chem. 15 (1970).
14. D. Strauss and P. C. Ford, J.C.S. Chem. Comm. 194 (1977).
15. J. D. Petersen and F. P. Jakse, Inorg. Chem. 18:1818 (1979).
16. L. H. Skibsted and P. C. Ford, J.C.S. Chem. Comm. 853 (1979).
17. M. Talebinasab-Sarvari and P. C. Ford, Inorg. Chem. submitted.

MECHANISM OF OCTAHEDRAL SUBSTITUTIONS ON TRANSITION METAL COMPLEXES. ATTEMPTS TO DISTINGUISH BETWEEN D AND I_d MECHANISMS

Smiljko Ašperger

Department of Chemistry, Faculty of Pharmacy and
Biochemistry, University of Zagreb
Zagreb, Croatia, Yugoslavia

Abstract - The evidence obtained from studying the induced aquations of $Co(NH_3)_5N_3^{2+}$ and of $Co(NH_3)_5(O_2CNH_2)^{2+}$ by nitrous acid, and of $Co(NH_3)_5(halide)^{2+}$ by Hg^{2+}, led to the conclusion that a common five-coordinate intermediate, $Co(NH_3)_5^{3+}$, was formed in all these reactions. Recent examinations of different reactions, including $KMnO_4$ induced aquation of $Co(NH_3)_5(DMSO)^{3+}$, however, suggest different intermediates in different reactions. The mechanism of several octahedral replacement reactions, studied in our laboratory, is discussed. These reactions are: aquations of trans-chloronitro- and trans-dichlorobisethylenediaminecobalt(III) ions, and of (dimethyl suphoxide)pentaamminecobalt(III) ions, in mixed aqueous-organic solvents; base-hydrolysis of $Co(NH_3)_5(DMSO)^{3+}$; anations of $Fe(CN)_5(H_2O)^{3-}$ ion; replacements in $Fe^{II}(CN)_5(ligand)^{n-}$ ions; replacements in the binuclear dimer of $Fe(CN)_5(H_2O)^{3-}$. It was shown that often there is no sharp distinction between D and I_d mechanism. The clear-cut D mechanism is much less frequent than it was originally thought to be the case. In the reaction systems mentioned the kinetic results are more consistent with I_d than with D mechanism.

It is well established that substitution reactions of octahedral complexes occur by a predominantly bond-breaking mechanism, but the details vary from one case to the other. In some cases and S_N1 (lim) (or D)(1) mechanism is claimed to operate, forming a five coordinate intermediate, which lives long enough to discriminate between entering ligands. In other cases a dissociative-interchange (or I_d)(1) mechanism operates, i.e. a dissociative type of exchange of positions between particles in the first and the second coordination sphere. For example, it has been postulated that induced aquations of $Co(NH_3)_5(O_2CNH_2)^{2+}$ by nitrous acid(2,3) and of(3) $Co(NH_3)_5X^{2+}$ (X = Cl^-, Br^-, I^-) by Hg^{2+} proceed through a common five-coordinate intermediate(4,5), $Co(NH_3)_5^{3+}$. On the other hand, spontaneous aquation of the $Co(NH_3)_5Cl^{2+}$ ion should proceed by an I_d mechanism(6). To avoid complications of ion-pairing, anionic $Co(CN)_5(OH_2)^{2-}$ has been used(7,8) and it has been claimed that replacements of water by azide or thiocyanate ions occur by the D mechanism. Another example(9) of the D mechanism (with no ion-pairing) was said(7,10) to involve activation by SO_3^{2-} in the systems trans-$Co(CN)_4(SO_3)X^{n-}$ (X = CN^-, NCS^-, NH_3, OH^-, N_3^-, NO_2^-, SO_3^{2-}). Some doubt has been thrown upon the claim that the above mentioned induced aquations follow a D mechanism through a common intermediate because it has also been shown that the assisted aquation of $Co(NH_3)_5Cl^{2+}$ by Hg(II) in sulphate media proceeds by paths involving Hg^{2+}, $HgSO_4$, $Hg(SO_4)_2^{2-}$ with intermediates having approximately 0, 15, and 30% efficiencies, respectively, for conversion to $Co(NH_3)_5SO_4^+$ that are independent of sulphate concentration(11,12). The efficiency of conversion of $Co(NH_3)_5^{3+}$ to $Co(NH_3)_5(SO_4)^+$ in the nitrosation reactions was dependent on sulphate concentration, which led to the conclusion(12) that the intermediates in the Hg(II)-assisted aquations and nitrosation reactions are, most probably, not the same. Similarly, it was observed(13) that Hg(II)-assisted aquation of $Co(NH_3)_5Cl^{2+}$ in nitrate media proceeds by paths involving Hg^{2+} and $HgNO_3^+$ with intermediates having approximately 2 and 98% efficiencies, respectively, for conversion to $Co(NH_3)_5(NO_3)^{2+}$, independent of nitrate concentration. It was therefore concluded(12) that neither intermediate can be the same as the one in the nitrosation reactions. The same authors(12) also found that the assisted aquations of $Co(NH_3)_5(DMSO)^{3+}$ (DMSO = dimethyl sulphoxide) by MnO_4^- and of $Co(NH_3)_5N_3^{2+}$ by nitrous acid do not proceed via the same intermediate. Plots of the product ratio $[Co(NH_3)_5Y^{(3-n)+}]/[Co(NH_3)_5(OH_2)^{3+}]$ against nucleophile concentration were linear for constant as well as variable μ. Smaller values of slopes obtained for $Co(NH_3)_5(DMSO)^{3+}$ than those for $Co(NH_3)_5N_3^{2+}$ indicate that their intermediates are different(12).

For the base hydrolysis of $(NH_3)_5CoX^{2+}$ (X = Cl^-, Br^-, I^-, NO_3^-) an S_N1CB mechanism has been proposed(14,15) on the basis of competition experiments with various entering ligands. But arguments have also been presented in favor of the assumption that the intermediate has some memory of the leaving group(16,17), and that an experimentally significant proportion of the base hydrolysis occured by I_d rather than D mechanism.

Our own results to be presented will show that I_d mechanism was prevailing in most of the replacements we investigated. Let us first recall the much investigated aquation of trans-$Co(en)_2Cl_2^+$ and of trans-$Co(en)_2(NO_2)Cl^+$ for which a bond-breaking mechanism was generally accepted(18). We have studied these aquations in a range of binary mixed solvents, including methanol, ethanol, dioxan, and acetone-water solutions extended to 10% (v/v) water content(19). Both complexes aquated completely in water-rich mixtures and incompletely in solutions containing higher percentages of organic components. The logarithms of rate constants correlated linearly with Grunwald-Winstein solvent Y values, with slopes of m=0.32 for the dichloro- and m=0.09 for the chloronitro-complex. The lower susceptibility of the chloronitro-complex to solvent ionizing power indicates that the remaining bonding to the leaving chloride in the transition state is larger than in the dichloro-complex. It was previously suggested that in the aquation of the chloronitro-complex ion the leaving chloride does not completely dissociate before the entering water molecule starts to associate; the term "solvent assisted dissociation" has been coined to describe this situation(18,20). Alternatively, it could be claimed that the aquations of both dichloro- and chloronitro-complexes are I_d processes, where "accidental bimolecularity" resulted in different bond strengths of the incoming water molecule to the central metal atom of the complexes in the transition state.

The next reaction for which we presented evidence in favour of I_d mechanism is the aquation of pentaammine(dimethylsulphoxide)cobalt(III) perchlorate in water-organic solvent mixtures (non-aqueous components were dioxan, acetone, acetonitrile, methanol, ethanol, and n-propanol)(21). The D mechanism is represented by equations (1) and (2), and the I_d mechanism by equation (3) ($M^{3+} = Co(NH_3)_5^{3+}$):

$$M(DMSO)^{3+} \underset{k_{-1}}{\overset{k_1}{\rightleftharpoons}} M^{3+} + DMSO \tag{1}$$

$$M^{3+} + H_2O \underset{k_{-2}}{\overset{k_2}{\rightleftharpoons}} M(H_2O)^{3+} \tag{2}$$

$$M(DMSO)^{3+} + H_2O \underset{k'_{-1}}{\overset{k'_1}{\rightleftharpoons}} M(H_2O)^{3+} + DMSO \tag{3}$$

The rate constant k'_1 for the I_d mechanism should be sensitive to the mole fraction of water provided the change of mole fraction of water in the bulk solvent is reflected by changes in the amount of water in the solvation shell of $Co(NH_3)_5(DMSO)^{3+}$ when the activated complex is formed. This is so because water should be considered as entering explicitly into the activated complex ("accidental bimolecularity"). Furthermore we have shown that the changes in aquation rate constant do not parallel changes in the heats of mixing of DMSO with the solvent components and hence that the interaction between DMSO and solvent components is not the controlling factor in the change of k_{aq}. We have also shown that the dependence of k_{aq} on the Grunwald-Winstein Y-factor is very small and hence that changes in the water concentration do not significantly affect the solvating power of the solvent for the activated complex as compared to the reactant complex. It appears that the importance of the mole fraction of water in the solvation shell of $Co(NH_3)_5(DMSO)^{3+}$ overrides the importance of other factors mentioned, and that in an I_d mechanism the rate constant can show a marked dependence on the mole fraction of water, as has been found experimentally(21). On the other hand the k_{aq} for the D mechanism should be insensitive to the mole fraction of water in the solvent medium because water does not enter the activated complex $[Co(NH_3)_5 \ldots DMSO)^{3+}]^{\ddagger}$.

The next group of octahedral replacement reactions extensively studied is that of the base hydrolysis of $Co(NH_3)_5X^{2+}$, for which an S_N1CB mechanism has been proposed(14). The competition ratios $[Co(NH_3)_5Y^{2+}]/[Co(NH_3)_5(OH)^{2+}][Y^-]$ obtained with base hydrolyses of pentaammine-complex ions in presence of anions Y^- showed little dependence on the nature of the leaving group(14), which would suggest a common intermediate of reduced coordination number. We determined(16) the competition ratios for base hydrolysis of $Co(NH_3)_5$ $(DMSO)^{3+}$ with $Y^- = N_3^-$, NCS^-, NO_2^-, AcO^-, and SO_4^{2-} as 12.8 ± 0.7, 9.9 ± 0.2, 7.8 ± 1.0, 1.8 ± 0.6, and 3.7, respectively, using the same base hydrolysis conditions as Buckingham et al.(3). Our data for DMSO leaving ligand show that the competion ratios are about 50% higher than those of Buckingham et al.(14) (which we could reproduce, however). The ion pairing seems to be of minor importance since the competition ratios DMSO as leaving ligand remained constant when the solvent composition was changed from aqueous to 50% (v/v) aqueous dioxane solution. It appears that the intermediate has appreciable memory of the leaving group, and that a clear-cut D mechanism is not operating in this instance.

Another group of reactions I would like to discuss, is the replacement in pentacyano(ligand)-ferrate(II) ions, which has been extensively studied(22-29). As pointed out(26-29) the D and I_d mechanisms can equally well explain the main kinetic observations. These can be summarized as follows: limiting reaction rates, at sufficiently large concentrations of entering ligand Y, were observed with all leaving ligands, X, except water, when the replacement in aqueous solution obeyed the second-order rate law: $d[Fe^{II}(CN)_5Y]/dt = k_y[Fe(CN)_5(OH_2)^{3-}][Y]$. Neglecting the reverse replacement of Y by X, because of large excess(27) of Y over X (>200-fold), and because of the relative stability of the final reaction product $Fe^{II}(CN)_5Y$, application of the steady-state approximation to the hypothetical intermediates $Fe(CN)_5^{3-}$ (D mechanism) and $Fe(CN)_5(OH_2)^{3-}$ (I_d mechanism) will give the same rate law with the observed rate constants (4) and (5).

$$k_{obs.}^{D} = \frac{k_1 k_2 [Y]}{k_{-1}[X] + k_2[Y]} \quad (4) \qquad k_{obs.}^{I_d} = \frac{k_1'[H_2O] k_2'[Y]}{k_{-1}'[X] + k_2'[Y]} \quad (5)$$

where k_1 and k_{-1}, and k_1' and k_{-1}' are the rate constants for the formation, and its reverse, of the hypothetical intermediates $Fe(CN)_5^{3-}$ and $Fe(CN)_5(OH_2)^{3-}$, respectively, and k_2 and k_2' the rate constants of formation of $Fe^{II}(CN)_5Y$ from these potential intermediates. Therefore, both the D and I_d mechanism can explain the fact that, at sufficiently large concentrations of Y, limiting reaction rates are observed. For $Fe(CN)_5(OH_2)^{3-}$, i.e. $X = H_2O$, $k_{-1}[H_2O] \gg k_2[Y]$, equation (4) for the D mechanism reduces to (6), in agreement with the observed second order rate law(13,24).

$$k_{obs.}^D = \frac{k_1 k_2 [Y]}{k_{-1}[H_2O]} \qquad (6)$$

If the I_d mechanism is operating, the aqua-complex ($X = H_2O$) will react according to (7),

$$Fe(CN)_5(OH_2)^{3-} + Y \underset{k_{-2}'}{\overset{k_2'}{\rightleftarrows}} Fe^{II}(CN)_5Y + H_2O \qquad (7)$$

and, neglecting the reverse reaction, $k_{obs.}^{I_d}$ is given by (8)

$$K_{obs.}^{I_d} = k_2' [Y] \qquad (8)$$

It follows tha D and I_d mechanism should be distinguishable if the replacement of water by Y in $Fe(CN)_5(OH_2)^{3-}$ is carried out in a solvent which does not co-ordinate, or co-ordinates relatively slowly. The choice of solvents for this purpose appears to be limited; ethylene glycol seems to be satisfactory. If the D mechanism is operating the slopes of the straight lines obtained on plotting $k_{obs.}^D$ against Y should decrease with increasing water concentration (equation (6)). Moreover, in pure ethylene glycol, at a sufficiently high concentration of Y, the second-order rate law should not hold (equation (4), $X = H_2O$). However, if the I_d mechanism is operating, equation (8) requires that the second-order rate law applies in pure and mixed solvents, and this was strictly borne out by our results (27). Namely, $k_{obs.}$ for the replacement of water in $Fe(CN)_5(OH_2)^{3-}$ with pyridine, 3-cyanopyridine, nicotinamide, 4-aminopyridine, and cyanide in ethylene glycol solvent containing $\geqslant 0.055$ mol dm^{-3} water was found to be practically independent of water concentration. Values of $\Delta H^{\neq}$ and $\Delta S^{\neq}$ for the replacement of water by Y in the aquacomplex are almost identical for water and ethylene glycol solvents. This support the assertion that Y replaces water in both solvents.

In the replacement $Fe(CN)_5(OH_2)^{3-} + Y \xrightarrow{k_f} Fe^{II}(CN)_5Y + H_2O$ it was observed(26) that k_f depends on the nature of Y which may be attributed to a change of k_2 (the rate constant of the reaction of intermediate $Fe(CN)_5^{3-}$ with Y). If this were true, then the reacion intermediate should discriminate between reagents, and the D mechanism would be ensured. On the other hand, it was pointed out (26,28) that uncharged ligands react at almost the same rates (although the basicity of these reagents is different), and that the differences in rates are due to differences in the charges on Y. This claim was also supported by other investigators(30). The kinetic results could still be interpreted in terms of a D mechanism, where k_2 is very near to diffusion - controlled limit(23,24) and is therefore very much influenced by interionic forces(31).

The alternative explanation is in terms of an I_d mechanism. The entering reagent does not contribute energetically to the transition state (accidental bimolecularity), but its chance of filling the position vacated by dissociation of water depends on its charge. Therefore, only uncharged entering ligands should react at roughly the same rates. Consequently, the I_d mechanism explains the kinetic data equally well as does the D mechanism. It appears that a possible way of distinguishing between these two mechanisms is by variation of solvent compositions, as described above.

We also recently presented arguments(28) in favour of the I_d mechanism for the replacements of 3-cyanopyridine (3CN-py) by pyridine (py) in $Fe(CN)_5(3CN-py)^{3-}$, and of nitrosobenzene (PhNO) by CN^- in $Fe(CN)_5(ONPh)^{3-}$ in solutions containing different water concentrations. The water contents were varied by addition of sorbitol, glucose and sucrose. The replacements were carried out in the presence of a ⩾200 fold excess of the entering ligand Y, and a 10 fold excess of the leaving ligand X, over the concentration of the starting complex. Under these conditions rate constants for the D mechanism (eq. 4), and for the I_d mechanism (eq. 5), reduce to

$$k_{obs.}^D \simeq k_1 \quad \text{and} \quad k_{obs.}^{I_d} \simeq k_1' [H_2O]$$

Consequently, in the case of an I_d mechanism $k_{obs.}^{I_d}$ should depend on the water activity, since water "accidentally" enters the activated complex. A large, practically linear, reduction in rate with decreasing water concentration was indeed observed, and the results were interpreted in terms of an I_d mechanism via a $Fe(CN)_5(OH_2)^{3-}$ reaction intermediate. The observed decrease in rate upon addition of saccharides is not caused by an increase in viscosity of these solutions. It was previously found(27) that the rate of replacement of water in $Fe(CN)_5(OH_2)^{3-}$ by various ligands is practically the same in aqueous solution and in aqueous solutions containing 50% glucose or sucrose. This is to be expected in a dissociative type process.

Another objection which might be raised is that k_1' (or k_1, if D mechanism operates) is a function of the nature of solvent, so that the decrease of $k_{obs.}$ might be caused by the decrease of k_1' (or k_1). Indeed, the $Fe(CN)_6^{4-}$ is a much stronger electron-pair donor than $Fe(CN)_6^{3-}$, as manifested by a large shift of the half-wave potential of this reversible one-electron reduction to the more negative $E_{1/2}$ values in organic solvents(32). Drastic changes in $E_{1/2}$ values were found, due to the interaction between the iron (II) species and the solvent molecules(32): e.g. in aqueous solution, methanol, ethanol, nitromethane, dimethylsulfoxide, N,N-dimethylformamide, and 1-methylpyrrolidinone-(2) the $E_{1/2}$ values are 0.95, 0.46, 0.29, -0.07, -0.27, -030, and -0.35 volts, respectively(32).

We found that the effect of the solvent on the rate of dissociation of the ligand X from $Fe^{II}(CN)_5X$ is very much dependent on the nature of the solvent. The more negative the $E_{1/2}$ of the mentioned redox system in a particular solvent, the larger the rate in this solvent. The increase in rate can be for several orders of magnitude, depending on the $E_{1/2}$. For example, we found that the rate constant of disappearance of 3×10^{-5} mol dm^{-3} $Fe(CN)_5(4CN-py)^{3-}$ in ethanol, in presence of an about 100 fold excess of oxygen (yielding most probably a μ-peroxo-complex) is 2.5×10^{-4} s^{-1} at 25°C. The rate constant of the same reaction in aqueous solution is smaller by a factor of about 10^3 (unpublished results). It follows that the organic solvents can only increase the rate of dissociation of $Fe^{II}(CN)_5X$ relative to that in water because there are more electrons on the metal centre in organic solutions. Thus, the reduction of the rate we observed when the water concentration was decreased by adding saccharides can be adequately interpreted in terms of an I_d mechanism. One would only expect that the reduction of water concentration by addition of saccharides will be much more effective than the addition of, e.g., ethylene glycol and glycerol, in the same concentrations, due to the accelerating effect of these solvents on the rate of dissociation of X. This is what we previously observed(27).

The last replacement reaction I would like to discuss is that of the binuclear dimer of $Fe(CN)_5(OH_2)^{3-}$, for which the structure $(CN)_4Fe(\mu-CN)(\mu-NC)Fe(CN)_4^{6-}$, involving two non-linear cyanobridges, has been proposed(33). However, this structure would require a deviation from linearity of the bridging cyanide for roughly 90° and is therefore unlikely(34). We believe that the binuclear complex is singly bridged, and we have proposed the following mechanism for its reaction with Y (eqs. 9 and 10):

$$(NC)_5Fe^{II}(\mu-NC)Fe^{II}(CN)_4(OH_2)^{6-} + H_2O \underset{k_{-1}}{\overset{k_1}{\rightleftharpoons}} 2Fe(CN)_5(OH_2)^{3-} \quad (9)$$

$$Fe(CN)_5(OH_2)^{3-} + Y \underset{k_{-2}}{\overset{k_2}{\rightleftharpoons}} Fe^{II}(CN)_5Y^{3-} + H_2O \quad (10)$$

The dissociation of the dimer is the rate-determining step in which the iron-nitrogen bond is broken and water coordinated, yielding $Fe(CN)_5(OH_2)^{3-}$. We determined ratios of rate constants (competition ratios) for the reactions of various ligand pairs (Y_1 and Y_2) with $Fe(CN)_5(OH_2)^{3-}$ and its binuclear dimer(34). The competition ratios at 25°C with Y_1, Y_2 defined as (a) pyridine(py), PhNO, (b) 3CN-py, 4CN-py, (c) 4CN-py, py, (d) 4CN-py, PhNO, and (e) 3CN-py, PhNO are: for the binuclear dimer, (a) 1.35, (b) 1.04, (c) 1.61, (d) 2.17, (e) 2.21; for $Fe(CN)_5(OH_2)^{3-}$, (a) 1.32, (b) 1.01, (c) 1.66, (d) 2.15, (e) 2.19. Almost equal competition ratios obtained with the dimer and the monomer in the presence of the same pairs Y_1, Y_2 suggest that the aqua-complex is the only intermediate in the replacement involving the binuclear complex, and that the binuclear complex is singly bridged(34).

The almost equal competition ratios obtained with the dimer and the monomer also suggest that the reactions (9) and (10) follow an I_d mechanism, as does the replacement of water by Y in the monomer.

Furthermore, the decrease in water content of aqueous solutions of the dimer by added saccharides reduced the reaction rate of the dimer with Y, as should be expected for the reaction (9), in case of an I_d mechanism. For example, when the molarity of water was brought from 55.5 to 27.7 by adding sorbitol or glucose, the rate constant of the reaction of the dimer with pyridine decreased from 2.4×10^{-2} to 1.0×10^{-2} s^{-1} at 25°C (unpublished results).

In addition, the ratio of rate constants k_{-1}/k_2, in the I_d replacements scheme (11) and (12),

$$Fe^{II}(CN)_5 Y_1 + H_2O \underset{k_{-1}}{\overset{k_1}{\rightleftharpoons}} Fe(CN)_5(H_2O)^{3-} + Y_1 \quad (11)$$

$$Fe(CN)_5(H_2O)^{3-} + Y_2 \overset{k_2}{\longrightarrow} Fe(CN)_5 Y_2 + H_2O \quad (12)$$

$$\frac{1}{k_{obs.}} = \frac{1}{k_1[H_2O]} + \frac{k_{-1}[Y_1]}{k_1[H_2O] k_2[Y_2]} \quad (13)$$

could be determined from linear plots(34) of $1/k_{obs.}$ against $1/[Y_2]$ (or against $[Y_1]$), (eq. 13). The ration k_{-1}/k_2 for the ligand pairs Y_1 and Y_2 defined, as previously, (a) pyridine, PhNO and (e) 3CN-py, PhNO are (a) 1.33, and (e) 2.19. These values are almost identical with the competition ratios (a) and (e) in the reaction of $Fe(CN)_5(H_2O)^{3-}$ with the same pairs Y_1 and Y_2. Thus, the replacements $Fe(CN)_5(H_2O)^{3-} + Y$, and $Fe(CN)_5 Y_1 + Y_2$ follow the same (I_d) mechanism(34).

In conclusion it may be said that, in the octahedral replacements discussed, often there is no sharp distinction between D and I_d mechanisms. The extent of discrimination of a potential five-coordinate intermediate for entering reagents varies considerably with the nature of this intermediate. Whereas, e.g., in $Fe^{II}(CN)_5 X + Y$ replacements discussed discrimination between reagents is mainly a function of the electrical charge of Y, the intermediates $Co(CN)_5^{2-}$

and $Co(NH_3)_4SO_3^+$ discriminate between reagents with no marked correlation with charge(35). It appears logical that the shorter the life of the five-coordinated intermediate (due to its reactivity) the smaller is its ability for discrimination and the larger the I_d character of the replacement. The high reactivity of most intermediates in replacements of transition metal complexes makes the clear-cut D mechanism much less frequent that it was originally thought to be.

An important contribution to the understanding of the mechanism of octahedral substitutions is expected from measurements of the volumes of activation ($\Delta V^{\ddagger}$). However, Langford recently pointed out(36) that, prior to the interpretation of data coming from $\Delta V^{\ddagger}$ measurements, it is necessary to evaluate the large contribution of the nonlabile ligands of the complex to the volume change. For example, as one ligand is lost in a dissociative reaction the nonlabile ligands strengthen bonds(36). This reduces the volume of the primary coordination sphere, followed by the collapse of the solvent onto this sphere(36). Unfortunately, this role of the nonlabile ligands is not yet adequately known.

REFERENCES

1. Symbolism employed by C. H. Langford and H. B. Gray, "Ligand Substitution Processes" W. A. Benjamin, Inc., New York (1965).
2. A. Haim and H. Taube, Inorg. Chem. 2:1199 (1963).
3. D. A. Buckingham, I. I. Olsen, A. M. Sargeson, and H. Satrapa, Inorg. Chem. 6:1027 (1967).
4. A. M. Sargeson, Pure Appl. Chem. 33:527 (1973).
5. W. G. Jackson and A. M. Sargeson, Inorg. Chem. 15:1986 (1976).
6. C. H. Langford and W. R. Muir, J. Am. Chem. Soc. 89:3141 (1967); J. W. Moore and R. G. Pearson, Inorg. Chem. 3: 1334 (1964).
7. J. E. Byod and W. K. Wilmarth, Inorg. Chim. Acta Rev. 5:7 (1971).
8. A. Haim and W. K. Wilmarth, Inorg. Chem. 1:573 (1962); R. Grassi, A. Haim, and W. K. Wilmarth, ibid. 6:237 (1967).
9. D. R. Strans and J. Yandell, Inorg. Chem. 9:751 (1970).
10. A. M. Sargeson, Pure Appl. Chem. 33:527 (1973); C. K. Poon, Coord. Chem. Rev. 10:1 (1973).
11. F. A. Posey and H. Taube, J. Am. Chem. Soc. 79:255 (1957).
12. W. L. Reynolds, S. Hafezi, A. Kessler, and S. Holly, Inorg. Chem. 18:2860 (1979).
13. W. L. Reynolds and E. R. Alton, Inorg. Chem. 17:3355 (1978).
14. D. A. Buckingham, I. I. Olsen, and A. M. Sargeson, J. Am. Chem. Soc. 88:5443 (1966); 89:5129 (1967); 90:6539, 6654 (1968), and refs. therein.
15. D. A. Buckingham, I. I. Creaser, and A. M. Sargeson, Inorg. Chem. 9:655 (1970).

16. M. Birus, W. L. Reynolds, M. Pribanic, and S. Asperger, Croat. Chem. Acta, 47:561 (1975).
17. W. L. Reynolds and S. Hafezi, Inorg. Chem. 17:1819 (1978).
18. F. Basolo and R. G. Pearson, "Mechanism of Inorganic Reactions", Wiley, New York, p. 239 (1967).
19. M. Pribanic, M. Birus, D. Pavlovic, and S. Asperger, J.C.S. Dalton, 2518 (1973).
20. T. P. Jones, W. E. Harris, and W. J. Walace, Can. J. Chem. 39:2371 (1951).
21. W. L. Reynolds, M. Birus, and S. Asperger, J.C.S. Dalton, 761 (1974); W. L. Reynolds, S. Asperger, and M. Birus, J.C.S. Chem. Comm. 823 (1973).
22. D. Pavlovic, I. Murati, and S. Asperger, J.C.S. Dalton, 602 (1973).
23. H. E. Toma and J. M. Malin, Inorg. Chem. 12:2080 (1973).
24. H. E. Toma, J. M. Malin, and E. Giesbrecht, Inorg. Chem. 12:2084 (1973).
25. Z. Bradic, D. Pavlovic, I. Murati, and S. Asperger, J.C.S. Dalton, 344 (1974).
26. Z. Bradic, M. Pribanic, and S. Asperger, J.C.S. Dalton, 353 (1975).
27. D. Pavlovic, D. Sutic, and S. Asperger, J.C.S. Dalton, 2406 (1976).
28. I. Murati, D. Pavlovic, D. Sutra, and S. Asperger, J.C.S. Dalton, 500 (1978).
29. M. A. Blesa, J. A. Olabe, and P. J. Aymonino, J.C.S. Dalton, 1196 (1976).
30. A. D. James and R. S. Murray, J.C.S. Dalton, 326 (1977).
31. D. N. Hague, "Fast Reactions", Wiley-Interscience, p. 12-14, New York (1971).
32. V. Gutmann, G. Gritzer, and K. Danksagmüller, Inorg. Chim. Acta, 17:81 (1976).
33. G. Emschwiller and C. K. Jørgensen, Chem. Phys. Letters, 5:561 (1970).
34. R. Juretic, D. Pavlovic, and S. Asperger, J.C.S. Dalton, 2029 (1979).
35. M. L. Tobe, "Inorganic Reaction Mechanism", p. 93, Nelson, London (1972).
36. C. H. Langford, Inorg. Chem. 18:3288 (1979).

USE OF ELECTRON PARAMAGNETIC RESONANCE SPECTROSCOPY TO STUDY THE INTERACTION BETWEEN COBALT SCHIFF BASE COMPLEXES AND PHOSPHINES OR PHOSPHITES IN SOLUTION[†]

J. Barrie Raynor and Gerard Labauze

Department of Chemistry, The University
Leicester, U.K.

Abstract — EPR measurements on frozen solutions of a wide range of phosphine or phosphite adducts of Co^{II} Schiff base complexes have given well resolved spectra exhibiting cobalt and phosphorus hyperfine structure. From analysis of the ^{31}P hyperfine tensor, the R-P-R bond angle can be deduced in the phosphine or phosphite, PR_3. The calculated spin densities in the cobalt and phosphorus orbitals allows a determination of the relative contributions of σ- and π-bonding in the Co-P bond and their influence upon the phosphine or phosphite bond angle. From a full analysis of the ^{31}P polarisation tensor, a correlation is found between its magnitude and the magnitude of the splitting of the cobalt d_{xz} and d_{yz} orbitals. In analysing the ^{59}Co spin Hamiltonian parameters, use is made of the theory of McGarvey and the approximations of Attanasio.

1. INTRODUCTION

Electron Paramagnetic Resonance is a useful tool to probe the ground state, stereochemistry and bonding in complexes of divalent cobalt. The basic distinction between high spin and low spin Co^{II} (d^7) complexes is readily seen when the EPR spectra of magnetically dilute crystals or frozen solutions is recorded. High spin Co^{II} is a ^4F state ion and only yields an EPR spectrum at temperatures close to that of liquid helium. A typical one is shown in Figure 1 and is characterised by wide-ranging g-values tipically extending up to ca 6. From the g-values, information may be deduced about the stereochemistry of the complex which is usually based on

---

[†] along this paper $1G = 10^{-4}$ T

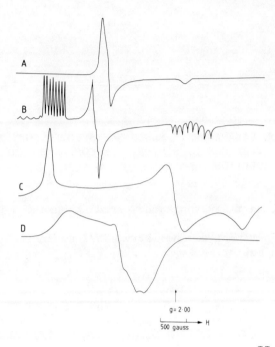

Fig. 1. Representative EPR spectra of high spin Co^{II} complexes recorded at 4.2 K. (a) [Co(Me$_6$tren)Cl]Cl, (b) Co(terpy)Cl$_2$ (c) Co(δ-pic)$_2$Cl$_2$ and (d) Co thermolysin. (From F.S. Kennedy, H. A. O. Hill, T. A. Kaden, and B. L. Vallee, Biochem. Biophys. Res. Comm. 48:1533 (1972)).

a tetrahedron, trigonalbipyramid or octahedron(1). Hyperfine coupling to cobalt is often seen, but gives no additional information.

The low spin alternative is a doublet state ion whose EPR spectrum is readily seen at 77 K. g-values typically lie in the region 1.7-3.5 although there are examples in which the g-values lie beyond these limits. The ground state of low-spin Co^{II} is notoriously varied and depends heavily upon the stereochemistry, the nature of the ligands and especially to any axial perturbation. In $C_{2v}(x)$ symmetry, "square planar" complexes, e.g. Co(acacen), have a d_{yz} ($^2A_2$) ground state which has two other doublet states close by (d_{z^2}, $^2A_1$ and d_{xz}, $^2B_1$) which interact by spin-orbit coupling since their energy separation is comparable to the spin-orbit coupling constant. A full analysis of the g-tensor has been carried out by McGarvey(2) which invokes also a nearby quartet state. Axial coordination, e.g. Co(acacen)NH$_3$ destabilizes the $d_{z^2}(^2A_1)$ orbital which becomes more σ antibonding than in the planar complexes and causes a change in ground state to $d_{z^2}(^2A_1)$ (Figure 2).

Very strong axial interactions ultimately stabilise a high spin state in, e.g. Co(3MeO-SALen)H_2O(3). Complexes with these ground states have been thoroughly reviewed by Daul, Schlapfer and von Zelewsky(4).

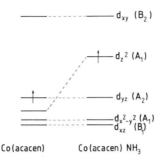

Fig. 2. The order of the d-orbital energy levels in low spin d^7 Co^{II} complexes, with the effect of added base in the axial position destabilising the d_{z^2} orbital.

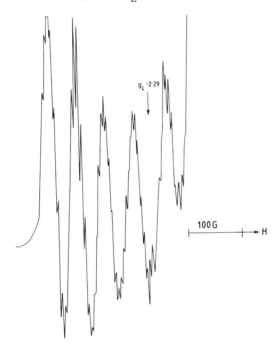

Fig. 3. Some of the perpendicular ^{59}Co features of $[Co(CN)_4Cl_2]^{4-}$ species in a KCl host lattice showing superhyperfine coupling to two equivalent chlorine nuclei. (From M. C. R. Symons and J. G. Wilkinson, J. Chem. Soc. Dalton, 1086 (1972)).

This paper concentrates on complexes having the $d_{z^2}(^2A_1)$ ground state in which solvent or base interaction along the z-axis is a very sensitive probe of structure and bonding. No "octahedral" complexes exist which have this ground state, the nearest being those complexes stabilised in diamagnetic host lattices related to $[Co(CN)_4X_2]^{4-}$. When KCl crystallises in the presence of $Co(CN)_5^{3-}$ ions, the complex ion replaces a K^+ and an appropriate number of Cl^- ions to form species such as $[Co(CN)_4(NC)_2]^{4-}$, $[Co(CN)_4Cl_2]^{4-}$, $[Co(CN)_5Cl]^{4-}$. These ions are characterised by hyperfine coupling to the axial nitrogen or chlorine atoms (Figure 3)(5). In alcoholic solution, the spectrum of $Co(CN)_5^{3-}$ is well resolved (Figure 4) and upon addition of nitrogen bases or phosphines, (shfs) to N or P is clearly seen on the parallel Co features (Figures 5 and 6). Analysis of the spin Hamiltonian parameters (Table I) shows the distribution of unpaired electron density and illustrates how influential axial ligands are. A noteworthy feature about these axial ligand (hfs) is that coupling to P is very large, thus making it a particularly sensitive probe to electron distribution in cobalt complexes. How this is achieved and the results obtained will be considered in detail below.

2. SCHIFF BASES WITH P BASES

A range of nine Schiff Base and related complexes of cobalt(II) with up to seven phosphines or phosphites interacting in the axial position have been studied in CH_2Cl_2 solution(6) (Table II). Their EPR spectra were recorded at 77 K and at about 200 K. The frozen solution spectra exhibited three well separated g-features at ca. g = 2.14-2.47, 2.11-2.18 and 2.02. The two g-features at higher field exhibited well resolved cobalt (hfs) and each line was split into two by further coupling to one phosphorus atom. Resolution on

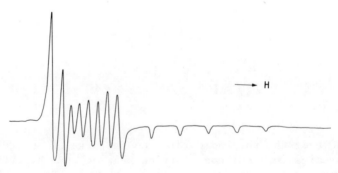

Fig. 4. EPR spectrum of a frozen solution of $[Co(CN)_5]^{3-}$ in Ethanol at 77 K.

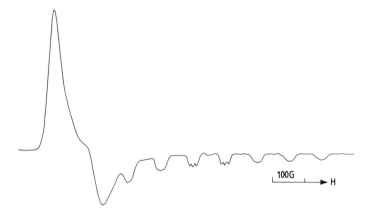

Fig. 5. EPR spectrum of a frozen solution of $[Co(CN)_5]^{3-}$ with excess of Et_2NH at 77 K.

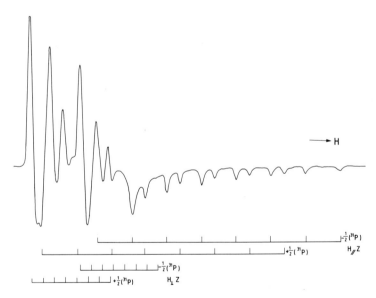

Fig. 6. EPR spectrum of a frozen solution of $[Co(CN)_5]^{3-}$ with excess of PPh_3 at 77 K.

the low field g-feature was usually poor, but hyperfine couplings could be calculated from the isotropic value. The observation of splitting of the cobalt or phosphite was coordinate in solution. Typical EPR spectra are shown in Figure 7.

The spectra were all interpreted in terms of a $^2A_1$ ground state ($\underline{a}\, d_{z^2} + \underline{b}\, d_{x^2-y^2}$). In C_{2v} symmetry, the d_{z^2} and $d_{x^2-y^2}$ orbitals both have the same symmetry. The coefficients are such that $a^2 + b^2 = 1$. The analysis of the g- and hyperfine tensors was carried out according to the theory of McGarvey(2) which takes into account the effect of admixture of excited quartet states. Because there are more unknowns than equations we approximate $C_3 = C_4 = C_5$ and put $C_6 = 0$ since it is espected to be small(7). In the McGarvey notation (2),

$$C_1 = \frac{\lambda\, \alpha_{3d}^2 \beta_{b_1}^2}{\Delta E(^2A_1 - ^2B_1)} \tag{1}$$

where α_{3d}^2 = spin density in d_{z^2} orbital and is calculated from P/P^{2+} where $P^{2+} = 0.0254$ being the value of P for an electron entirely in a 3d orbital on Co^{2+} $\lambda = 533$ cm^{-1} and $\beta_{b_1}^2$ takes into account electron delocalisation to the ligands. The coefficients C_2-C_6 are similar

Table I. Spin Hamiltonian Parameters for $Co(CN)_4B_2$ complexes

	$Co(CN)_4Cl_2^{2-}$ [a]	$Co(CN)_5^{2-}$ [b]	$Co(CN)_5NHEt_2^{3-}$ [c]	$Co(CN)_5P\Phi_3^{3-}$ [c]
$g_\perp$	2.019	2.012	2.002	1.993
$g_\parallel$	2.287	2.181	2.315	2.169
A (Co), cm^{-1}	+.0127	+.0083	+.0094	+.0076
A (Co), cm^{-1}	+.0067	−.0031	−.0007	−.0030
A (ligand), G [d]	13	37.6	11.5	150
A (ligand), G	5	31.3	−	120
A_{iso}(ligand), G	7.6	33.4	−	130
A_{aniso}(ligand), G	5.3	4.2	−	18
s, %	0.4	2.5	−	2.7
p, %	5.2	7.8	−	8.9
P, cm^{-1}	.01117	.0178	.0215	.0162
3d, %	52	79	83	72

[a] M. C. R. Symons and J. C. Wilkinson, J. Chem. Soc. A, 2069 (1971).
[b] R. J. Booth and W. C. Lin, J. Chem. Phys. 61:1226 (1974).
[c] J. B. Raynor and R. L. Nye, J. Chem. Soc. Dalton, 504 (1976); Inorg. Chim. Acta, 22:L28 (1977).
[d] A in cm^{-1} = $4.6693 \cdot 10^{-5} \cdot g \cdot A$ in G.

Table II. Some Schiff Base Complexes of Cobalt(II)

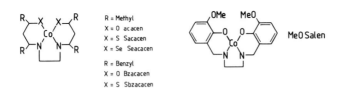

R = Methyl
X = O acacen
X = S Sacacen
X = Se Seacacen

R = Benzyl
X = O Bzacacen
X = S Sbzacacen

MeO Salen

R = Methyl
Sacac

R = Phenyl
Phsacac

Saphen

BASES: P(OMe)$_3$ PBu$_3$
        P(OEt)$_3$ PPh$_3$
        P(OPh)$_3$ dppe
        P(OEt)$_2$Ph

and are mainly dependent upon the energy gaps ΔE between $^2A_1$ and $^2B_2$, $^4B_1$, $^4B_2$, $^4A_2$ and $^2A_2$ respectively. The equations are solved for the following parameters; a,b,P,k,C_1,C_2 and C_3.

3. SPIN DENSITY ON COBALT

For all the adducts, the coefficient a is ca. 0.99 showing that the ground state is essentially d_{z^2}. The spin density in this orbital α^2_{3d} varies considerably with P base and with equatorial ligating atom. For any one phosphine, α^2_{3d} decreases in the order acacen > Sacacen > Seacacen. The absence of **hfs** to N of acacen or Se of Seacacen shows that there is very little spin delocalisation via σ-bonding from the x and y component of the d_{z^2} orbital. Since the total spin density on Co + P is <1, then the balance of spin

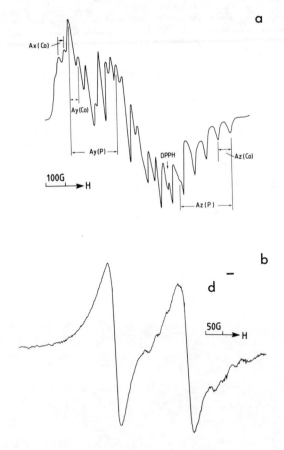

Fig. 7. EPR Spectrum of (a) a frozen solution of Co(Seacacen).P(OMe)$_3$ at 77 K and (b) at 200 K.

density must be delocalised onto the Schiff base via π-bonding (Figure 8). This can only be achieved in complexes where the cobalt is above the plane of the ring and the complex is saddle shaped. This mechanism may well help to stabilise the shape of these molecules since π bonding typically increases from <5% in acacen to 20% in Sacacen and Seacacen. Furthermore, it accounts for why only one base is added, not two. Two bases may only be added in strictly planar systems where rigidity prevents bending (as in Co porphyrins) and where better electron delocalisation can be achieved to two P atoms. Representative results are given for adduct of Co(Seacacen) in Table III.

4. SPIN DENSITY ON PHOSPHORUS

The experimental hyperfine coupling to P on each g-feature need correcting for indirect dipolar coupling using the point dipolar approximation(8). The principal value of this tensor was 2G. The corrected anisotropic hyperfine tensor was strongly asymmetric and was decomposed into two axially symmetric tensors. The choice of which two, of the three possible tensors, was governed by the obvious need for one principal direction to be along the Co-P bond (the z-axis). The principal direction of the second tensor could be along either the x- or the y- axes. Each was considered in turn. When the second tensor was along the y-axis, the principal value was large and positive (2-4x that of the first tensor). This was rejected because there is no way in which positive spin density can get into a p (or d) orbital on phosphorus along the y-axis. When the second tensor was along the x-axis, the principal value was negative and could readily be explained by spin polarisation.

Spin polarisation of phosphorus p_x and p_y π- electrons by an unpaired electron in a Co-P σ- orbital must be contrasted with the much more common spin polarisation of σ electrons (e.g. αC-H) by an unpaired electron in a C p_z (π) orbital as frequently found in organic radicals. In our molecules, the d_z2 (α-spin) unpaired electron

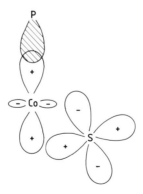

Fig. 8. π-Bonding between Co and equatorial S atoms.

Table III. Spin Hamiltonian Parameters for Co(Seacacen) with various bases.

	P(OMe)$_3$	P(OEt)$_3$	P(OPh)$_3$	PBu$_3$	PPh$_3$	dppe
g_x	2.285	2.281	2.433	2.332	2.447	2.411
g_y	2.170	2.173	2.170	2.180	2.180	2.183
g_z	2.028	2.026	2.024	2.031	2.024	2.024
$-A_x(Co), 10^{-4} cm^{-1}$	18	18	18	18	18	18
$-A_y(Co), 10^{-4} cm^{-1}$	33	34	28	30	24	26
$A_z(Co), 10^{-4} cm^{-1}$	64	62	68	61	71	67
$P, 10^{-4} cm^{-1}$	173	170	221	178	234	211
K	.096	.104	.139	.112	.139	.133
c_1	.025	.025	.030	.027	.033	.032
c_2	.042	.043	.061	.048	.062	.058
c_3	.122	.118	.124	.131	.125	.123
α_2^2	.68	.67	.87	.70	.92	.83
α_3^2	.033	.031	.034	.032	.036	.034
$\Delta\,(^2B_1-^2A_1), cm^{-1}$	14670	14080	15390	13720	14920	13890
$\Delta\,(^2B_1-^2A_1), cm^{-1}$	8550	8380	7660	7780	7930	7600
$A_x(P), G$	196	195	214	131	99	105
$A_y(P), G$	246	242	261	165	152	173
$A_z(P), G$	278	277	293	196	175	100
A_{zz}, G	19	21	19	18	13	9
$-A_{xx}, G$	33	30	31	22	34	45
$c_3^2 s(P), \%$	6.5	6.5	7.0	4.5	3.9	4.2
$c_3^2 p(P), \%$	9.4	10.4	9.4	8.9	6.4	4.5
θ	103°	104°58'	103°41'	106°34'	104°14'	101°5

polarises each of the pairs of electrons in the $3d_{xz}(Co) - 3p_x(P)$ and $3d_{yz}(Co) - 3p_y(P)$ π bonds to give some β-spin in the $3p_x$ and $3p_y$ orbitals on phosphorus (see Figure 9). Likewise, there will be some polaristion of the non bonding $2p_x$ and $2p_y$ electron pairs on phosphorus but this is likely to be a much smaller effect. If the d_{xz} and d_{yz} cobalt orbitals are equal in energy, then they will be polarised equally and transmission of apparent spin density onto $3p_x$ and $3p_y$ phosphorus orbitals will be equal and effectively cancel each other out. The result will be an axially symmetric phosphorus hyperfine tensor. On the other hand, if the d_{xy} and d_{yz} cobalt orbitals are different in energy, then polarisation of the electron pairs in these orbitals will be to different extents with the result that the apparent spin density in the $3p_x$ and $3p_y$ phosphorus orbitals will be different. This difference in apparent spin density in the x and y directions will contribute to the anisotropic hyperfine tensor such as to reduce its symmetry from axial. Using the example of Co(saphen).P(OMe)$_3$, the corrected ^{31}P anisotropic hyperfine tensor is (in Gauss)

```
         x    y    z
        -48   4   44
```

which decomposes into either

```
    (a)        -13   -13   26    tensor 1
           and -34    17   17    tensor 2

or  (b)        -31   -31   62    tensor 1
           and -17    34  -17    tensor 2
```

In (a) the principal value of tensor 1, $A_{zz}26G$ (z), is positive and represents 12.9% spin density in the phosphorus $3p_z$ orbital, whilst the principal value A_{xx} of tensor 2, -34G (x), is negative and arises from the imbalance (difference) in the polarisation of the $3p_x$ and $3p_y$ orbitals. In the alternative pair of decomposed tensors (b), the principal value, A_{zz}, 62G (z) is suspiciously high and would represent 31% spin density in the phosphorus $3p_z$ orbital. Furthermore, the principal value of the second tensor, A_{yy}, 34G (y) is positive. Since there is no mechanism for positive spin density to get into the p_x orbitals on phosphorus, and this value is far too high anyway ($\equiv$ 17% spin density in $3p_y$ orbital), then we reject (b) as a possibility.

The magnitude of the spin polarisation tensor, A_{xx}, must be related to the rhombic distortion in the molecule, since it disappears in axially symmetrical molecules. The amount of rhombic distortion may be calculated from the cobalt hyperfine and g- tensors. Some of the deduced parameters in solving the McGarvey(2) equations are related to the energy of excited states. In particular, the excited doublet state ^2E (in C_{4v} symmetry) is split in rhombic symmetry to $^2B_1(d_{xz})$ and $^2B_2(d_{yz})$ ($C_{2v}(Z)$). The energy above the ground

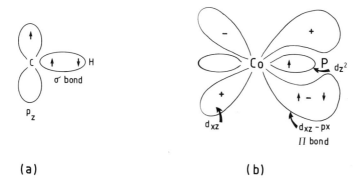

Fig. 9. (a) Spin polarisation of (a) a C-H σ-bond by a π-electron in p_z on C and (b) a Co-P π bond by a σ-electron in d_{z^2} on Co

state of these two states $^2B_1$ and $^2B_2$ are given by the coefficients C_1 and C_2 using equations similar to (1). The energy difference between $^2B_1$ and $^2B_2$ we call Δ. We have plotted A_{xx} vΔ for all our results and these points are shown in Figure 10. The line drawn is the least squares best fit and uses a cubic expression. The correlation coefficient for the data is 0.84 which is very good. The relationship is qualitatively acceptable since the greater the splitting of the $^4B_1$ and $^4B_2$ states (approximately proportional to the $^2B_1$ and $^2B_2$ states), the greater the difference in the polarisation of the ^{31}P in the x and y directions, and so $|A_{xx}|$ becomes larger.

From n, the p:s ratio of spin density on P, the internal angle θ of the phosphine or phosphite can be calculated using the equation

$$\cos \theta = \frac{1.5}{2n+3} - \frac{1}{2}$$

Differences between θ for any one phosphine or phosphite when coordinated to different Schiff bases are small and suggests that there is little steric interaction between the ligand and the down-turned Schiff base. The calculated angles θ are slightly larger than those found from X-ray diffraction studies on typical transition metal complexes. For example, our bond angles in PPh_3 adducts (average $\theta = 104°\ 57'$) compared with $102°\ 36'$ in $PPh_3Cr(CO)_5$, and for $P(OPh)_3$ adducts (average $\theta = 102°\ 34'$) compared with $100°$ in $P(OPh)_3Cr(CO)_5$.

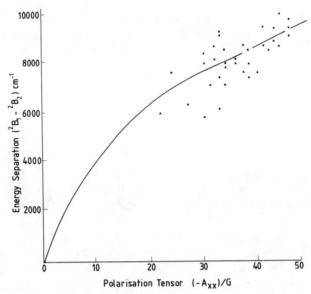

Fig. 10. Plot relating the axial distortion Δ against the ^{31}P polarising tensor ($-A_{xx}$).

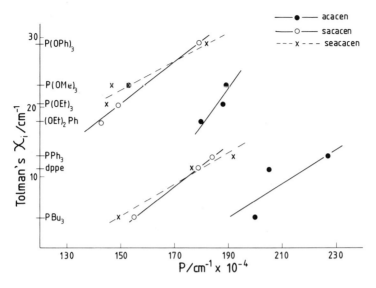

Fig. 11. Relationship between the Tolman χ_i parameter and the cobalt P parameter for a range of cobalt Schiff base adducts

We suggest this is due to the inevitably longer Co-P bond length because of the unpaired electron in the d_{z^2} orbital, which results in weaker repulsion between the Co-P bonding electrons and the P-O or P-C bonding electrons thus allowing the O-$\hat{\text{P}}$-O or C-$\hat{\text{P}}$-C bond angles to increase.

Tolman's χ_i measures σ-effects only(9). If we plot χ_i for our phosphines against the cobalt P parameter, then we find (Figure 11) that the PBu_3, dppe and PPh_3 adducts lie on a series of parallel lines for each Schiff base in a predictable fashion, and that the $P(OEt)_2Ph$, $P(OEt)_3$, $P(OMe)_3$, $P(OPh)_3$ adducts lie on another series of parallel lines, at values of P lower than expected, i.e. shifted to lower P values relative to the phosphines. We observe that the phosphite complexes all have larger C_S^2 and C_P^2 than phosphine complexes and also they have smaller P values. This is explained by the stronger π interactions which shorten the Co-P bond and place more spin density on the phosphorus with a corresponding lower spin density on cobalt. We believe that the shift to lower P in Figure 11 for the phosphite adducts reflects the greatly increased π bonding in these complexes.

The trends within each group reflect the known increased σ-donating power (inductive effect) in the series Bu > dppe > Ph and OEt > OMe > OPh. Thus the inductive effect in PBu₃ pushes spin density in the d_{z^2} orbital away from the cobalt onto the equatorial ligands and lowers P.

In conclusion, we see that EPR spectroscopy acts as a very sensitive probe to monitor the bonding interaction between a solute and a base in a non coordinating solvent.

REFERENCES

1. See e.g. A. Desideri, L. Morpurgo, J. B. Raynor, and G. Rotilio, Biopys. Chem. 8:267 (1978); F. S. Kennedy, H. A. O. Hill, T. A. Kaden, and B. L. Vallee, Biochem-Biophys. Res. Comm. 48:1533 (1972).
2. B. R. McGarvey, Can. J. Chem. 53:2498 (1975).
3. M. A. Hitchman, Inorg. Chim. Acta, 26:237 (1978).
4. C. Daul, C. W. Schlapfer, and A. von Zelewsky, Structure and Bonding, 36:130 (1979).
5. M. C. R. Symons and J. G. Wilkinson, J. Chem. Soc. Dalton, 1086 (1972).
6. G. Labauze and J. B. Raynor, J. Chem. Soc. Dalton, in the press (1980).
7. D. Attanasio, G. Dessy, V. Fares, and G. Pennesi, Chem. Phys. Letters, (1980).
8. B. A. Goodman and J. B. Raynor, Adv. Inorg. Chem. and Radiochem. 13:135 (1970).
9. C. A. Tolman, J. Am. Chem. Soc. 92:2953 (1970).

AN NMR STUDY OF SOLVENT INTERACTIONS IN A PARAMAGNETIC SYSTEM

R. M. Golding, R. O. Pascual, and C. Suvanprakorn

The University of New South Wales
P.O. Box 1
Kensington, N.S.W. 2033. Australia

Abstract - The temperature dependence of the proton nmr spectra of dithiocarbamato iron(III) complexes is markedly solvent dependent. A study is made of the temperature dependence of the nmr shifts for the N-CH$_2$ protons in tris(N,N-dibutyldithiocarbamato)iron(III) in acetone, benzene, carbon disulfide, chloroform, dimethylformamide, pyridine and some mixed solvents. This contribution shall outline first how the nmr shifts may be interpreted in terms of the Fermi contact interaction and the dipolar term in the multipole expansion of the interaction of the electron orbital angular momentum and the electron spin dipol-nuclear spin angular momentum. This analysis yields a direct measure of the effect of the solvent system on the environment of the transition metal ion. The results are analysed in terms of the crystal field environment of the transition metal ion with contributions from (a) the dithiocarbamate ligand (b) the solvent molecules and (c) the interaction of the effective dipole moment of the polar solvent molecule with the transition metal ion complex. The model yields not only an explanation for the unusual nmr results but gives an insight into the solvent-solute interactions in such systems.

1. INTRODUCTION

Several years ago we reported(1) that the reduction potential,s $E_{\frac{1}{2}}$, for several iron(III) dithiocarbamates in acetone-water mixtures was markedly dependent with solvent variations at least as great as 0.3 V between $E_{\frac{1}{2}}$ values for an iron(III) dithiocarbamate in acetone and in water. We interpreted these results by considering, in a simple way, the interaction of a polar molecule in the solvent with the iron(III) dithiocarbamate.

It is known also that the nmr spectra of these compounds are not only markedly temperature dependent (2 - 5) but at least some are solvent dependent(6). For example, the magnitude of the nmr shift for the N - CH_2 protons in tris(N,N-diethyldithiocarbamato) iron(III) increases with increasing temperature(2) when the compound is in a chloroform solution whereas the magnitude of the nmr shift decreases with increasing temperature when the compound is in a dimethylformamide solution.

In this paper we shall extend our earlier interpretation of the redox results to the nmr data for the N - CH_2 protons in tris(N,N-diethyldithiocarbamato) iron(III). We shall show that the solvent dependence of the nmr shifts can be interpreted as arising from solvent interactions with the iron(III) dithiocarbamate system. Although the solvent interactions are small compared with the electronic interactions within the transition metal iron complex the effect is marked since in these cases for the d^5 iron system there are low lying electronic states where the energy separation is sensitive to small changes in the crystal field environment of the transition metal ion.

2. THEORY

The crystal field environment of the d^5 system in the series of iron(III) dithiocarbamate complexes may be expressed in the form of a potential with octahedral symmetry with a small trigonal component. Therefore, we may write for our purpose in this paper the effective crystal field potential, V(r), in terms of a series of spherical harmonics for the transition of metal ion as:

$$V(\underline{r}) = \frac{2}{5}(21\pi)^{\frac{1}{2}} \left[\frac{\sqrt{7}}{2\sqrt{3}} Y_{40}(\theta,\phi) + \frac{\sqrt{5}}{2\sqrt{6}} Y_{44}(\theta,\phi) + \frac{\sqrt{5}}{2\sqrt{6}} Y_{4-4}(\theta,\phi) \right] \Delta$$

$$-14(\pi/5)^{\frac{1}{2}} Y_{20}(\theta,\phi) \delta \qquad (1)$$

The form of the first term in equation (1) is such that for the t_2^5 ($^2T_2$) state the crystal field interaction of octahedral symmetry affects the energy level by -2Δ and the second term splits the level by 3δ.

The crystal field parameters Δ and δ may be considered as arising from three terms:

(i) a dominant contribution from the ligands, Δ_ℓ and δ_ℓ
(ii) a contribution from the solvent molecules, Δ_s and δ_s
(iii) a contribution from the interaction of the effective dipole moment of the polar solvent molecules, Δ_d and δ_d.

Hence $\Delta = \Delta_\ell + \Delta_s + \Delta_d$

and $\delta = \delta_\ell + \delta_s + \delta_d$

From the nmr results of the temperature dependence of the proton magnetic resonance shifts for the $N-CH_2-$ protons in tris(N,N-diethyldithiocarbamato) iron(III) in a range of solvents and solvent mixtures we shall explore these solvent interactions.

In interpreting the nmr shift, ΔB, obtained from the nmr spectra of paramagnetic molecules it is usual to consider two terms, the Fermi contact interaction where the contribution may be expressed(7) as

$$\Delta B = a<S_z>/g_N\mu_N \tag{2}$$

where $<S_z>$ is the time-averaged value of the z component of the electron spin and a is the hyperfine interaction constant measured in cycles per second. The dipolar term in the multipole expansion of the interaction of the electron orbital angular momentum and the electron spin dipolar-nuclear spin angular momentum which may be expressed(8) in terms of the magnetic susceptibility components, $\chi_{\alpha\alpha}$, is

$$\frac{\Delta B}{B} = -\frac{\mu_0}{4\pi}\left\{\left[\chi_{xx} - \tfrac{1}{2}(\chi_{xx} + \chi_{yy})\right](3\cos^2\theta-1) + \tfrac{3}{2}(\chi_{xx} - \chi_{yy})\sin^2\theta\cos 2\phi\right\}/3R^3 \tag{3}$$

(g_N is the Landé g-value for the nucleus, μ_N the nuclear magneton, μ_0 the permeability of a vacuum; the polar coordinates of the nmr nucleus in relation to the paramagnetic centre are (R,θ,ϕ), where the z - axis is the principal axis of the complex).

Although equations (2) and (3) are approximations a combination of (2) and (3) will mirror rather closely the temperature dependence of the nmr shifts although care needs to be exercised in a detailed interpretation of the results(9). To proceed, therefore, we need to establish the form of (2) and (3) applicable to the analysis of the nmr study of the solvent interaction with the iron(III) dithiocarbamate.

A previous proton nmr study(2) of tris(N,N diethyldithiocarbamato)iron(III) indicated that the transition metal ion is an intermediate crystal field environment such that the $^6A_1$ state and the lowest lying $^2T_2$ state for the d^5 ion are close in energy. Hence we need to consider both states.

First we shall examine the form of the nmr shift for the $^6A_1$ state. Studies of iron(III) dithiocarbamate complexes with a $^6A_1$ ground state well below the $^2T_2$ state show that the nmr shift, ΔB, is inversely proportional to the absolute temperature, T(10). This indicates that the origin of the nmr shift in this case arises through the contact term.

For the $^2T_2$ level we need to consider the contribution to the nmr shift by the term proportional to $\chi_\parallel - \chi_\perp$ and the Fermi interaction. Since in these compounds the effective crystal field environment of the iron atom is approximately of octahedral symmetry with a small most likely trigonal component(11,12) we shall assume that the form of the Fermi contact interaction is as given in reference (1). The magnetic susceptibility components $\chi_\parallel$ and $\chi_\perp$ can be determined from the appropriate equations in ref.(13). (We shall neglect bonding in this paper determining $\chi_\parallel$ and $\chi_\perp$).

Combining the above results yield the following expression

$$\frac{\Delta B}{B} = \alpha y_2 + \beta y_6 + \gamma(\chi_\parallel - \chi_\perp) \tag{4}$$

where $$y_2 = \frac{(x + 16/3)\exp(x) - (16/3)\exp(-x/2)}{\exp(x) + 2\exp(-x/2) + 3\exp(-E/kT)}$$

and $$y_6 = \frac{105x \exp(-E/kT)}{\exp(x) + 2\exp(-x/2) + 3\exp(-E/kT)}$$

where $x = \zeta/kT$, ζ the spin-orbit coupling constant. E is the energy separation of the $^2T_2$ and the $^6A_1$ states: when E is negative the ground state is the $^6A_1$ state. The energy separation, E, is related to Δ of equation (1) as a first approximation by the expression.

$$E = 2\Delta - 15B - 10C \tag{5}$$

where B and C are the usual Racah parameters(14).

We would expect the electronic and atomic structure of the complex at least near the transition metal ion not to be affected significantly by the solvent. Hence in analysing the nmr results we shall treat the solvent interaction as a perturbation where the dominant crystal field interaction will arise from the ligands. (As we shall show later the results confirm this approach). Therefore, for a specific nmr nucleus in an iron(III) dithiocarbamate the parameters α, β and γ in equation (4) would not change significantly for the N-CH$_2$ protons in tris(N,N-diethyldithiocarbamato) iron(III) from solvent to solvent and we shall treat these parameters as constant for a specific proton in an iron(III) dithiocarbamate. This is analogous to the recent method used to analyse the nmr shifts in a series of lanthanide complexes(15). Hence the temperature dependent nmr results for the N-CH$_2$ protons in tris(N,N-diethyldithiocarbamato) iron(III) in all solvents may be analysed as a single set of data using equation (4) where the energy separation of the $^6A_1$ and $^2T_2$ states, E, and the distortion parameter δ are allowed to vary from solvent to solvent. The best fit yields

3. RESULTS AND DISCUSSION

The 60 MHz proton nmr shift, ΔB, for the $N - CH_2$ protons in tris(N,N-diethyldithiocarbamato)iron(III) was determined in a number of solvents and in some solvent mixtures over a wide temperature range. (The diamagnetic contribution to the nmr shift was obtained by measuring the nmr shift of the $N - CH_2$ protons in the diamagnetic cobalt analogue). In this paper six different solvents were used: acetone, benzene, carbon disulfide, chloroform, dimethylformamide and pyridine. Also three chloroform-dimethylformamide solvent mixtures were used: 0.75, 0.49 and 0.24 mole fraction of chloroform.

The ΔB values for all the nine solvent systems were fitted to equation (4) where the variables E and δ were chosen for each solvent to yield the best fit to all the data. The results are given in Table I.

Figure 1 illustrates the fit with the data for five solvents - chloroform, benzene, carbon disulfide, pyridine and dimethylformamide.

The first significant factor about the results is that the E values are remarkedly similar and correspond to a Δ value of about

Table I. The E and δ values found to give the best fit to the experimental data using equation (4). The Q values of Dance and Miller for some of the solvents are given:

Solvent system	E	δ	Q_1	Q_2
acetone	- 499	- 222	4.65	0
benzene	- 499	- 371	2.43	0
carbon disulfide	- 510	- 242	-	-
chloroform	- 426	- 823	2.91	1.81
dimethylformamide	- 503	- 161	5.14	0
pyridine	- 503	- 225	4.31	0
0.75 mole fraction of chloroform in diemethylformamide	- 472	- 503	-	-
0.49 mole fraction fo chloroform in dimethylformamide	- 492	- 331	-	-
0.24 mole fraction of chloroform in dimethylformamide	- 496	- 245	-	-

(The best fit was achieved when the E values were about the same and about $- 500$ cm^{-1})

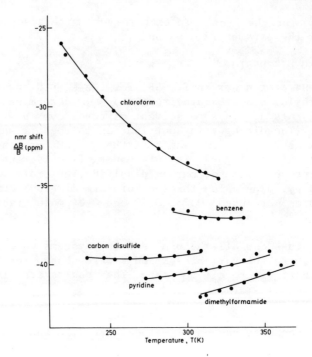

Fig. 1. Temperature dependence of the CH_2 isotropic shifts of tris (N,N-diethylditiocarbamate)iron(III) in various solvents. The dots indicate the experimental data and the curves the best fit of equation (4). The acetone results lie between the pyridine and dimethylformamide results.

27,000 cm^{-1}. (The Racah parameters for the Fe^{3+} ion are about 1075 cm^{-1} and 3900 cm^{-1} for B and C respectively). The variation in E corresponds to less than 0.4% variation in the octahedral symmetric component of the crystal field potential. In addition the distortion parameter is small for all cases and hence not only is the trigonal crystal field component to the crystal field potential small but the variation from solvent to solvent is small. This support our approach of treating the solvent interaction as a perturbation.

In order to analyse these results further we need to deduce the value for the distortion parameter arising from the ligand in an attempt to examine the variation of the specific components to δ, namely, δ_s and δ_d from solvent to solvent.

Dance and Miller(16) have found an empirical parameterization of solute-solvent interactions expressed as a sum of products of solute and solvent parameters P_i and Q_i respectively which we may express for our case as

$$\delta = P_0 + P_1Q_1 + P_2Q_2$$

Q_1 is interpreted as the electric field experienced by the solute in a solvent cavity and Q_2 a measure of hydrogen bonding. They analysed 225 data spanning 24 solvents and 16 solutes/phenomena in 7 chemical classes. The Q values of Dance and Miller for the appropriate solvents are given in Table I.

A linear regression analysis of the δ values in Table I for the solvents with given Q values yields

$$\delta = -553 + 74.7Q_1 - 269.0Q_2 \tag{6}$$

Hence from the equation (6) we may equate δ_ℓ to -553 cm^{-1} as this would correspond to the case with no solvent solute interactions ($Q_1 = Q_2 = 0$).

Next we need to explore whether we can ascertain the signs of δ_s and δ_d and their relative magnitudes. If the solvent molecules are treated as negative point changes at a distance b from the transition metal ion then δ_s would be proportional to $1/b^3$ and on this simple model we would expect δ_s to be positive. On the other hand δ_d is proportional to d/b^4 where d is the effective dipole moment of the solvent molecule. But in this case δ_d based on this simple model would be negative. Choosing the δ_ℓ value as -553 cm^{-1} it follows from the results in Table I that for all solvents except chloroform $\delta_s + \delta_d$ must be positive. Since we expect δ_s to be positive and δ_d negative we may infer that for all solvents other than chloroform the dominant solvent interaction must arise through the δ_s term. In the case for chloroform it appears that δ_d is the dominant term.

We may gain further insight into this difference if we return to our simple model. When b is large we would expect that $|\delta_s|>|\delta_d|$ but as b becomes smaller there may be a region when $|\delta_s|<|\delta_d|$. From the results it would suggest that the chloroform-transition metal ion distance is relatively small and that the chloroform molecules is able to penetrate closer to the transition metal ion than any other of the solvents used. This is supported by X-ray studies[17] which show that when iron(III) dithiocarbamates are crystallized from chloroform solution in at least some of the compounds chloroform is complexed, the chloroform protons being bonded to the sulfur atoms of the dithiocarbamate ligands.

Finally, a plot of the distortion parameter as a function of the mole fraction of chloroform in dimethylformamide clearly shows that there is not a linear relation between the mole fraction of chloroform and the distortion parameter.

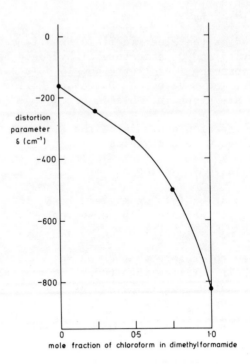

Fig. 2. Dependence of the distortion parameter δ on the mole fraction of chloroform in dimethylformamide.

As seen from Figure 2 the results imply that the dimethylformamide molecules at least partially shield the chloroform molecules from the iron(III) dithiocarbamate solute.

In this analysis of the nmr data we have that the results for a variety of solvents and over a wide temperature range may be interpreted as arising from small changes in the crystal field environment of the iron atom due to two solvent interactions - a term which is an intrinsic property of the solvent and a second term arising from a solute-solvent interaction. Although the application of the model has been simplified the results nevertheless give an insight into the effect of the solvent on the nmr shifts of these iron dithiocarbamate complexes.

REFERENCES

1. R. M. Golding, L. L. Kok, K. Lehtonen, and R. K. Nigam, Aust. J. Chem. 28:1915 (1975).
2. R. M. Golding, W. C. Tennant, C. R. Kanekar, and A. H. White J. Chem. Phys. 45:2688 (1966).

3. R. M. Golding, W. C. Tennant, J. M. P. Bailey, and A. Hudson, J. Chem. Phys. 48:764 (1968).
4. R. M. Golding, B. D. Lukeman, and E. Sinn, J. Chem. Phys. 56:4147 (1972).
5. R. M. Golding, Pure Appl. Chem. 32:123 (1972).
6. W. D. Perry and R. S. Drago, J. Am. Chem. Soc. 93:2183 (1971).
7. R. M. Golding, Molec. Phys. 8:561 (1964).
8. R. J. Kurland and B. R. McGarvey, J. Magn. Reson. 2:286 (1970)
9. R. M. Golding and L. C. Stubbs, J. Magn. Reson. 33:627 (1979).
10. R. M. Golding, "Applied Wave Machanics", Van Nostrand, London (1969).
11. R. M. Golding and H. J. Whitfield, Trans. Faraday Soc. 62:1713 (1966).
12. R. M. Golding, Molec. Phys. 12:13 (1967).
13. M. Das, R. M. Golding, and S. E. Livingstone, Transition Met. Chem. 3:112 (1978).
14. J. S. Griffith, "The theory of Transition Metal Ions", Cambridge University Press, London (1961).
15. H. A. Bergen and R. M. Golding, Austr. J. Chem. 30:2361 (1977).
16. I. G. Dance and I. R. Miller, to be published.
17. G. L. Raston and A. H. White, J. Chem. Soc. 2405 (1975).

PROTONATION AND COMPLEXATION EQUILIBRIA OF MACROMOLECULAR BIOLIGANDS

IN AQUEOUS AND MIXED SOLVENT SOLUTIONS. THE SOLVENT EFFECT

Kalman Burger and Bela Noszal

L. Eötvös University, Department of Inorganic and
Analytical Chemistry
H-1443 Budapest, P.O. Box 123, Hungary

Abstract - The experimental and computational methods of the determination of the effect of solvents on the protonation and metal complexation equilibria of macromolecular bioligands is discussed. The effect of the solvent dependence of the secondary structure (e.g. conformation) of bioligands on these processes is especially emphasized.

1. INTRODUCTION

Coordination chemical equilibria of macromolecular polyfunctional ligands in solution are extremely sensitive to changes in the composition of the solvent(1). Besides the changes in the dielectric properties of the system and the different solvation of the functional groups in solvents of different composition the solvent dependence of the conformation (secondary structure) of the macromolecules may exert also a decisive effect on the protonation and metal complex formation equilibria. The change in the conformation of the macromolecular ligands alters their coordination chemical behaviour not only by changing their space requirement but also by the alteration of the H-bonding interactions in the system. The latter changes are reflected in the protonation equilibria and complex formation reactions.

2. MICROCONSTANTS - GROUP CONSTANTS

The interpretation of the solvent induced changes in the coordination equilibria was made difficult by the overlap of the multistep protonation (or complexation) equilibria of the polyfunctional ligands. The formation of protonation (or complexation) isomers, i.e. |species with equal proton (or metal) - ligand ratios but of different structure| made the assignment of the conventionally determined macro-equilibrium constants to the corresponding functional groups very difficult or even impossible. The extremely high number of the microconstants(2,3) which would be needed to characterize correctly these groups made their experimental determination in the macromolecular system impossible. A ligand containing $\underline{n}$ basic groups can be completely characterized by $2^{\underline{n}-1} \cdot \underline{n}$ protonation microconstants. The protonation equilibria and the corresponding 12 microconstants, of a molecule containing only three basic groups are shown e.g. in Figure 1. The macromolecular bioligands contain always more functional groups than 3 or 4. In no system could more than eight microconstants be determined so far(4). To overcome this difficulty, group constants(5) have been derived from the microconstants to characterize the functional groups in overlapping processes.

The differences between the values of microconstants assigned to the protonation of the same functional group (e.g. K^A, K^A_B and K^A_{BC} in Figure 1 representing the protonation of group A) are due to the different protonation of the other basic groups in the system

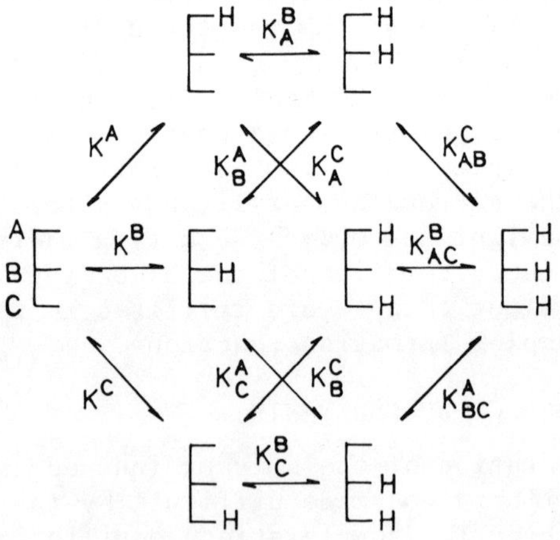

Fig. 1. The protonation equilibria of a molecule containing three basic groups A, B, C.

which influences their electrophylic or nucleophylic nature. If these groups are, however, separated from group A by more than three atoms their effect on the electron density of group A becomes negligible(6). Thus $K^A = K_B^A = K_{BC}^A$ in Figure 1 and accordingly $K^B = K_A^B = K_C^B$ and $K^C = K_B^C = K_{AB}^C$. In this way the microconstants characterizing the protonation of the same groups can be considered equal, and called group constants (K_A, K_B and K_C). On the basis of these considerations the correlation between the macroconstants (β values) and the microconstants

$$\beta_1 = K^A + K^B + K^C$$

$$\beta_2 = K^A K_A^B + K^A K_A^C + K^B K_B^C = K^B K_B^A + K^C K_C^A + K^C K_C^B$$

$$\beta_3 = K^A K_A^B K_{AB}^C = K^B K_B^C K_{BC}^A = K^C K_C^B K_{BC}^A$$

known from the litterature can be given e.g. for the equilibria in Figure 1:

$$\beta_1 = K_A + K_B + K_C$$

$$\beta_2 = K_A K_B + K_A K_C + K_B K_C$$

$$\beta_3 = K_A K_B K_C$$

Since the number of β values known from the measurements (i.e. the number of equations) and the number of the group constants is equal, the latter ones can be computed by a suitable method from the former ones. As a matter of course, the primary experimental data can be also used directly for the determination of the group constants(7).

In polypeptides usually more than three carbon atoms separate the two nearest donor groups (if the peptide bonds having rather low donor properties are neglected), the only exceptions are the C terminal aspartic and glutamic acids in which three C-atoms separate two carboxylates. The smallest in-chain distance between two donor groups is to be found in the sequence asp – asp, where six C-atoms separate the two carboxyls. Since conjugation does not help the electron shift between the nearest donor groups through the peptide chain the mutual influence of the different groups on the electron densities can be neglected.

Donor groups separated by several other atoms may, however, come near to each other by steric reasons causing a change in their coordination chemical behaviour. The formation of an H-bond, e.g. between two such groups increases the protonation constant of one of the pillar atoms and decreases that of the other(8). Such interactions decrease the number of species in the system. The species

$\begin{bmatrix}-H\\ \\ \\ \end{bmatrix}$ and $\begin{bmatrix}\\-H\\ \\ \end{bmatrix}$ furthermore $\begin{bmatrix}-H\\-H\\ \\ \end{bmatrix}$ and $\begin{bmatrix}-H\\ \\-H\\ \end{bmatrix}$ in Figure 1 become equal due to the formation of an H-bond between groups A and B. The new picture is represented in Figure 2 indicating also the group constants characterizing the processes (K_f represents the protonation resulting in the formation of the H-bonds, K_d the protonation resulting in its cleavage and K_C the protonation of group C). The correlation of these group constants and the primary complex products (β) can be given as follows:

$$\beta_1 = K_f + K_C$$
$$\beta_2 = K_f K_C + K_f K_d$$
$$\beta_3 = K_f K_C K_d$$

It is seen that the number of equations is equal to the number of unknowns. So these group constants can be also easily calculated.

One can conclude from all these that the group constant concept is suitable for the characterization of overlapping coordination equilibria in macromolecular systems in all cases when the effect of electron shift between the nearest donor groups on the electron density of the donor atoms can be neglected. However hydrogen bond or metal bridge formation between these groups does not prevent the application of the concept. Thus group constants are suitable for the quantitative characterization of protonation and complexation equilibria of polypeptides.

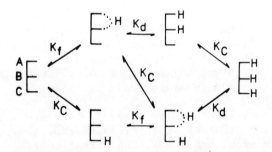

Fig. 2. The protonation equilibria of a molecule containing three basic groups A, B, C, when H-bond formation between A and B groups occurs.

3. THE EFFECT OF SOLVENT ON POTENTIOMETRIC DATA

For the study of the solvent effect, comparable equilibrium constants have to be determined in water and in solution made with non-aqueous solvents or solvent mixtures. Potentiometric (usually pH-metric) equilibrium measurements are used for this purpose in polyfunctional systems. The solvent effect makes the application of potentiometry somewhat difficult. The substitution of water by organic solvents results in changes of the autoprotolysis constant of the solvent changing the pH-scale. The lower relative permittivity of the system favours association processes which have to be considered, e.g., in the determination of the ionic strength of the solution. Diffusion potentials at the liquid junctions connecting the galvanic cell with the reference electrode may falsify the measured data.

To overcome these difficulties and to get data which can be directly correlated with the concentration of free hydrogen ions (or in metal complexation processes to free metal ions) the measuring equipment has to be carefully calibrated. In the study of protonation equilibria this should be done e.g. using acidimetric titration curves of acids and bases strong enough to be completely dissociated in the given solvent (usually $HClO_4$ is titrated with NaOH). The $[H^+]$ and electromotive force (E, mV) values derived from these titration curves obtained in solution of analogous composition with those used in the equilibrium studies but not containing the ligand (e.g. the polypeptide) are used for the calculation of E_o, j_a, j_b and K_w constants in the equation:

$$E = E_o + 0.059 \log [H^+] + j_a[H^+] + j_b K_w [H^+]^{-1}$$

where j_a and j_b are the constants for the correction of the diffusion potential in the acidic and basic range of titration, K_w is the autoprotolysis constant of the solvent which takes into consideration the change in the pH scale caused by the solvent. For these calculations a simple computer program is used, which also compares mV values calculated according to the equation with the measured data proving in this way the accuracy of the constants used and indicating the millivolt and concentration ranges in which the experimental work could furnish data suitable for the equilibrium studies.

4. THE EFFECT OF SOLVENT ON THE PROTONATION EQUILIBRIA OF POLYPEPTIDES.

In macromolecular system (e.g. in polypeptides, proteins, etc.) a dominant part of the solvent effect is due to the solvent dependence of the conformation of the molecule. To separate this effect from that of the differences in solvation of functional groups, alteration of the relative permittivity of the system, etc., equilibrium constants of analogous processes measured in macromolecular systems and in analogous model systems of low molecular weight have to be compared. The protonation group constants, determined as discussed above, of two corticotropin(9) fragments (of the N terminal 4 amino acid containing $ACTH_{1-4}$ and of the N terminal 32 amino acid containing $ACTH_{1-32}$ polypeptides) measured in water and in trifluoroethanol - water and propylen glycole - water solvent mixtures are shown in Table I(10).

Since $ACTH_{1-4}$ does not undergo structural changes due to the solvent effect its protonation constants reflect the variation in the basicity of its functional groups (carboxylate, amino and phenolyc hydroxyls) on changing the solvent:

1) Terminal carboxylate groups show higher protonation constants in the solvent mixtures (lg K = 3.88 and 3.89, while lg K = 3.20 was obtained in water) which can be attributed to the lower relative permittivity of solvent mixtures favouring the formation of neutral particles ($RCOO^- + H^+ \rightleftharpoons R\ COOH$).

Table I. Protonation group constants (logarithmic values) of ACTH fragments in aqueous solutions and in solvent mixtures

Functional group	Aqueous solution		Trifluoroethanol-water		Propylene glycol-water	
	$ACTH_{1-4}$	$ACTH_{1-32}$	$ACTH_{1-4}$	$ACTH_{1-32}$	$ACTH_{1-4}$	$ACTH_{1-32}$
Terminal COO^-	3.20	3.54	3.88	4.08	3.84	4.00
Glutamic acid + COO^- Aspartic acid		3.59 4.10 4.25 5.03		4.17 4.73 4.80 5.47		4.21 4.46 5.02 5.29
Histidine N		6.43		6.13		6.06
Terminal NH_2	7.27	7.47	6.77	6.77	6.87	6.82
Lysine $-NH_2$		9.70 10.13 10.45 10.82		8.41 9.49 9.75 9.82		8.56 9.73 9.84 9.84
Tyrosine OH	10.70	10.94 10.98	9.95	9.84 9.87	10.04	9.84 9.84

2) The constants lg K = 6.77 and 6.87 belonging to the terminal amino group are lower than those obtained in water (lg K = 7.27), indicating changes in solvation conditions of unprotonated and protonated amino groups favouring the unprotonated state. In fact, the solvent mixture favours the electrically neutral state of these groups too.

3) The development of reduced constants of phenolic hydroxyls (as compared to aqueous media) is probably due to different effects in the two solvent mixtures. In solutions containing propylene glycol, the interaction of the phenolate oxygen with propylene glycol, being suitable for forming an H-bond of chelate nature with the deprotonated phenolate, is responsible for this phenomenon, while in solutions containing trifluoroethanol the specific solvation of the proton influenced by it causes the lower protonation group constants. (The autoprotolysis constant of the mixture containing trifluoroethanol is $-\lg K_w = 11.04$).

In $ACTH_{1-32}$, the solvent-dependence of the protonation group constants differing from those mentioned above can be explained by modifications in the secondary structure. It is known that the decrease in the water activity of corticotropin solutions results in the transformation of the unordered globule structure of the polypeptide in aqueous solution into the ordered α helix structure(11,12). This brings some donor groups nearer to each other and removes other, furthering the formation of some H-bonds and hindering that of others.

In both solvent mixtures, the lowest value of the constants assigned to the ε-amino groups of $ACTH_{1-32}$ shows a stronger decrease as that compared with the corresponding value for aqueous medium, while the constants belonging to the five most basic groups show a much smaller deviation from each other and from the mean value than in water. According to this, the helical arrangement occurring on decreasing the concentration of water favours the development of one hydrogen bond producing an extremely low stability constant, however, it hinders the formation of further hydrogen bonds with the partecipation of ε-amino or phenolic OH groups leading to a globule structure characteristic of $ACTH_{1-32}$ in aqueous solutions.

The secondary structure (conformation, etc.) of a molecule may completely reverse the solvent effect. For instance all data in Table I show that the replacement of 50 v/v % of the water with organic solvents resulted in an increase of the protonation constants of the carboxylates and in a decrease of those of the amino and phenolate groups in all ACTH containing systems.

Table II presents the protonation macroconstants of BPTI, the basic tripsin inhibitor polypeptide (Kunitz base)(13), measured in water(14) and in a 50 v/v % water - propylene glycol solvent mixture(15). The method for computation of the corresponding group constants in the system containing also polynuclear species is currently under study. The macroconstants show already that in this system the substitution of water by propylene glycol resulted in the decrease of the protonation constants of the carboxylates and the increase of those of the amino groups.

The basic differences in the solvent effect shown by these two polypeptide systems may be due to the dimerisation of BPTI in the solutions investigated. According to the equilibrium measurements these solutions contain differently protonated monomeric and dimeric species. The composition of the protonated species (bound proton: peptide ratios), the protonation equilibrium constants and the dimer formation constants are shown in Table II.

Table II. Protonation macroconstants (logarithmic values) of BPTI in aqueous solution and in 50 v/v % water – propylene glycol mixture

Functional groups	Proton : Peptide ratio	Successive protonation constants in water	in solvent mixture
Phenolic OH	2 : 1	22.28	
	4 : 1	21.16	
$\varepsilon-NH_2$	6 : 1	19.88	
	7 : 1	9.47	
	8 : 1	8.97	10.33
Terminal NH_2	9 : 1	7.90	8.50
$-COO^-$ + Dimer formation	19 : 2	8.27^a	–
	20 : 2	–	13.61^b
$-COO^-$	20 : 2	4.36	–
	22 : 2	8.71	8.28
	24 : 2	7.41	6.07
	27 : 2	9.94	6.42

$$^aK = \frac{[H_{19}P_2]}{[H_9P][H_{10}P]} \quad ; \quad ^bK = \frac{[H_{20}P_2]}{[H_9P][H_{11}P]}$$

where P = BPTI

According to these data, when starting from the monomeric species containing protonated amino and deprotonated carboxylate groups, the decrease in pH results in the dimer formation accompanied in water with the uptake of one or, in the solvent mixture, two protons. The further decrease in pH results in the protonation of the remaining carboxylates in the dimer. With the protonation of the last carboxylate group the dimer dissociates. Thus, it becomes clear that the deprotonation of the terminal carboxyl group and the protonation of one or two other carboxylates and that of the amino groups are essential for the formation of the BPTI dimer in solution. It is however impossible to state on the basis of equilibrium measurements alone the exact position of the hydrogen bonds in the dimer.

The partial substitution of water in the solvent by propylene glycol favours the dimerisation process. This is reflected in an increase of the dimer formation constant and in the fact that the association of monomeric species containing unprotonated carboxylates and protonated amino groups is accompanied in water with the protonation of one carboxylate and in the solvent mixture with that of two. The dimer formation increases the protonation constants of these carboxylates showing that the dimerization of BPTI is due to the formation of intermolecular H-bonds in which these carboxylate-oxygens act as pillar atoms. Thus the pH dependence of the dimerization is determined by the protonation - deprotonation equilibria of these groups. The increased acidity of the other carboxyls in the solvent mixture may be due to the steric effect of the macromolecular dimer.

5. METAL COMPLEXATION EQUILIBRIA OF POLYPEPTIDES

Equilibrium studies of the metal complex formation of biopolymers, e.g. polypeptides or proteins can be based on the potentiometric determination of the free (uncomplexed) metal ion activity in solutions of different pH containing the peptide and the metal ion in different concentrations(16,17). For the measurement of the uncomplexed metal ion concentration in the presence of the metal coordinated by the bioligand, metal electrodes (e.g., silver or mercury), amalgam electrodes (e.g., zinc, cadmium, bismuth), and ion-selective membrane or ion exchange electrodes (e.g., calcium) can be used. The coordination of metal ions to donor atoms protonated at the pH of the solutions to be investigated can be followed by the pH metric equilibrium study of the deprotonation equilibria caused by this process.

The potentiometric equilibrium studies could be supplemented e.g. by UV, IR, ^{1}H and ^{13}C-NMR measurements. In this way specific interactions of the different functional groups could be followed. The comparison of the equilibrium constants measured in the macromolecular system with analogous constants of smaller model ligands could also help in the assignement of the functional groups to the corresponding coordination processes. In systems containing donor groups of different basicities one might determine the binding sites in the macromolecular ligand on the basis of the pH dependence of the equilibrium processes.

Such complex investigations in non-aqueous solvents and especially in water - organic solvent mixtures could lead to the determination of the solvent effect. Such investigations are currently made in our laboratory. In most of the cases the solvent dependence of metal complexation equilibria is in strong correlation with that of the protonation reactions.

REFERENCES

1. K. Burger, "Protonation and Complexation of Macromolecular Polypeptides" in "Metal Ions in Biological Systems", Vol. 9, M. Dekker, New York, p. 213 (1979).
2. N. Bjerrum, Z. Physik. Chem. 106:209 (1923).
3. J. E. Sarneski and C. N. Reilley, "The Determination of Proton Binding Sites by NMR Titrations" in "Essays on Analytical Chemistry". In Memory of Anders Ringbom, ed., E. Wänninen, Pergamon Press, Oxford, New York, Toronto, Sydney, Paris, Frankfurt, p. 35 (1977).
4. D. L. Rabenstein, J. Am. Chem. Soc. 95:2797 (1973).
5. B. Noszal and K. Burger, Acta Chim. Acad. Sci. Hung. 100:275 (1979).
6. J. L. Sudmeier and C. N. Reilley, Anal. Chem. 36:1698 (1964).
7. B. Noszal, to be published.
8. K. Burger, F. Gaizer, B. Noszal, M. Pekli, and G. Takacsi Nagy, Bioinorg. Chem. 7:335 (1977).
9. B. Riniker, P. Sieber, and W. Rittel, Nature New Biol. 235:114 (1972).
10. B. Noszal and K. Burger, Inorg. Chim. Acta, to be published.
11. M. Löw, L. Kisfaludy, and S. Fermandjian, Acta Biochim. Biophys. Acad. Sci. Hung. 10:229 (1975).
12. D. Greff, F. Toma, S. Fermandjian, M. Löw, and L. Kisfaludy Biochim. Biophys. Acta, 439:219 (1976).
13. R. Huber, D. Kukla, O. Epp, and H. Formanek, Naturwissenschaften, 57:389 (1970).
14. K. Burger, I. Zay, F. Gaizer, and B. Noszal, Inorg. Chim. Acta, 34:L239 (1979).
15. K. Burger, I. Zay, and F. Gaizer, to be published.
16. K. Burger, F. Gaizer, I. Zay, M. Pekli, and B. Noszal, J. Inorg. Nucl. Chem. 40:725 (1978).
17. B. Noszal and K. Burger, Inorg. Chim. Acta, 35:L387 (1979).

INTRAMOLECULAR HYDROPHOBIC AND AROMATIC-RING STACKING INTERACTIONS

IN TERNARY COMPLEXES IN SOLUTION

Helmut Sigel

Institute of Inorganic Chemistry, University of Basel
Spitalstrasse 51, CH-4056 Basel, Switzerland

Abstract — The occurrence of intramolecular ligand-ligand interactions in ternary complexes in solution leads to an intramolecular equilibrium between two isomeric complexes, i.e. between an 'open' and a 'closed' species. The source of the ligand-ligand interaction in the 'closed' isomer may be (aside from covalent bond formation, ionic interactions and hydrogen bond formation) a hydrophobic or aromatic-ring stacking interaction between suitable groups of the two different ligands already coordinated to the same metal ion, e.g. between the indole moiety of tryptophanate and the purine moiety of adenosine 5'-triphosphate. More such example are listed and discussed.

1. INTRODUCTION

Ternary or mixed ligand complexes are formed by complexation of a metal ion (M) with two different kinds of ligands (A or B). Such species, M(A)(B), play a role, e.g., in analytical chemistry, in catalytic processes, and in biological systems; it is therefore not surprising that they are receiving more and more attention(1,2).

A common way to quantify the stability of such ternary complexes is by calculating the values of $\Delta \log K_M$ (equation 1)(3-6),

$$\begin{aligned}
\Delta \log K_M &= \log K^{M(A)}_{M(A)(B)} - \log K^{M}_{M(B)} \\
&= \log K^{M(B)}_{M(B)(A)} - \log K^{M}_{M(A)} \\
&= \log \beta^{M}_{M(A)(B)} - \log K^{M}_{M(A)} - \log K^{M}_{M(B)}
\end{aligned} \qquad (1)$$

i.e. by comparing the coordination tendency of, e.g., the ligand B towards M(A) (equation 2) relative to M(aq) (equation 3):

$$M(A) + B \rightleftharpoons M(A)(B)$$

$$K^{M(A)}_{M(A)(B)} = [M(A)(B)]/([M(A)][B]) \qquad (2)$$

$$M + B \rightleftharpoons M(B)$$

$$K^{M}_{M(B)} = [M(B)]/([M][B]) \qquad (3)$$

The value of $\Delta \log K_M$ is also the logarithm of the equilibrium constant due to equilibrium 4.

$$M(A) + M(B) \rightleftharpoons M(A)(B) + M \qquad (4)$$

In accordance with the statistical expectation(5) and the general rule, $K^{M}_{M(L)} > K^{M(L)}_{M(L)_2}$, negative values for $\Delta \log K_M$ (equation 1) are usually observed, but it must be pointed out that certain ligand combinations leads to positive values for $\Delta \log K_M$(2,3,5-9).

In study of ternary complexes one has to distinguish between two possibilities:

(i) ternary complexes in which the two kinds of ligands 'simply coordinate to the same metal ion, i.e. only indirect effects are possible (see e.g., refs 2-5,8),
(ii) those complexes in which also a direct intramolecular ligand-ligand interaction occurs. We shall deal now with the second possibility which allows to influence the stability and structure through the formation of intramolecular 'bonds' between the two different ligands A and B already coordinated to the same metal ion(2).

Covalent bond formation(2) is well-known, e.g., coordinated pyruvate and glycinate form a Schiff base within the coordination sphere of a metal ion(10). Ionic bonds(2) may be formed between oppositely charged side chains: several amino acids are predestinated for such electrostatic interactions(6,11). Hydrogen bond formation also seems possible(12) and is known to exist in solid mixed ligand complexes(13). Hydrophobic and aromatic-ring stacking interactions between suitable groups of the coordinated ligands are further, very subtle possibilities to influence the structure of mixed ligand complexes as well as the position of equilibrium 4; these latter two types of interactions will be in the focus of the remainder of this article.

It is evident that a possible intramolecular ligand-ligand interaction of the mentioned type has not to occur to hundred

percent; such an interaction may well occur only in a certain number
of complex species present. In other words, one must be aware that
two isomers of the ternary complex M(A)(B) must be considered, i.e.
an 'open' and a 'closed' form. This then leads to an intramolecular
equilibrium between these two isomers, which may be schematically
represented as shown in equilibrium 5.

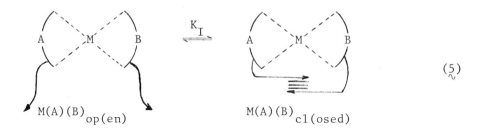

(5)

$$K_I = [M(A)(B)_{cl}] / [M(A)(B)_{op}] \quad (6)$$

The dimensionless constant K_I (equation 6) of equilibrium 5 is
of great interest, because knowledge of this constant allows to
calculate the individual percentages of the two isomers.

Based on a reasoning similar to the one used by Mariam and
Martin(14) for the problem of the extent of chelate formation in
binary complexes(14,15), values of K_I may be determined in the
following way(2,16,17). If the two isomers of equilibrium 5 exist,
then equilibrium 2 may be rewritten:

$$M(A) + B \rightleftharpoons (M(A)(B)_{cl} + M(A)(B)_{op})$$

$$K^{M(A)}_{M(A)(B)} = K_{exp} = ([M(A)(B)_{cl}] + [M(A)(B)_{op}])/([M(A)][B]) \quad (7)$$

For conditions where only the open isomer is formed, i.e. without
an intramolecular interaction, it holds:

$$K_{op} = [M(A)(B)_{op}] / ([M(A)][B]) \quad (8)$$

A combination of equations 6, 7 and 8(16), together with the $\Delta \log K_M$ description(17), results in equation 9.

$$K_I = \frac{K_{exp}}{K_{op}} - 1 = \frac{10^{\Delta \log K_{M/exp}}}{10^{\Delta \log K_{M/op}}} - 1 \quad (9)$$

With equation 9 values for K_I may now be calculated, provided the
values of K_{exp} (experimentally accessible) and K_{op} (often estimated)
or of $\Delta \log K_{M/exp}$ and $\Delta \log K_{M/op}$ are known. However, it must be
emphasized that often the result can be considered only as an estimation because it is derived from differences between constants
which are connected with an experimental error. One may also attempt

to determine K_I directly(18), but this is usually rather difficult because the position of the intramolecular equilibrium 5 is independent of the total concentration of the ternary complex present.

2. RESULTS AND DISCUSSION

Intramolecular aromatic-ring stacking between two suitable ligands within a ternary complex in solution was first shown in 1974 to occur in the ternary Cu^{2+} complexes formed by 2,2'-bipyridyl (Bpy) and adenosine 5'-triphosphate (ATP^{4-}) or inosine 5-triphosphate (ITP^{4-})(19-21); the evidence came from spectrophotometric, NMR and stability studies. In the meanwhile several more such examples, including other metal ions, have been studied(16,18,21-28). X-ray structure determinations of solids of such ternary complexes were first published in 1977(29,30) usually an intermolecular stacking between the aromatic moieties of the coordinated ligands was found (29-32). The first example(33) showing an intramolecular stacking (adenine-bipyridyl overlap) also in the solid was described in 1978 for $[Cu(5'-AMP \cdot H)(2,2'bipyridyl)(H_2O)]_2(NO_3)_2$. It is evident that in crystals the kind of stacking is determined by the lattice; it was pointed out already earlier(32,34) that such stacking interactions contribute significantly towards the stability of the crystal lattice.

The following mixed ligand system, which was studied in solution recently in some detail(35,36), involves Cu^{2+}, 1,10-phenanthroline (Phen) and phenylcarboxylates, i.e.

$\langle\text{phenyl}\rangle-(CH_2)_n-COO^-$ where n = 0 to 5

The results of these measurements, together with those of the corresponding complexes formed with $HCOO^-$ and CH_3COO^-, are given in Table I. For the complexes $Cu(Phen)(HCOO)^+$ and $Cu(Phen)(CH_3COO)^+$ small positive values for $\Delta \log K_{Cu}$ are observed, i.e. equilibrium 4 is slightly displaced to its right hand side. This is in agreement with the general observation that the combination of a heteroaromatic N base with an O donor ligand favors the ternary complexes in systems with Mn^{2+}, Co^{2+}, Ni^{2+}, Cu^{2+} or Zn^{2+}(2-9). However as expected, for the systems with the phenylcarboxylates (n = 0 to 5), which allow the formation of an intramolecular stack with $Cu(Phen)^{2+}$, $\Delta \log K_{Cu}$ is even more positive. Most important: from the data in Table I it is nicely seen that the extent of intramolecular stacking depends upon the flexibility of the coordinated ligands. There seems to be an optimal fit for the stack in $Cu(Phen)(2-phenylacetate)^+$; a tentative structure is shown in Figure 1. In $Cu(Phen)(benzoate)^+$ the overlap of the aromatic rings is certainly not very pronounced and in the ternary complexes with the phenylcarboxylates having a longer chain (n = 2 to 5) the increased flexibility reduces the ligand-ligand interactions. Furthermore, as one would expect, with an

TERNARY COMPLEXES IN SOLUTION   153

Table I. Results[a] regarding the position of the intramolecular equilibrium

$$Cu(Phen)(R-COO)^+_{op} \underset{}{\overset{K_I}{\rightleftharpoons}} Cu(Phen)(R-COO)^+_{cl}$$

for the aromatic-ring stacking in the complex of

⌬—$(CH_2)_n-COO^-/Cu^{2+}/$1,10-phenanthroline

R-COO$^-$	$\Delta \log K_{Cu/exp}$ [b]	$\Delta\Delta \log K_{Cu}$ [c]	K_I [e]	% Cu(Phen)(R-COO)$^+_{cl}$ [f]
HCOO$^-$	0.06	0.05[d]		
CH$_3$COO$^-$	0.04			
n = 0	0.14	0.09±0.03	0.23±0.08	19±5
1	0.46	0.41±0.05	1.57±0.30	61±5
2	0.28	0.23±0.03	0.70±0.12	41±4
3	0.20	0.15±0.03	0.41±0.10	29±5
4	0.22	0.17±0.02	0.48±0.07	32±3
5	0.20	0.15±0.03	0.41±0.10	29±5
β-N.ac.[g]	0.70	0.65±0.04	3.47±0.41	78±3

[a]The data refer to measurements in 50% (v/v) aqueous dioxane at I = 0.1 (NaClO$_4$) and 25°C; they are taken form refs 35 and 36.
[b]See equation 1 and 4.
[c]$\Delta\Delta \log K_{Cu} = \Delta \log K_{Cu/exp} - \Delta \log K_{Cu/op}$ (see[d]), i.e. this value is the decisive parameter of equation 9. The given error limits are estimated.
[d]This value is used as an estimate for the 'open' from (equation 5), i.e. $\Delta \log K_{Cu/op} = 0.05$ (see equation 9).
[e]K_I (equation 5 and 6) was calculated with equation 9.
[f]Percentage of the 'closed' isomer in equation 5 (see also Figure 1).
[g]β-N.ac. = 2-(β-naphthyl)acetate.

increasing size of the involved aromatic-ring systems the extent of stacking increases: the replacement of 2-phenylacetate by 2-(β-naphthyl)acetate (β-N.ac; see Table I) promotes the interaction --; about 80 percent of Cu(Phen)[2-(β-naphthyl)acetate]$^+$ exist in 50% aqueous dioxane in the 'closed' isomeric form(35,36).

Other examples which have been studied in detail are the ternary M(Phen)(ATP)$^{2-}$ complexes, where M^{2+} = Mg^{2+} or Ca^{2+}(2,18,26). The stability of these complexes is also rather large, i.e. $\Delta \log K_M \simeq +0.5$(26). This stability enhancement results from the intramolecular stacking interaction between the purine moiety of ATP and the aromatic-ring system of 1,10-phenanthroline; and indeed the

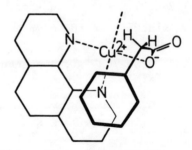

Fig. 1. Tentative and simplfied structure for the ternary stacked Cu(1,10-phenanthroline)(2-phenylacetate)$^+$ complex.

stacked isomer dominates in these systems: it is present to about 90 percent(2) or more(18). As already pointed out, with decreasing size of the involved aromatic-ring systems and increasing flexibility of the ternary complex the extent of stacking decreases. The results of Figure 2 are in agreement herewith: the percentage of the stacked isomer in M(ATP)(tryptophanate)$^{3-}$ complexes is smaller(2,18, 22). It should be noted in this connection that the occurrence of an indole-purine interaction in these ternary complexes has been confirmed in several independent studies by different methods(18, 22,23,25,28).

Intramolecular hydrophobic interactions are closely related to the described aromatic-ring stacking interactions. The first such example was published in 1976(37): a hydrophobic interaction was shown to occur between the Cu^{2+} and Zn^{2+} complexes of 2,2'-bipyridyl and 1,10-phenanthroline with the trimethylsilyl moiety of the also coordinated 3-(trimethylsilyl)-1-propanesulfonate or 3-(trimethylsilyl)propionate(37,38), as well as with alkyl groups of other aliphatic carboxylates(35,38,39); then the attention was focussed on systems with amino acids(17,40).

Though weak, such hydrophobic interactions are also observed between suitable groups in the absence of metal ions(17,37,38,41). For example, ^{1}H-NMR shift studies(17) showed that the methyl resonance of leucine is increasingly shifted upfield with increasing concentrations of 2,2'-bipyridyl or 1,10-phenanthroline, thus indicating that a hydrophobic adduct between these aromatic systems and the isopropyl moiety of leucine is formed. The addition of Zn^{2+} to such a system leads to a much larger upfield shift thus giving clear evidence that the hydrophobic interaction between the isopropyl group of leucine and the aromatic ligands is significantly promoted by the formation of a metal ion bridge between these ligands: about 30 percent of Zn(Phen)(leucinate)$^+$ exist in the

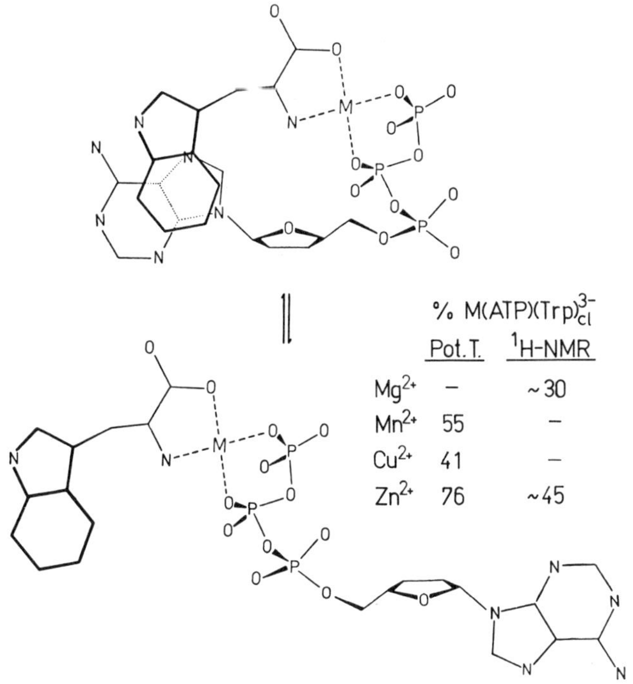

TERNARY COMPLEXES IN SOLUTION

Fig. 2. Intramolecular stacking equilibrium (equation 5) between the stacked and the open isomer of M(ATP)(tryptophanate)$^{3-}$. The inserted estimations of the percentages of the 'closed' isomer were calculated(2) with equation 9 and the stability data obtained from potentiometric pH-titrations(22), or were determined from ^{1}H-NMR shift measurements(18) (the average percentage of the results obtained from the shifts of the resonances of H-8, H-2 and H-1' is given).

	% M(ATP)(Trp)$^{3-}_{cl}$	
	Pot.T.	^{1}H-NMR
Mg^{2+}	–	~30
Mn^{2+}	55	–
Cu^{2+}	41	–
Zn^{2+}	76	~45

'closed' form(17). Estimations of K_I (equations 5 and 9)(2,17), based on published stability data(42), for M(phenylalaninate)(norvalinate) complexes, where M^2 = Co^{2+}, Ni^{2+} or Cu^{2+}, give 21, 5 and 11 percent, respectively, for the isomer with an intramolecular interaction.

To give a sounder basis about this kind of hydrophobic interaction some ^{1}H-NMR shift results(17) are summarized in Figure 3. In addition it must be pointed out in this connection that in an association where the aliphatic side chain of an amino acid is located above or below the plane of an aromatic ring the signals of the aliphatic protons should be shifted upfield, relative to the signals obtained for the free amino acid, owing to the ring current of the aromatic system(43). In contrast, protonation or coordination of a metal ion shifts the signals of protons close to the binding site

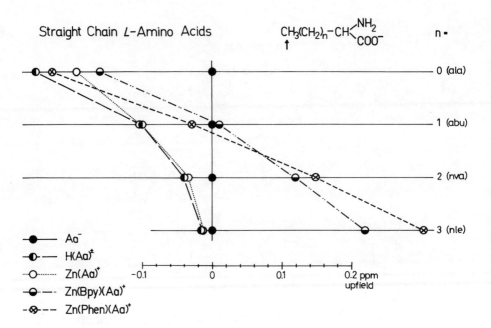

Fig. 3. Chemical shift of the terminal methyl group (always the midpoint of the multiplets was measured) of the shown amino acids (L-alaninate, L-α-aminobutyrate, L-norvalinate, and L-norleucinate) for the protonated (◐) and Zn^{2+} (○), $Zn(2,2'-bipyridyl)^{2+}$ (◓) or $Zn(1,10-phenanthroline)^{2+}$ (⊗) coordinated amino acids, relative to the resonance position of the amino acid anions (●). The plotted values are from Table V of ref 17.

in the ligand downfield. NMR is therefore the ideal method to trace the coordination of the diamagnetic Zn^{2+} and especially to monitor the ligand-ligand interaction within a ternary complex containing one ligand with an aromatic group and another one with an aliphatic side chain. The results of Figure 3 show that the hydrophobic interaction is dependent upon the length (and possibly the volume) of the aliphatic side chain of the amino acid. Obviously, the longer the side chain of the aliphatic amino acid, the larger is the upfield shift of the terminal methyl group, resulting from the interaction between these groups and the aromatic-ring system of 1,10-phenanthroline or 2,2'-bipyridyl within the ternary complexes; it should be noted that the complexation of Zn^{2+} alone (as well as protonation) leads, as expected, to a downfield shift(17).

3. CONCLUSION

The energies involved with aromatic-ring stacking and hydrophobic interactions are small(17). Scheraga(41) calculated on a theoretical basis for several types of interaction values of ΔG^o between -1.3 to -6.3 kJ mol^{-1}. A difference in complex stability ($\Delta\Delta$ log K_M values; see e.g. Table I) of 0.1 log unit between a system with no intramolecular interaction and one in which some interaction occurs corresponds to -0.6 kJ mol^{-1}, while 0.3 and 1.0 log unit correspond to -1.7 and -5.7 kJ mol^{-1}; these values correspond further to K_I values (equation 5) of 0.26, 1.0 and 9.0, respectively, i.e. to 21, 50 and 90 percent formation of the 'closed' isomer. It is evident that even though these energies are small, the resulting effect on the position of an equilibrium (e.g. equation 4) or on the formation of a certain structure (equation 5) may be significant.

The most important point of the described intramolecular ligand-ligand interactions in ternary complexes is probably not so much that the formation of certain mixed ligand complexes is favored by these interactions, but rather that such interactions are responsible for the creation of distinct structures; with regard to the specificity and selectivity observed in nature this seems to be the most fascinating point.

<u>Acknowledgment</u>. The support of our work on "Ternary complexes in Solution" by the Swiss National Science Foundation is gratefully acknowledged.

REFERENCES

1. H. Sigel, ed., "Mixed-Ligand Complexes", Vol. 2 of the Series "Metal Ions in Biological Systems", Marcel Dekker, Inc., New York and Basel (1973).
2. H. Sigel, Plenary Lecture, Conference Volume of the XXth Internat. Conf. on Coord. Chem.(Dec. 1979, Calcutta), D. Banerjea, ed., published by I.U.P.A.C through Pergamon Press, Oxford.
3. H. Sigel, Chimia, 21:489 (1967).
4. R. B. Martin and R. Prados, J. Inorg. Nucl. Chem. 36:1665 (1974).
5. H. Sigel, Angew. Chem. 87:391 (1975); Angew. Chem. Internat. Edit. 14:394 (1975).
6. G. Brookes and L. D. Pettit, J.C.S. Dalton Trans. 1918 (1977).
7. H. Sigel, "Structural Aspects of Mixed-Ligand Complex Formation in Solution", p. 63 or ref. 1.
8. H. Sigel, B. E. Fischer, and B. Prijs, J. Am. Chem. Soc. 99:4489 (1977).
9. H. Sigel, Inorg. Chem. 19:1411 (1980).
10. D. L. Leussing, "The Formation of Schiff Bases in the

Coordination Sphere of Metal Ions" in Vol. 5 of "Metal Ions in Biological Systems", H. Sigel ed., Marcel Dekker, Inc., New York and Basel, p.1 (1976).

11. T. Sakurai, O. Yamauchi, and A. Nakahara, Bull. Chem. Soc. Jpn. 51:3203 (1978).
12. K. H. Scheller, T. H. J. Abel, P. E. Polanyi, P. K. Wenk, B. E. Fischer, and H. Sigel, Eur. J. Biochem. in press (1980).
13. L. G. Marzilli and T. J. Kistenmacher, Accounts Chem. Res. 10:146 (1977).
14. Y. H. Mariam and R. B. Martin, Inorg. Chim. Acta, 35:23 (1979).
15. H. Sigel, K. H. Scheller, V. M. Rheinberger, and B. E. Fischer, J.C.S. Dalton Trans. 1022 (1980).
16. J. B. Orenberg, B. E. Fischer, and H. Sigel, J. inorg. Nucl. Chem. 42:785 (1980).
17. B. E. Fischer and H. Sigel, J. Am. Chem. Soc. 102:2998 (1980).
18. P. R. Mitchell, B. Prijs, and H. Sigel, Helv. Chim. Acta, 62:1723 (1979).
19. C. F. Naumann, B. Prijs, and H. Sigel, Eur. J. Biochem. 41:209 (1974).
20. C. F. Naumann and H. Sigel, J. Am. Chem. Soc. 96:2750 (1974).
21. P. Chaudhuri and H. Sigel, J. Am. Chem. Soc. 99:3142 (1977).
22. H. Sigel and C. F. Naumann, J. Am. Chem. Soc. 98:730 (1976).
23. R. Basosi, E. Gaggelli, and E. Tiezzi, J. Chem. Research (S), 278 (1977).
24. Y. Fukuda, P. R. Mitchell, and H. Sigel, Helv. Chim. Acta, 61:638 (1978).
25. J.-J. Toulmé, Bioinorg. Chem. 8:319 (1978).
26. P. R. Mitchell and H. Sigel, J. Am. Chem. Soc. 100:1564 (1978).
27. E. Farkas, B. E. Fisher, R. Griesser, V. M. Rheinberger, and H. Sigel, Z. Naturforsch. 34:208 (1979).
28. R. Basosi, E. Gaggelli, and G. Valensin, J. Inorg. Biochem. 10:101 (1979).
29. B. E. Fischer and R. Bau, J.C.S. Chem. Commun. 272 (1977).
30. K. Aoki, J.C.S. Chem. Commun. 600 (1977).
31. R. W. Gellert and R. Bau, "X-Ray Structural Studies of Metal-Nucleoside ond Metal-Nucleotide Complexes" in Vol. 8 of "Metal Ions in Biological Systems", H. Sigel, ed., Marcel Dekker, Inc., New York and Basel, p. 1 (1979).
32. P. Orioli, R. Cini, D. Donati and S. Mangani, Nature, 283:691 (1980).
33. K. Aoki, J. Am. Chem. Soc. 100:7106 (1978).
34. C. -Y. Wei, B. E. Fischer, and R. Bau, J.C.S. Chem. Commun. 1053 (1978).

35. K. H. Scheller, Ph. D. Thesis, University of Basel, 1978.
36. U. K. Häring, K. H. Scheller, and H. Sigel, details to be published.
37. P. R. Mitchell and H. Sigel, Angew. Chem. 88:585 (1976); Angew. Chem. Internat. Edit. 15:548 (1976).
38. P. R. Mitchell, J.C.S. Dalton, 771 (1979).
39. K. H. Scheller and H. Sigel, to be published.
40. B. E. Fischer and H. Sigel, XIXth Internat. Conf. Coord. Chem. Prague, Proceedings II, p. 42a (1978).
41. H. A. Scheraga, Accounts Chem. Res. 12:7 (1979).
42. A. Gergely, I. Sovago, I. Nagypal, and R. Kiraly, Inorg. Chim. Acta, 6:435 (1972).
43. L. M. Jackman and S. Sternhell, "Applications of Nuclear Magnetic Resonance Spectroscopy in Organic Chemistry", 2nd ed., Pergamon Press, Oxford, p. 94 (1969).

STUDY ON SOME DIOXYGEN CARRIERS - NEW MODELS OF NATURAL SYSTEMS

B. Jeżowska-Trzebiatowska and A.A. Vogt

Institute of Chemistry
University of Wrocław
Wrocław, Poland

Abstract - Some systems capable of activation and transportation of dioxygen and nitric oxide in solution have been investigated. The structure of these complexes has been related to their reactivity.

In recent years the rapid development of interest in metal complexes having the ability of uptaking the molecular oxygen has been observed (1-9). It is believed that the elucidation of the properties and structures of these low-molecular activators and oxygen carriers will allow more complete understanding of the role, function, structure and mechanism of action of natural systems which involve the oxygen into biological processes.

A significant influence on the effectiveness of both natural and synthetic activators or oxygen carriers is exerted by the medium. This is related to the possibility of interaction processes between solvent molecules, ions or other solute molecules and the active centres of natural or synthetic systems. Sometime these environmental ligands are necessary for the correct functioning of the system and sometimes their coordination results in the loss of transport abilities or activation of the O_2 molecule. On the other hand, for aqueous solution, owing to the presence of various acid-base equilibria related to the ligand, even a slight change in pH of the medium may result in considerable changes in effectiveness of the system.

One should not neglect other factors influencing this effectiveness, including the polarity and steric hindrance of the solvent used. In the light of the above statements one can understand the role of hydrophobic proteins, e.g. globin, which protect the active centres of haemic carriers concealed in special "pockets" against any undesirable effect of the medium.

Our studies on models of dioxygen carriers and enzymatic systems engaging this gas, resulted into the discovery and recognition of three synthetic systems having the ability of uptaking oxygen reversibly and of uptaking nitric oxide and carbon dioxide. Methods for their synthesis and characteristics as gas carriers are shown in Figure 1.

The first is the system Co-imidazole-aminoacid, acting in aqueous solutions over the physiological pH range 6.5-8.5 as the dioxygen and nitric oxide natural carriers (10-14).

Our studies were focused on the following α-aminoacids: glycine, alanine, α-aminobutyricacid, valine, norvaline, iso-leucine, serine, threonine, homoserine, methionine, asparagine, glutamine, arginine HCl, ornithine, lysine, cytruline, proline, hydroxyproline, sarcosine.

The system Co-imidazole-aminoacid is active for many cycles (more than 100) and is one of the most effective carriers of both gases. Moreover it is characterized by a favourable temperature range (0-50°C) which determines the existence of the free and oxygenated species at 1 atmosphere of oxygen pressure, as indicated by the values of the equilibrium constants (14). Oxygenated species of these complexes are also relatively resistant (especially in the case of proline and sarcosine) to the spontaneous deactivating intramolecular oxidation processes.

The second system is constituted by cobalt (II) complexes with a bidentate Schiff base and monoamine. They are capable to transport oxygen and carbon dioxide in non-aqueous solutions, both in polar solvents (alcohols, dimethylformamide, formamide, dimethylsulfoxide, tetrahydrofurane, dioxane, etc.) and in nonpolar solvents as benzene, cyclohexane (15-17). Among the bidentate Schiff bases we have studied the following compounds, N-n-propylsalicylidenimine (salprop), N-n-butylsalicylidenimine (salbut), N-isobutylsalicylidenimine (salisobut), N-benzylsalicylidenimine (salben), N-3-morpholinpropyl-1-salicylidenimine (salmorph). This system is less effective and convenient because of the presence of volatile amines and organic solvents than the system Co-imidazole-aminoacid. In addition its oxygenated form is more sensitive to the deactivating processes of intramolecular activation.

The third investigated system is Co-primary monoamine complexes. It is active in non-aqueous solvents, to uptake reversibly the molecular oxygen and irreversibly the nitric oxide. In our studies the following monoamines were used: n-propylamine, n-butylamine,

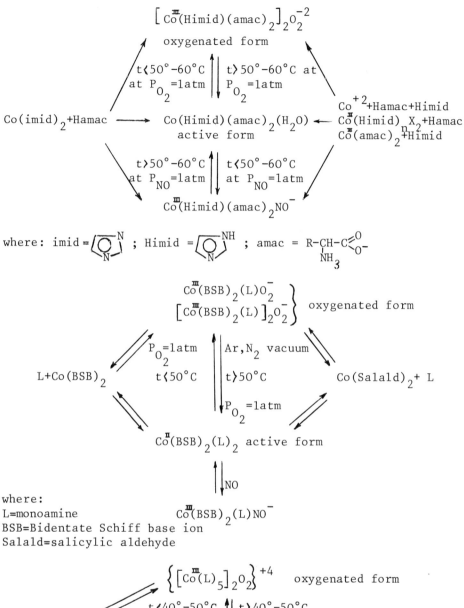

Fig. 1. Some synthetic models of dioxygen and nitric oxide carriers.

isobutylamine, benzylamine, 3-morpholinpropylamine. This system is characterized by the lowest efficiency, mainly because it is the most sensitive to the intramolecular oxidation processes.

The problem of efficiency of the gas carriers depends, besides on the sensitivity to irreversible oxidation processes and medium influence, also upon the molar ratio metal ion: ligand. A comparison of the examined complexes made by volumetric and spectroscopic methods revealed the greatest efficiency of the system Co-imidazole--aminoacid. For this system the maximal efficiency of the O_2 and NO transport may be achieved at Co: aminoacid:imidazole = 1:4-5:2 molar ratio. Both other systems are less efficient. For Co:BSB:amine and Co:monoamine, the ratios are 1:2:30 and 1:40, respectively, that is directly connected with the equilibria present in solution.

In the system Co-BSB-amine, under argon atmosphere, there is the equilibrium between the tetracoordinated initial complex $Co(BSB)_2$ and the hexacoordinated complex $Co(BSB)_2(amine)_2$. This equilibrium was examined by investigating the concentration (Figure 2) and the temperature dependence of the extinction values and band positions of the electronic spectra over the range 600-1700 nm (15).

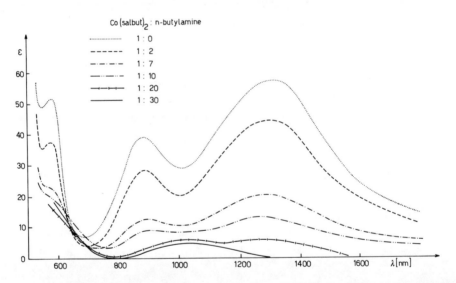

Fig. 2. Dependence of the electronic spectrum of the Co(II)-BSB-monoamine system on $Co(BSB)_2$: monoamine molar ratio in toluene solution at 20°C.

The spectroscopic data show that an increase in concentration of amine, or a decrease in temperature, shift the equilibrium towards the hexacoordinated complex, which exhibits only a weak band at about 1000 nm assigned, for the O_h symmetry, to the $^4T_{1g} \rightarrow {}^4T_{2g}$ transition.

Tetracoordinated complexes were characterized by the spectrum containing two specific, relatively strong bands located at 900 nm and 1300 nm, respectively. The full shift of this equilibrium towards the hexacoordinated complex occurs at 20±2°C with a Co(BSB)$_2$: amine ratio of 1:25-30.

Simultaneously, the dependence of these equilibria upon the nature of the ligands was observed, and, as it follows from the data shown on Table I, it is mainly connected with the steric effects, rather than with the basicicy of the axial and coupled amines in the Schiff base.

Table I. Thermodynamic parameters for the equilibrium
$$Co(II)(BSB)_2 + 2\ \text{monoamine} \rightleftharpoons Co(II)(BSB)_2(\text{monoamine})_2$$

R for "trans" monoamine	n-propyl	n-butyl	iso-butyl	benzyl	N-3-morpholyn propyl	piperidyl
R' for monoamine coupled in Schiff bases	R'=R	R'=R	R'=R	R'=R	R'=R	R'≠R
$\Delta H\left(\dfrac{\text{cal}}{\text{mol}}\right)$ [a]	-0.69	-0.72	-0.38	-0.90	-0.81	
$K_{20\pm2°C}$	30	50	34	214	115	17
pK_a	10.53	10.60	10.43	9.62	--	11.12

[a] Along this paper 1 cal = 4.187 J.

This effect explains the lower equilibrium constant in complexes with salicylidenebutylamine and piperidine. Moreover, this effect, in the case of secondary amines, is strong enough as to prevent the formation of hexacoordinated complex in reaction of any of the complexes with bidentate Schiff base. The system Co-BSB-secondary amine cannot be considered as dioxygen carrier.

Similar equilibria between tetra and hexacoordinated complexes exist also for the oxygen-free form of the Co-monoamine complexes whose spectra are dependent upon the amine concentration versus cobalt ions and upon the temperature (17). The hexacoordinated complexes formed in these systems are less stable than in the Co-BSB-amine system, and the quantitative approach to the problem is difficult because the system was investigated in a strong coordinating solvent as dimethylformamide. Also in this case dependence on steric hindrance was observed.

A somewhat different equilibrium is observed for the oxygen - free form of the system Co-aminoacid-imidazole (14). Temperature dependence of its spectra in the range 300-1400 nm may suggest the presence of equilibrium between two octahedral complexes. These are the hexacoordinated complex $Co(Himid)-imidazole(aminoacid)_2(H_2O)$, of lower symmetry (more intense band at 500 nm) responsible for the transfer of O_2 and NO, and the higher symmetrical hexacoordinated complex $Co(aminoacid)_2(H_2O)_2$ which has the lower extinction coefficients band at 500 nm.

After oxygen uptaking by hexacoordinated high-spin "active" complexes in solutions of the systems Co-BSB-amine and Co-amine, the additional new equilibria are established depending on the polarity of the solvent. In polar solvents an equilibrium between dimeric complex containing the dioxygen molecule coordinated as μ-peroxo O_2^{-2} bridge, and a monomeric complex containing dioxygen molecule as superoxo group does exist. In non-polar solvents only the dimeric complexes are formed. The presence of monomeric complexes in solution of polar solvents (DMF) is supported by the EPR spectra (15).

The monomeric complexes give in the EPR spectra a very characteristic signal with well resolved anisotropic hyperfine structure (Figure 3I).

Eight lines of this structure derives from the contact interaction of the unpaired electron with the nuclear spin = 7/2 of one cobalt ion ^{59}Co. These spectra are the evidence that the density of the unpaired electron is located mainly on the oxygen molecule as supported by the low value of the hyperfine splitting constant $A_{\parallel} \simeq 26.6$ G $A_{\perp} \cong 16.6$ G. As a comparison values of the $A_{iso} = 50-120$ G are observed for low-spin Co(II) complexes (18), where the unpaired electron is placed mainly on the cobalt. For liquid solutions of the oxygenated form of both systems we have obtained also the 8-lines isotropic spectra (Figure 3II).

A similar isotropic spectrum was obtained for the system Co-imidazole-aminoacid, however, not for solution of the oxygenated form, but in the solid precipitated from acetone solution (Figure 3III). Appearance of hyperfine structure results in this case from dilution of monomeric complex by the dimeric diamagnetic complex. Similar g_{iso} and A_{iso} parameters were obtained in EPR spectrum of solutions of oxidation products by Ce^{+4} ion in acid medium of oxygenated form of Co-imidazole-aminoacid complexes i.e. dimeric cobalt complexes with μ-peroxo bridge (Figure 3IV). These products are the paramagnetic complexes containing the irreversibly uptaken O_2 molecules, as the μ-superoxo bridge.

Unpaired electron on that bridge interacts symmetrically with the nuclear spin 7/2 of both cobalt (III) atoms to give the 15-line hyperfine structure. Complexes with the μ-superoxo bridges were not

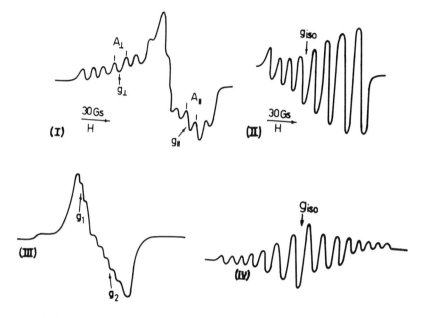

Fig. 3. (I) EPR spectrum of the frozen solution of oxygenated Co-monoamine in DMF at -150°C. $g_{\parallel}$ = 2.080-2.082, $A_{\parallel}$ = 22.1-22.7 G, $g_{\perp}$ = 2.006-2.007, $A_{\perp}$ = 15.9-16.8 G.

(II) EPR spectrum of the liquid solution of oxygenated Co-monoamine in DMF at -10°C. g_{iso} = 2.031-2.032, A_{iso} = 13.0-13.1 G.

(III) EPR spectrum of the solid of oxygenated Co-imidazole-aminoacid system at -120°C. g_1 = 2.04, g_2 = 2.00, A = 13.0 G.

(IV) EPR spectrum of the liquid solution (1N in HNO_3) of the oxidized (by means of Ce^{+4} ions) form of oxygenated Co-imidazole-aminoacid systems at 40°C. g_{iso} = 2.031-2.034, A_{iso} = 12.3-12.4 G.

obtained in the case of both the remaining systems, because they are stable only in aqueous medium at high concentration of H^+ ions. They are also unknown for natural systems. In the electronic spectra of complexes with μ-peroxo bridge, there are two intensive charge transfer bands in the range 300-420 nm. These bands may be assigned to the charge transfer transition dioxygen → cobalt (III) ion. This was proved by the lack of these bands in spectra of active form of complexes and by their appearance and decay during formation and transformation of the oxygenated complexes. These bands may be assigned to two transitions $\pi_a^* \to d_z^2$ and $\pi_b^* \to d_z^2$, where π_a^* and π_b^* orbitals are the nondegenerated components formed at the moment of the angular coordination of the dioxygen from its two degenerated π^* acceptor orbitals.

Mutual intensity of both bands is thus strongly dependent upon the dihedral angle Co-O-O-Co (19). The more this angle differs from the right angle, the greater is the difference observed in the intensity of these bands; when the deviation reaches 90° the intensity of a band could be weak enough to be overlapped by the stronger band. This is connected with various degree of contribution of both components of the split π-antibonding orbitals into the strong interaction with d_z^2 orbitals. Such model was used in the interpretation of differences in mutual intensity of both bands in the electronic spectra of oxygenated form of the dioxygen carriers. This differences are illustrated in Figure 4 taking as an example the Co-imidazole-aminoacid system.

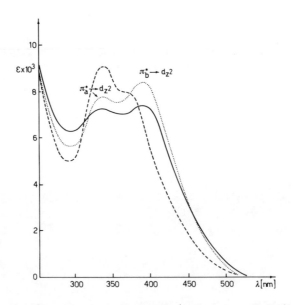

Fig. 4. Electronic spectra of oxygenated form of Co(II)-imidazole--aminoacid system in aqueous solution at 0°C.

SOME DIOXYGEN CARRIERS

The steric effects of the coordinated ligands are responsible of this behavior, as far as they determine the values of the dihedral angle. The interactions with a solvent should also be not neglected. This is outlined by the considerable differences in the spectra we found for the oxygenated form of the Co-amine solvent system, depending on the polarity and steric hindrance in the solvent (Figure 5). These factors, because of the possibilities of rotation around the oxygen-oxygen bond, might effect significantly the dihedral angle.

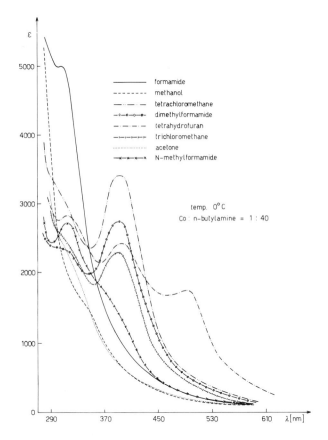

Fig. 5. Electronic spectra of the oxygenated form of Co-n-butylamine-solvent system at 0°C.

There is also another possible explanation for the differences in the above mentioned spectra. This is the monomer-dimer equilibrium existing in polar solvents. Then, one of the bands should be assigned to the monomeric form of complex $\pi_u^* \rightarrow d_z^2$. This band could be overlapped by the bands resulting from the dimeric form $\pi_a^* \rightarrow d_z^2$, $\pi_b^* \rightarrow d_z^2$. In the most polar solvent, such as formamide, for which the highest monomer concentration was found, the band at higher energy is considerably dominating. This concept is also supported by certain kinetic data concerning the intramolecular oxidation processes. However on the light of this theory there is no strict agreement between the polarity of solvent and the nature of the spectrum predicted by this theory.

Bidentate Schiff base complexes exhibit interesting properties of catalytic oxidation of hydrazine with molecular oxygen, with high heat effects in many cycles (16-17) according to the Scheme 1.

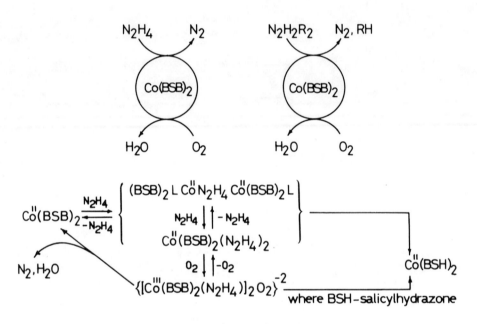

Scheme 1

The system Co(II)-(BSB)$_2$-N$_2$H$_4$ may be treated as a specific case of the Co(II)(BSB)$_2$-monoamine system. The reversible oxygen uptaking should be expected. However, the thermodynamic unstability and readiness for oxydation characteristic for hydrazine, as well as the lack of special hindrance and its ability for bridge formation, are the cause for the different behaviour of the system Co(II)(BSB)$_2$-N$_2$H$_4$ in the presence of oxygen. The lifetime of the oxygen adduct is very short (1 s at room temperature) and it decomposes with evolution of nitrogen and water as well as hydrocarbons in the case of alkyl or aryl derivatives of hydrazine. Coordination of hydrazine to cobalt has been determined by spectroscopic investigations of equilibrium Co(BSB)$_2$-hydrazine in solution. Under oxygen-free atmosphere, enhancement of the hydrazine concentration, or temperature lowering resulted in decreasing of intensity of Co(BSB)$_2$ bands at 1250 and

925 nm like in the Co-BSB-monoamine system. At Co:N_2H_4 = 1:2 molar ratio, only one band of low intensity, at 975 nm, characteristic for hexacoordination Co(II) complexes were observed.

These facts allowed to state the coordination of hydrazine to cobalt, but not to find out exactly, whether we deal with the equilibrium 1 or 2:

(1) $Co(BSB)_2 + 2N_2H_4 \underset{\longleftarrow}{\overset{K_1}{\longrightarrow}} Co(BSB)_2(N_2H_4)_2$

(2) $2Co(BSB)_2 + N_2H_4 \underset{\longleftarrow}{\overset{K_2}{\longrightarrow}} (BSB)_2CoN_2H_4Co(BSB)_2$

since the relation lnK = f (T) was not linear for none of them. Most likely they appear simultaneously; for the Co:N_2H_4 = 1:0.5 and 1:1 ratios the equilibrium (1) was predominant, while for 1:2 and 1:5 molar ratio was predominant the equilibrium (2). Coordination of hydrazine results in enhancement of its sensitivity to oxidation. Decrease of the electron density on nitrogen results in weakening of the N-H bond; the presence of a proton acceptor, in this case the oxygen uptaken as O_2^{2-} bridge, facilitates the releasing of the atomic hydrogen. The EPR spectra of the system with uptaken oxygen present no signals, thus revealing the lack of equilibrium between dimeric and monomeric complex, characteristic for the system $Co(BSB)_2$-monoamine-O_2. This means that in solution exist only dimeric species. No EPR signals are observed also at decomposition of the ternary complex with uptaken oxygen, thus allowing to suppose that the decomposition with evolution of dinitrogen has no radical character. Chromatographic examination of gaseous products of the discussed reaction indicates the total 4-electrons reduction of oxygen. It does not exclude, however, that the oxygen reduction may be dielectronic with formation of hydrogen peroxide, which subsequently, under these conditions, may oxidize hydrazine. In this case the diimide should be formed as intermediate as well.

Significant influence of the medium may be observed on the catalytic activity as well as on the effectiveness of dioxygen transport. A particularly significant effect of the medium was found in the Co-imidazole-amino acid system. Its highest effectiveness is achieved in the pH range 6.5-9. A considerable decrease in pH of the dissolved oxygenated form of Co-imidazole-aminoacid system, achieved by dropping it into an aqueous (1M) oxy-acid solution at room temperature, results in the immediate decomposition of this form with evolution of oxygen. At -3 up to 0°C unstable intermediates are formed. Their decomposition is considerably accelerated by a slight increase of temperature or by the presence of a catalyst (activated carbon). The spectra of these intermediates, which contain a very characteristic d-d band at about 500 nm, are indicative of the presence of the μ-hydroperoxo bridge O_2H^-. As a result of its further protonation hydrogen peroxide is evolved whereas the cobalt ions undergo irreversible oxidation to Co(III) ions.

A considerable increase in pH to 10-12 of the solution of oxygenated form of this system also leads, not to complete deactivation of the system but to a decrease in its effectiveness. This is accompanied by characteristic changes in the electronic spectra of this species. Instead of the two charge transfer bands shown above, related to the $\pi_{a,b}^* \rightarrow d_z^2$ transition of metal orbitals in the 300-400 nm region, a new, poorly shaped band appears at about 360-370 nm. On the other hand, the second band lies slightly below 300 nm. This spectrum is highly characteristic of the double-bridge cobalt complexes of a μ-peroxo-μ-hydroxo type, e.g. simple amino acid complexes with two bridges (Figure 6). Thus, in media with high hydroxide ion concentrations, one can observe the formation of an additional μ-hydroxo bridge. This bridge makes the system very rigid, thus explaining its much lower effectiveness in oxygen carrying. Furthermore, the OH ion stabilizes the third oxidation state of cobalt, facilitating the deactivation process of intramolecular oxidation.

The formation of an additional OH^- bridge is facilitated when two cis-cis positions are readily available in the dimeric μ-peroxo complex. For the discussed complexes these might be the positions freed by the two carboxylic groups, readily dissociating in highly alkaline media. The involvement of the medium in processes taking place in solution of the oxygenated form of the dioxygen carrying system is particularly evident in the case of irreversible spontaneous transformation of this form into inactive products - simple Co(III) complexes. As shown by spectroscopic studies, in this case the process proceeds in two steps. In the first step - more rapid - a dimeric complex containing an additional μ-hydroxo bridge is formed. In the slower second step, the double bridge complex is transformed into simple monomeric hydroxo-amino acid Co(III) complexes. Simultaneously, imidazole is liberated from the complex. These processes proceed in a quite different way for the two remaining systems where no intermediate is found to be formed.

The examples reported represent a strong evidence of the considerable sensitivity to the medium of molecular oxygen carrying complexes, particularly those acting in water. A slight change in the medium may lead to a sudden decrease in system activity. In this respect, the natural systems surpass considerably most synthetic analogues. This is related, above all, to a considerable "rigidity" of the natural systems where the active centres are built into large organic molecules whose role consists not only in stabilizing particular forms involved in the processes, but also in screening these centres against harmful ions or molecules provided by the medium. In the case of low-molecular synthetic analogues their structures makes usually possible a direct contact with the ions and molecules, with simultaneous sensitivity to various types of rearrangements.

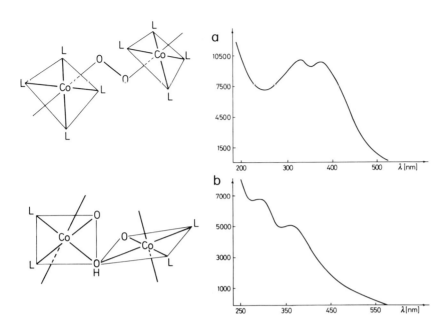

Fig. 6. (a) Electronic spectrum of $(aminoacid)_2(Himid)Co(III)O_2^{-2}$
Co(III)(Himid)(amino-acid)$_2$ in aqueous solution at 0°C.
(b) Electronic spectrum of $[(aminoacid)_2(Himid)$

$$Co(III) \begin{array}{c} O_2^{-2} \\ \diagup \\ \diagdown \\ OH \end{array} Co(III)(Himid)]^- (aminoacid)_2 \text{ or}$$

$$[(aminoacid)_2 Co(III) \begin{array}{c} O_2^- \\ \diagup \\ \diagdown \\ OH \end{array} Co(III)(aminoacid)_2]^- \text{ in}$$

aqueous solution, at pH 11.

REFERENCES

1. L. H. Vogt, H. Faigenbaum, S. Wimberly, Chem. Rev. 63:269 (1963).
2. F. Bayer, P. Schretzman, Structure and Bonding, 2:181 (1967).
3. S. Fallab, Angew. Chem. Int. Ed., 6:496 (1967).
4. A. G. Sykes, J. A. Weil, Progr. Inorg. Chem., 13:1 (1970).
5. J. S. Valentine, Chem. Rev., 73:235 (1973).
6. V. J. Choy, C. J. O'Conner, Coord. Chem. Rev., 9:145 (1972).
7. R. W. Erskine, B. O. Field, Structure and Bonding, 28:1 (1976).
8. G. Mc Lendon, A. E. Martell, Coord. Chem. Rev., 19:1 (1976).
9. R. D. Jones, D. A. Summerville, F. Basolo, Chem. Rev., 79:139 (1979).

10. B. Jeżowska-Trzebiatowska, A. A. Vogt, H. Kozłowski,
    A. Jezierski, Bull. Acad. Polon. Sci., ser. sci., chim.,
    XX 3:187 (1972).
11. B. Jeżowska-Trzebiatowska, K. Gerega, A. A. Vogt, Inorg.
    Chim. Acta, 31:183 (1978).
12. B. Jezowska-Trzebiatowska, K. Gerega, G. Formicka-Kozłowska,
    Inorg. Chim. Acta, 40:187 (1980).
13. B. Jeżowska-Trzebiatowska, K. Gerega, G. Formicka-Kozłowska,
    Inorg. Nucl. Chem. Letters, in press.
14. B. Jeżowska-Trzebiatowska, A. A. Vogt, Inorg. Chim. Acta,
    in press.
15. B. Jeżowska-Trzebiatowska, A. A. Vogt, A. B. Vogt,
    B. Froncek, Inorg. Chim. Acta, in press.
16. B. Jeżowska-Trzebiatowska, A. A. Vogt, P. Chmielewski,
    Inorg. Chim. Acta Letters, 45:L 107 (1980).
17. B. Jeżowska-Trzebiatowska, A. A. Vogt, P. Chmielewski,
    Proc. Xth Summer School - Symposium on Coordination
    Chemistry, Karpacz, 1979.
18. C. Paul, C. W. Schläpfer, A. von Zalewsky, Structure and
    Bonding, 36:129 (1979).
19. A. B. P. Lever, H. B. Gray, Acc. Chem. Res., 11:348 (1978).

DIRECT NMR STUDIES OF IONIC SOLVATION

Pierre Laszlo, André Cornélis, Alfred Delville,
Christian Detellier, André Gerstmans, and Armel Stockis

Institut de Chimie et de Biochimie (B6), Université de
Liège, Sart-Tilman, B-4000 Liège, Belgium

Abstract -The nuclear magnetic resonance of quadrupolar ions provides a wealth of information about solvation thermodynamics and dynamics. In this contribution, we consider the local ordering induced by an ion, surrounding itself with neutrals, solute or solvent molecules, in competition with its counterion. Examples include determination of the solvent electron-accepting, or hydrogen bond donor, abilities, anionic activation, and hydrophobic interactions involving Q^+ cations, preferential solvation, and cation chelation, including that by calciproteins. Our approach is to describe the first solvation shell of an ion such as Na^+ in specific terms, i.e. there are solvent molecules truly coordinated to the ion. Beyond this coordination sphere, we treat the solvent as a continuum, whose effects are estimated through bulk parameters such as the macroscopic viscosity η or the dielectric constant ε.

1. INTRODUCTION

We shall report here on a large scale study of ion-molecule interactions in solution. It was launched to investigate local ordering around ionic charges, according to the following classification:
- A) competition between solvent molecules for clustering around an ion;
- B) competition between solvent molecules and neutral solutes for clustering around an ion;
- C) competition between solvent molecules and the counter-ion for coordination to an ion.

In recent years, new techniques have been brought to bear on these questions. Among them nuclear magnetic resonance of quadrupolar nuclei belonging to the ion itself plays a prominent role.

Presence of an electric quadrupole moment makes such a nucleus interact with local electrostatic field gradients q. These are set up in a permanent way if the shell consisting of all nearest neighbors to the ion has less than tetrahedral (T_d) or octahedral (O_h) symmetry. The magnitude of the interaction is written $(e^2qQ)/\hbar$, were Q is a constant which measures the magnitude of the electric quadrupole moment for the nucleus. We shall denote by χ this quantity $(e^2qQ)/\hbar$; χ is the quadrupolar coupling constant.

From the above, χ increases whenever the local symmetry around an ion is decreased or destroyed. Provided that the nucleus monitored has enough of a quadrupole moment Q, the corresponding longitudinal ($R_1 = T_1^{-1}$) or transverse ($R_2 = T_2^{-1}$) relaxation rates - or, nearly equivalently for most of the cases to be discussed in this presentation, the observed linewidth at half-height $\Delta\nu_{1/2} \simeq (\pi T_2)^{-1}$ - will be enhanced correspondingly.

We have performed a number of investigations using ^{7}Li, ^{23}Na, ^{39}K, ^{87}Rb alkali metal nuclei(1); and also ^{59}Co. These are the main techniques we shall refer to in this account.

Rather than being technique-oriented, however, our main goal is to contribute to molecular biology from a physico-chemical point of view. Therefore, we shall attempt to generalize from each of our results, obtained on rather simple model systems, to much more complex biomolecules such as enzymes.

2. ORDERING OF SOLVENT MOLECULES BY A CHARGED SOLUTE

To quote from a very recent theoretical study of electrolyte "Long ago it was realized that the presence of free ions in solution implied the existence of enormous local electric fields, a situation in which the solvent in the neighborhood of the ions might be in a state quite different from that of the bulk solvent. This would have enormous chemical consequences since the screening of the electrical forces between charges in solution is mediated by the solvent"(2). Among the consequences, free ions such as Na$^+$ surround themselves with clusters of irrotationally bound solvent molecules. Let us make this statement more quantitative: ordering of solvent molecules around an ion keeps them bound for a residence time τ_B significantly longer than the correlation time τ_R descriptive of their random reorientation in the pure solvent. For structure-forming Na$^+$ sodium ions, and with various types of solvents such as water or methanol, the ratio τ_B/τ_R is in the range 2-5, depending upon the solvent and upon the type of measurement(3-5). This corresponds to an activation energy for bound → free solvent exchange between ca. 250 and 1,000 cal mol^{-1}.[†]

The long-lasting solvates can profitably be viewed as true complexes. On this basis, if two solvents L_A and L_B compete for coordination to the sodium cation, ^{23}Na chemical shifts and linewidths will be, in the presence of preferential solvation, non-linear functions of the composition of the binary ($L_A + L_B$) mixture. We observe indeed such curves, deviating strongly from ideality, when the ^{23}Na nmr observables for Na$^+$ClO$_4^-$ are plotted against composition of binary solvent mixtures of THF (as the reference solvent) and various amines: unidentate amines studied(6) include aniline, pyridine, pyrrolidine, piperidine, propylamine, and i-propylamine; bidentate amines studied(7) are ethylenediamine, diethylenetriamine, 1,3-diamino propane, and cadaverin.

The key to our analysis(6,7) stems from the realization that the well-authenticated tetracoordination of the sodium cation by solvent molecules of these types and sizes makes the question of preferential solvation in ($L_A + L_B$) binary mixtures isomorphous with that of the binding of ligands to four equivalent sites in a biomolecule, as exemplified by the binding of O_2 to hemoglobin. In the latter case, the positive cooperativity, termed the Bohr effect, can be approached with the Hill-Adair formalism(8). The Hill formalism, hitherto neglected in studies of preferential solvation, provides a better understanding than the existing and somewhat less powerful formalisms. It shows(6) <u>equality</u> of the intrinsic equilibrium constants <u>k_i</u>(i = 1 - 4) for the successive steps upon displacement of THF (= L_A) from sodium coordination by one of these <u>unidentate</u> amines (see Scheme 1 in the next page). Furthermore, <u>this</u> <u>study</u>(6,7) provided an unexpected fringe benefit: the intrinsic preferential solvation constant k are simply proportional to the measured ^{23}Na nmr chemical shifts δ in the same solvents (see Table I). It may appear a real surprise that a thermodynamic quantity (k) is related to a chemical shift, i.e. to an electronic distribution. However, there is precedence for such a finding: the ^{23}Na chemical shifts are linearly correlated to another thermodynamic quantity, the Gutmann donicity of the solvent, when a salt such as NaClO$_4$ is dissolved in a series of oxygen- and nitrogen-donor solvents(9). The strength of the predominantly electrostatic chemical bond between Na$^+$ and nitrogen increases with nitrogen basicity, which is measured indirectly by the δ values.

By the contrast to the linear and unit slope Hill plots obtained with unidentate amines, with bidentate amines Hill plots are curved and display considerable non-cooperativity. Detailed analysis of the results shows(7) that solvation of Na$^+$ by the polyamines

---

[†] along this paper 1 cal = 4.187 J

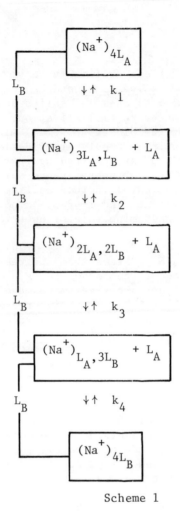

Scheme 1

follows the sequence: cadaverin < 1,3-diaminopropane << ethylendiamine << diethylenetriamine. Entry of the first and of the second diamine molecule into the sodium coordination shell are independent and equiprobable steps: $k_1 = k_3$ and $k_2 = k_4$, within the accuracy of the measurements. For instance, for ethylenediamine, $(k_1,k_3) = 1.0-1.5$ and $(k_2,k_4) = 83-102$: attachment of the second nitrogen is considerably easier, by two orders of magnitude. Hence, we determine the chelate effect by measurement on the bidentate ligands alone, without recourse to comparison with the corresponding unidentate ligands. This serves to avoid the methodological difficulties which have poisoned most of the previous studies of the chelate effect: the choice of proper concentration units, and of appropriate standard states.

Table I. ^{23}Na chemical shifts δ for NaClO$_4$ dissolved in the pure amines, and equilibrium constants for sequential displacement of THF by amine molecules in the Na$^+$ coordination shell (Scheme 1).

Solvent	δ (± 0.05 ppm)	k (± 10 %)
aniline	-3.30	0.46
pyridine	-0.42	1.3
piperidine	0.87	1.5
pyrrolidine	3.74	2.8
propylamine	6.68	4.4
i-propylamine	6.60	3.0

Chelation of cations is indeed a property of cation-binding proteins, which play an essential role for storage and transport of physiologically important ions such as Na$^+$, K$^+$, Ca^{++}, and Mg^{++}. In the words of Williams "Two conditions are required for strong chelation. The groups X must be held on a frame and the frame must be flexible. In proteins there are aspartate and glutamate carboxylate groups. (...).There is a requirement for an additional stabilizing feature from the protein fold to bring the anionic charges much closer together. This effect is seen in extra-cellular calcium-binding proteins (...): the fold helps to throw together several carboxylate groups to form a potential calcium-binding site and calcium then helps to stabilize and tighten this fold (...). In the binding of metal ions to these proteins a common situation is that (i) the protein folds somewhat loosely to give a potential chelating centre for a metal and (ii) a metal ion binds at this centre and, while doing so, tightens the protein fold. The cooperative nature of the chelation of the metal and the fold energy can produce a high degree of selectivity of metal-binding energy depending critically upon the metal-ion size, stereochemical demand and the fold energy"(10). A number of studies of our Liège group have indeed probed chelation of alkali and alkaline earth ions by calciproteins belonging to the [parvalbumin-troponin C-calmodulin] family(11-15).

3. ORDERING OF NEUTRAL SOLUTES BY IONS

Analogous to preferential solvation, there is competition between two kinds of neutrals: solvent, and solute molecules, for clustering around an ion. Since the previous section dealt with solvent sorting by the Na$^+$ cation, we shall present here results on solute ordering by this same ion.

We discovered a few years ago(16), at about the same time as an American group(17), that Na$^+$ ions induce the self-ordering of 5'-GMP in aqueous solution. 5'-GMP is the abbreviated name of a nucleotide, 5'-guanosine monophosphate, in which a guanine base is linked to a ribose ring at the 1'-position; this sugar ring is in turn connected to a phosphate group at the 5'-position. Alone among the nucleotides, guanosine monophosphate arranges itself into highly ordered aggregates, in aqueous solutions(18). The 5'-isomer (5'-GMP) not only forms gels at acidic pHs(18), but also self-orders at neutral and slightly basic pH(19). These aggregates formed at neutral pH have a surprising rigidity: from ^{1}H nmr studies, energy barriers of more than 15 kcal mol^{-1} prevent fast exchange of monomer nucleotides between sites(17). 5'-GMP aggregation is related to the exceptionally high rigidity also displayed by poly(riboguanylic acid) [poly(G)] in neutral aqueous solution(20-23), due to multistrand formation(24), with a four-stranded form characterized by X-ray in the solid(25,26).

Indeed, a key factor in the self-ordering of 5'-GMP is the hydrogen-bonding of guanines into centrosymmetric planar tetramers held together by eight hydrogen bonds(27).

We have studied the self-assembly of the sodium salt of 5'-GMP (5'-GMP^{--}, 2 Na$^+$) under conditions plausible for a pre-biotic environment: nucleotide formation has been documented under abiotic conditions(28-30), and 5'-GMP self-associates at the slightly basic pHs (7.8-8.4) characteristic of the present and, in all likelihood, also of the primeval ocean(31-33).

^{1}H chemical shifts show at concentrations below 0.2 M a normal base stacking process, with an apparent mean equilibrium constant of 3.8 M^{-1} at 299 K. Above a concentration of approximately 0.3 M at this temperature, an altogether different mode of self-ordering sets in. The multiplicity of H-8 resonances in the ^{1}H nmr spectra and that of phosphate resonances in the ^{31}P nmr spectra are consistent with the coexistence of two kinds of ordered structures, likely to be dimers of tetramers, and tetramers of tetramers, respectively. ^{23}Na chemical shifts and linewidths are also concentration-dependent: below a critical concentration, they increase normally with the fraction of ion-paired sodium, just as in model systems (5'-adenosine mono- and triphosphate, 2'-guanosine monophosphate, D-ribose 5-phosphate) with binding constants of about 2 M^{-1}, at 300 K.

Above a critical concentration(16), the ^{23}Na resonance undergoes a pronounced upfield shift(34), together with a considerable line broadening: these are diagnostic of a sharp disorder → order transition, which we have studied both by changing the concentration at constant temperature, or by varying the temperature at constant concentration(35). This ordering transition is highly cooperative, with a Hill coefficient(8) of about 6(35). The critical concentrations are determined either from the ^{23}Na chemical shifts or from linewidths reduced to unit viscosity: a phenomenological phase-separation model yields $\Delta H° = -17 \pm 2$ Kcal mol^{-1}, and $\Delta S° = -51 \pm 6$

cal mol^{-1} K^{-1}. The self-assembly is enthalpy-driven, rather than determined predominantly by hydrophobic forces, like the normal stacking interactions. The binding of Na$^+$ contributes directly(35) to the buildup of octamers and hexadecamers from hydrogen-bonded tetramers: these tetramers present a central cavity delimited by the four O$_6$ oxygens from the four guanines. This cavity has the appropriate size for inclusion of the Na$^+$ cation, which accounts for the considerable upfield shift of the ^{23}Na resonance upon 5'-GMP self-assembly(34). Besides this first class of Na$^+$-binding sites, the aggregates can bind Na$^+$ ions unto a second category of sites: the charged phosphate groups are brought into close proximity by the formation of aggregates of tetramers, and the Na$^+$ ions bridge phosphate groups, thus shielding these doubly charged anions from one another, in a similar way as the above described cation-binding by calciproteins.

In fact, self-assembly of 5'-GMP depends critically on the caption present: Pinnavia et al. using ^{1}H nmr(36) and Laszlo et al. (34,37-39) using ^{23}Na and ^{39}K nmr showed a potassium-selective interaction of alkali metal cations with the central cavity in the 5'-GMP tetramers. When sodium and potassium ions are present in conjunction, an organized structure $[(G_4, K^+, G_4), 4\ Na^+]^{11-}$ forms(40). Potassium ions bind selectively to the inner (cavity) site, where they are sandwiched in between two GMP tetramers(40). These are maintained at a spacing close to 3.3 Å which is characteristic of base stacking (41), while the K$^+$...O$_6$ oxygen distances are about 3 Å, i.e. only slightly greater than the K$^+$...O van der Waals contacts of 2.7-2.8 Å reported for a number of antibiotic ionophores(42).

A short comment is in order here: in water, the potassium cation is destructuring having slight negative solvation by many accounts(3). However, here, K$^+$ ions nucleate the self-ordering of 5'-GMP molecules because, entropy-wise, the tetramers are preformed so that there is a net entropy gain upon release of the water solvent molecules from K$^+$; and enthalpy-wise, the balance between K$^+$ hydration or K$^+$ binding to eight O$_6$ oxygens is probably close to zero, with an additional stabilization from reinforced hydrogen-bonding of the guanines due to polarization of the base by metal binding at O$_6$.

In these rather well-defined structures, sodium ions bind selectively to the outer sites, screening the electrostatic repulsion of the negatively charged phosphate groups(40). This results in the formation of a species with the (very likely) stoichiometry (G_4,K^+,G_4), 4 Na$^+$. They do so by true site-binding, whereas K$^+$ or Rb$^+$ ions undergo only atmospheric condensation, on the outer sites. Li$^+$ ions interact only very weakly(43), they are too small for the size of the cavity in the tetramers. Likewise, Cs$^+$ ions do not encapsulate within this cavity: too big.

K^+ and NH_4^+ ions are iso-steric to a good approximation. Hence, one could expect a competition for the inner (cavity) sites. We were surprised to discover that ammonium ions have a different function: they appear to effect duplication of 5'-GMP octamers into hexadecamers, by bridging phosphate groups belonging to distinct octameric units(44). Ammonium cations compete neither with the potassium cations for occupation of the inner sites, nor with the sodium cations for occupation of the outer sites, in the postulated $(G_8, K^+, 4\ Na^+)$ species. This finding of a dimerization from G_8 to G_{16}, brought about by ammonium ions - and the primitive atmosphere was ammonia-rich(28-30)-, is thus in harmony with the postulated co-existence between G_8 and G_{16} species for 5'-GMP, Na_2, especially at the higher concentrations(35). If biomolecules such as enzymes can discriminate between NH_4^+ (1.45 Å) and the iso-steric K^+ (1.33 Å) ions, it is truly remarkable that aggregates formed spontaneously by 5'-GMP, a relatively small molecule, are also capable of making the K^+/NH_4^+ distinction. Since the aggregates interact strongly with ammonium ions, we have investigated their interaction with amino acids(44): the system discriminates between glycine and alanine, two archaic amino acids formed in abundance under prebiotic conditions from the primitive atmospheric constituents(45,46). Glycine destroys the 5'-GMP clusters, perhaps because the amino acid competes, using its charged-NH_3^+ headgroup, with potassium ions for occupation of the core positions in the tetramers(44). No measurable effects, by contrast, are found with L-alanine(44).

Take as a whole, these results on the 5'-GMP self-assembly point to the considerable importance of ion binding to the stability of a supra-molecular structure devoid of covalent bonding. Likewise, the native conformation of proteins benefits to a very significant extent from the presence of salt bridges, bringing together selected groups(47).

4. WEAKENING OF SALT BRIDGES AND ANIONIC ACTIVATION

Alkylation of carbanions is a frequent and important reaction in organic chemistry. A special class of carbanions is the ambient enolate anions derived from β-dicarbonyl molecules, and which are capable of undergoing alkylation either at the central or at one of the terminal oxygens:

DIRECT NMR STUDIES OF IONIC SOLVATION

We have studied such a system: the sodium salt of ethyl acetylacetate $E^- Na^+$, with a view to correlate its chemical reactivity with the structure of the ion pairs present in solution(48).

For this purpose, the rate constant for alkylation by ethyl iodide is measured in THF; it is compared with the ^{23}Na chemical shift and linewidth for the substrate, in the same solvent, at the very same concentrations and temperatures.

The rate constant displays, in the presence of a specific Na^+ complexant such as the 15-crown-5 cyclic polyether, or the (2.2.1) cryptand, the phenomenon termed <u>anionic activation</u>: the reactivity is greatly enhanced, and reaches a maximum, with a plateau at ca. 5-10 equivalents of the crown ether or at 1-2 equivalents of the cryptand.

This is as expected. Consider the following schematic transition state:

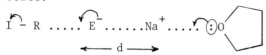

Clearly, the electrostatic stabilization of the enolate by the sodium counter-ion, in the low-dielectric solvent THF, decreases the reactivity of the anion. Were it possible to withdraw, even if very slightly, both partners of this ion pair from one another, i.e. lengthen <u>d</u> somewhat, and the anion reactivity will be increased. This is exactly what occurs when, instead of the THF molecule, back-solvation of the cation is effected by a yet more powerful sodium-complexant, such as a crown ether or a cryptand.

One might want to go one step further, and claim that the above-quoted reactivity plateau corresponds to the reactivity of the so-called "naked anion"(49), once the cation has been safely stowed away in the cavity of the crown ether (or cryptand). Even though such an intuitive concept is appealing, our spectroscopic results have shown that it does not apply to such a system.

Indeed, if one plots the ^{23}Na linewidth, reduced to unit viscosity, as a function of the molar ratio of sodium complexant C to the $E^- Na^+$ salt, a plateau is also reached with 5-10 equivalents when C is taken as 15-crown-5. However, rather than the expected monotonic decrease of the linewidth, when the cation is pulled away by the crown ether from the anion, one witness a sharp initial increase upon addition of minute quantities of the polyether, followed by a more gradual decrease to a value of the order of 500 Hz $(cP)^{-1}$.†

---

† $1 cP = 1 mPa \cdot s$

What these values, in the range of 500-1,000 Hz $(cP)^{-1}$, indicate is that in THF solution one is dealing exclusively with tight ion pairs and their aggregates. The complex titration curve just described, together with the concentration dependence of the ^{23}Na linewidth for E^-Na^+ alone, are uniquely described by the coexistence of the following species:

$$E^-Na^+ \qquad (E^-Na^+)_2 \qquad E^-Na^+C \qquad (E^-Na^+)_2C$$

The spectroscopic ^{23}Na data are analyzed to yield reliably the three corresponding equilibrium constants. It is thus possible to calculate back the relative amounts of each of these entities as a function of the $(C)/(E^-Na^+)$ molar ratio. A structure-reactivity correlation is successful - and outstandingly so- for only one of these: the rate constant for alkylation of E^-Na^+ by C_2H_5I is strictly proportional to the concentration of (E^-Na^+C). Therefore, at least to first approximation, this is the <u>single</u> kinetically active species in the system. The monomeric tight ion pairs E^-Na^+, their dimers $(E^-Na^+)_2$, the crowned dimers which presumably have a structure of the type $(E^-Na^+E^-Na^+C)$, are all inert under our conditions. This serves to show that the idea of a "naked anion"(49) is underly naive: instead we are dealing with an anion E^- which continues to benefit from stabilization, albeit reduced, from presence of the neighboring Na^+ cation which, in turn, is bound also to the complexant molecule C, either a crown ether or a cryptand.

This study provides another example of the competition between solvent molecules (THF) and neutral solutes (crown ethers or cryptands) for ion-binding. What about the reverse process, of enhanced ion pairing, brought about by partial withdrawal of ionic solvation? The next, and last, section will consider it.

5. EXCLUSION OF WATER MOLECULES FROM CHARGE SCREENING

The example chosen here is that of quaternary ammonium cations (Q^+). During the last 10 years, their salts have enjoyed extensive use as phase transfer catalysts(50). In the presence of Q^+, anions A^- are solubilized into the organic phase while retaining high nucleophilic activity in non polar solvents. In practice, the best results are achieved with symmetrical ammonium ions NR_4^+ in which the alkyl chains R have at least four carbon atoms(51). Hence, our study has focussed on such symmetrical cations Q^+.

The questions to be answered are the following:
1. granted that there is significant cation-anion interaction in the organic phase, yet phase transfer catalysis appears to occur right at the interface between the aqueous and the organic phases (50): what is the magnitude of the Q^+A^- interaction in water solution?

Once this first question is answered, a second question immediately arises:

2. in water solution, is the ion pairing between Q^+ and A^- best described as entropy-determined (the so-called hydrophobic effect, responsible e.g. for water solubilization of benzene in the presence of tetrabutyl ammonium bromide)(52,53) or as enthalphy-determined (enhancement of Coulombic Q^+A^- attraction provided by the low dielectric microenvironment of the long alkyl chains R)? Whatever its origin, the phenomenon amounts physically to rearrangement of water molecules around alkyl groups, due to the presence of an adjacent electrical charge(54).

Our approach is through the use of cobalt-59 nmr applied to Q^+A^- systems in which the anion is cobalt-centered(55). The rationale is that ^{59}Co nmr chemical shifts encompass some 20,000 ppm (!), due to the presence in Co^{III} atoms of low-lying excited states. Therefore, they ought to be sensitive to the presence of even minute amounts of Q^+A^- ion pairs in water, even if these ion pairs are of the solvent-separated type.

The variations in ^{59}Co chemical shifts for salts of the type $Co(CN)_6^{---}, 3M^+$ are shown as a function of salt concentration in Figure 1. Whereas, in the presence of H^+ or of alkali metal counterions (Li^+, Na^+, K^+, Rb^+, Cs^+), the concentration dependence is nil or slight, use of Q^+ counter-ions leads to considerable vatiations, increasing as expected with the number of carbon atoms in the alkyl chains. An approximate treatment of the data, neglecting further equilibria and the change of ion activity with the ionic strength, yields an equilibrium constant $K_1 \simeq 35$ M^{-1} for the ion pairing:

$$^+N(C_4H_9)_4 + Co(CN)_6^{---} \underset{}{\overset{K_1}{\rightleftharpoons}} Q^+, A^-$$
$$(Q^+) \qquad (A^-)$$

The solvent dependence of the extent of ion pairing is shown in Figure 2: as the dielectric constant is reduced, with respect to that of water, there is less and less change in the ^{59}Co chemical shift with concentration. One interpretation is that in low dielectric solvents, such as ethanol or dichloro-1,2-ethane, ion pairing is very extensive ($K_1 \rightarrow \infty$) and the small concentration dependence is to be attributed to an equilibrium between contact and solvent-separated ion pairs. In the high dielectric solvent N-methyl formamide ($\varepsilon > 180$), there is also negligible concentration dependence of the ^{59}Co chemical shift: dissociated ion pairs predominate, so that again concentration changes do not affect the chemical shifts.

We are engaged in the quantitative analysis of these results, in order to provide detailed answers to the above-mentioned questions

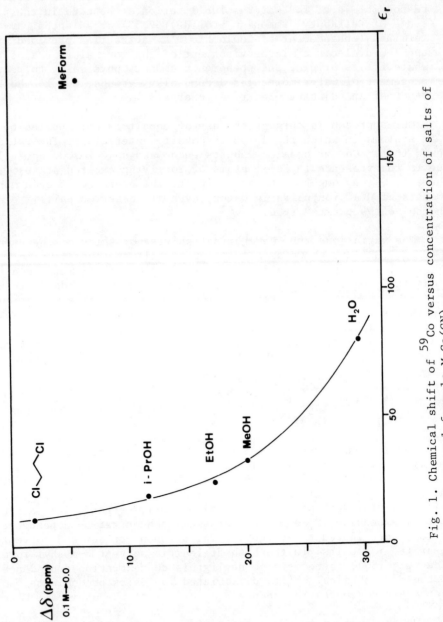

Fig. 1. Chemical shift of ^{59}Co versus concentration of salts of general formula $M_3Co(CN)_6$.

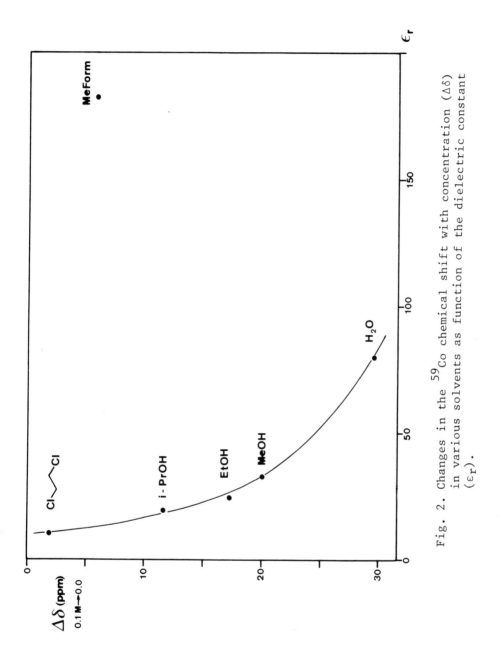

Fig. 2. Changes in the ^{59}Co chemical shift with concentration ($\Delta\delta$) in various solvents as function of the dielectric constant (ε_r).

In any case, we have here a system, and an extremely sensitive tool for examining it, to model enhancement of ion pairing by a low-dielectric microenvironment This should help to understand, not only phase-transfer catalysis, but also ionic transport across lipophilic membranes, neurotransmission which is mediated by Q^+A^- salts such as acetylcholine, and enzymatic action in which it is a fundamental factor.

6. CONCLUSION

In conformity with the general title of this Symposium, on "Solute-Solute-Solvent Interactions", we have provided here examples from work performed in our laboratory on solute-solvent, solute-solute, and solute-solute-solvent interactions. These interactions involving charged groups are fundamental to a better knowledge of important phenomena such as anion activation, ion-binding by proteins, phase-transfer catalysis, and the mechanism of enzymatic action.

REFERENCES

1. For reviews, see:
    B. Lindman ans S. Forsén, "The Alkali Metals", in "NMR and the Periodic Table", R. K. Harris and B. E. Mann, eds., Academic Press, London, p. 129 (1978).
    P. Laszlo, Nachrichten Chem. Tech. Lab. 27:710 (1979).
    P. Laszlo, Bull. Magn. Resonance, in press.
2. J. B. Hubbard, P. Colonomos, and P. G. Wolynes, J. Chem. Phys. 71:2652 (1979).
3. B. P. Fabricand, S. S. Goldberg, R. Leifer, and S. G. Ungar, Mol. Phys. 7:425 (1964).
4. L. Lanatowicz and Z. Pajak, Chem. Phys. Lett. 38:166 (1976).
5. Z. Pajak and L. Lanatowicz, Acta Phys. Pol. A53:555 (1978).
6. A. Delville, C. Detellier, A. Gerstmans, and P. Laszlo, J. Am. Chem. Soc. submitted for publication.
7. A. Delville, C. Detellier, A. Gerstmans, and P. Laszlo, J. Am. Chem. Soc. submitted for publication.
8. A. V. Hill, J. Physiol. (London), 40:IV (1910).
9. R. H. Erlich and A. I. Popov, J. Am. Chem. Soc. 93:5620 (1971).
10. R. J. P. Williams, Biol. Rev. 54:389 (1979).
11. J. Grandjean, P. Laszlo, and C. Gerday, FEBS Letters, 81: 376 (1977).
12. J. Grandjean and P. Laszlo, in "Ions and Protons Involved in Fast Dynamic Phenomena", P. Laszlo, ed., Elsevier, Amsterdam, p. 373 (1978).
13. C. Gerday, J. Grandjean, and P. Laszlo, FEBS Letters, 105: 384 (1979).
14. A. Delville, J. Grandjean, P. Laszlo, C. Gerday, Z. Grabarek, and W. Drabikowski, Eur. J. Biochem. 105:289 (1980).

15. A. Delville, J. Grandjean, P. Laszlo, C. Gerday, H. Brzeska, and W. Drabikowski, Eur. J. Biochem. submitted for publication.
16. A. Paris and P. Laszlo, A.C.S. Symposium Series, N. 34, H. A. Resing and Ch. G. Wade, eds., Washington, p. 418 (1976).
17. T. J. Pinnavaia, H. T. Miles, and E. D. Becker, J. Am. Chem. Soc. 97:7198 (1975).
18. W. Guschlbauer and J. F. Chantot, in "Proceedings of the International Conference" held in Dymaczewo near Poznan, 13-17.9.1976, Poznan, p. 96.
19. H. T. Miles and J. Frazie, Biochem. Biophys. Res. Comm. 49:199 (1972).
20. D. M. Gray and F. J. Bollum, Biopolymers, 13:2087 (1974).
21. M. Hattri, J. Frazier, and H. T. Miles, Biochemistry, 14: 5033 (1975).
22. T. Golas, M. Fikus, Z. Kazimerczuk, and D. Shugar, Eur. J. Biochem. 65:183 (1976).
23. F. B. Howard, J. Frazier, and H. T. Miles, Biopolymers, 16:791 (1977).
24. A. Yamada, K. Akasaka, and H. Hatano, Biopolymers, 17:749 (1978).
25. S. Arnott, R. Chandrasekaran, and C. M. Martilla, Biochem. J. 141:537 (1974).
26. S. B. Zimmerman, G. H. Cohn, and D. J. Davies, J. Mol. Biol. 92:181 (1975).
27. S. B. Zimmerman, J. Mol. Biol. 106:663 (1976).
28. M. Calvin, Angew. Chem. Int. Ed. Engl.12:12L (1975).
29. M. Calvin, "Chemical Evolution. Molecular Evolution towards the Origin of Living Systems on the Earth and Elsewhere", Clarendon Press, Oxford (1969).
30. C. C. Price, ed., "Synthesis of Life", Dowden, Hutchinson, and Ross, Stroudsburg, Pa. (1974).
31. T. H. Jukes, R. Holmquist, and H. Moise, Science, 189:50 (1975).
32. F. T. Mackenzie, in "Chemical Oceanography", 2nd ed., Vol. 1, J. P. Riley and G. Skirrow, eds., Academic Press, London, p. 309 (1975).
33. J. D. Burton, Chem. Ind. (London), 550 (1977).
34. P. Laszlo and A. Paris, Compt. Rend. Acad. Sci. Paris, 286D:717 (1978).
35. M. Borzo, C. Detellier, P. Laszlo, and A. Paris, J. Am. Chem. Soc. 102:1124 (1980).
36. T. J. Pinnavia, C. L. Marshall, C. M. Mettler, C. L. Fisk, H. T. Miles, and E. D. Becker, J. Am. Chem. Soc. 100: 3625 (1978).
37. C. Detellier, A. Paris, and P. Laszlo, Compt. Rend. Ac. Sci. Paris, 286D:781 (1978).
38. M. Borzo and P. Laszlo, Compt. Rend. Ac. Sci. Paris, 287C: 475 (1978).

39. A. Delville, C. Detellier, and P. Laszlo, J. Magn. Res., 34:301 (1979).
40. C. Detellier and P. Laszlo, J. Am. Chem. Soc. 102:1135 (1980).
41. A. W. Hanson, Acta Cryst. 17:559 (1964).
42. B. T. Kilbourn, J. D. Dunitz, L. A. R. Pioda, and W. Simon, J. Mol. Biol. 30:559 (1967); M. Pinkerton, L. K. Steinrauf, and P. Dawkins, Biochem. Biophys. Res. Comm. 35: 512 (1969); M. Dobler, J. D. Dunitz, and J. Krajewski, J. Mol. Biol. 42:603 (1969).
43. P. Laszlo and P. Rinaldi, unpublished results.
44. C. Detellier and P. Laszlo, Helv. Chim. Acta, 62:1559 (1979).
45. R. H. Lemmon, Chem. Rev. 70:95 (1979).
46. A. Brack and L. E. Orgel, Nature (London), 256:383 (1975).
47. M. F. Perutz, Science, 201:1187 (1978).
48. A. Cornélis and P. Laszlo, J. Chem. Res. (S) 462 (1978); (M) 5457 (1978).
49. C. L. Liotta, E. E. Grisdale, and H. P. Hopkins, Jr., Tet. Letters (48), 4205 (1975).
50. G. W. Gokel and W. P. Weber, J. Chem. Ed. 55:350; 429 (1978).
51. A. W. Herriott and D. Picker, J. Am. Chem. Soc. 97:2345 (1975).
52. E. M. Arnett, M. Ho, and L. L. Schaleger, J. Am. Chem. Soc. 92:7039 (1970).
53. S. J. Rehfeld, J. Am. Chem. Soc. 95:4489 (1973).
54. R. Zana, J. Phys. Chem. 81: 1817 (1977).
55. P. Laszlo and A. Stockis, to be published.

NMR STUDIES OF CALCIUM AND MAGNESIUM IN BIOLOGICAL SYSTEMS

Sture Forsén, Thomas Andersson, Torbjörn Drakenberg,
Eva Thulin, and Tadeusz Wieloch

Physical Chemistry 2 and Physical Chemistry 3
Chemical Centre, P.O.B. 740, Lund University
S-220 07 Lund 7, Sweden

Abstract - The NMR spectroscopic technique has developed to a state where studies of $^{25}Mg^{2+}$ and $^{43}Ca^{2+}$ in the millimolar concentration range are feasible. This development has in turn opened new possibilities in the study of calcium and magnesium binding proteins. In the article three proteins are chosen as examples: troponin C, calmodulin and phospholipase A_2. The type of information that can be obtained is: (i) binding constants in the range 1 to 10^4 M^{-1}; (ii) the competition of different cations for the calcium or magnesium binding site; (iii) the effects of other ligands (drugs etc.) on the protein; (iv) pK_a values for the groups involved in the Mg^{2+} or Ca^{2+} binding; (v) the dynamics of the binding site (correlation time). ^{43}Ca NMR studies of proteins with calcium binding constants exceeding 10^5 M^{-1} are generally not possible. In such situations the use of ^{113}Cd or ^{111}Cd NMR is often a powerful alternative. The Cd isotopes have both spin $I = 1/2$ and NMR signals from the individual ions when bound to the protein may be obtained. The chemical shift of cadmium is very sensitive to the nature of the ligands. Additional information may be obtained from competition experiments, chemical modifications, pH variations etc.

1. INTRODUCTION

Calcium and magnesium play an important role in biological systems. Calcium ions and to a lesser extent magnesium ions are structure formers that may trigger conformation changes in biological macromolecules. The calcium ion is a good cross-linking agent and is known to stabilize cell-cell interactions and subunit interactions in multisubunit proteins; also the binding of some proteins, for example prothrombin, to phospholipid bilayers appears to be mediated by calcium ions. About 70 different calcium binding proteins have been described(1) and the crystal structure of about 6 of these are known. With a few notable exceptions, staphylococc nuclease and phosopholipase A_2, the calcium ion is not observed to be bound at the active site. The calcium ion is liganded exclusively by oxygens: carboxyl groups, carbonyls from peptide bonds and sometimes hydroxyl groups or water molecules.

Much less is known about magnesium binding proteins. Magnesium ion is known to activate a number of enzymes many of which are involved in the transfer of phosphate groups. Also magnesium ions are required for the biological activity of t-RNA. Magnesium is found to be coordinated to both oxygen and nitrogen ligands (chlorophyll!).

The calcium ion has long been recognized as an important regulator of cellular activities(2). The detailed mechanism of its action is still largely unknown although it presently appears that many of the regulatory effects are brought about by a small thermostable calcium binding protein named calmodulin(3).

2. THE NMR APPROACH

There is a general lack of physical probes to study Ca^{2+} and Mg^{2+} in macromolecular complexes - both ions have a closed shell of electrons and no convenient optical spectroscopic properties. Both calcium and magnesium have however isotopes with magnetic nuclei, ^{25}Mg and ^{43}Ca, that makes NMR studies possible. Their NMR properties are summarized in Table I.

Table I. NMR properties of ^{25}Mg and ^{43}Ca

	^{25}Mg	^{43}Ca
Spin(I)	5/2	7/2
NMR freq. (MHz) at:		
B_o = 2.35 T	6.1	6.7
B_o = 6.0 T	15.6	17.2
Natural abundance (%)	10.1	0.15
Relative sensitivity[†] equal no. of nuclei	$3 \cdot 10^{-3}$	$6 \cdot 10^{-3}$
Effective relative sens. "receptivity"	$3 \cdot 10^{-4}$	$9 \cdot 10^{-6}$
Quadrupole moment (10^{-24} cm^2)	0.22	0.06

[†] ^1H = 1.000

NMR studies of ^{25}Mg and ^{43}Ca in biological systems is faced with a number of problems. In the first place the receptivity of both nuclei is low as may be inferred from Table I and the use of isotope enriched material is usually quite necessary in order to perform NMR studies at concentrations at least approaching those in living systems.

Secondly both ^{25}Mg and ^{43}Ca are nuclei with spins $I > \frac{1}{2}$ and thus have electric quadrupole moments that largely determines their NMR properties. In particular nuclear relaxation through interactions between the nuclear quadrupole moment and fluctuating electric field gradients at the place of the nucleus is a very efficient mechanism. Direct observations of NMR signals from quadrupolar nuclei firmly bonded to a large slowly tumbling macromolecule will therefore in general be exceedingly difficult. It is however still possible to obtain biochemically relevant information when there is a sufficiently fast chemical exchange between the macromolecular binding site(s) and the free solvated ion (4-6). (We shall shortly substantiate what is to be understood by "sufficiently fast"). This approach, which may be termed the indirect method, has for a number of years been used with satisfactory results in the study of Cl^- (^{35}Cl and/or ^{37}Cl) and Na^+ (^{23}Na) in biological systems(5). The basic theory of the indirect method has been extensively discussed elsewhere(5,6) and we will here restrict ourselves to a very brief and simplified account. This is perhaps best made by a reference to Figure 1A. The observed NMR relaxation rates, R_{obs}, are under circumstances where only a small fraction, p_B, of the cations are bonded to the macromolecule given by the Swift-Connick type expression

$$R_{obs} = R_{free} + \frac{p_B}{R_B^{-1} + k_{off}^{-1}} \tag{1}$$

where R_{free} is the relaxation rate of the "free" ion in aqueous solution, R_B is the relaxation rate of the bonded ion and k_{off} is the off-rate of the ion from the binding site. The difference $R_{obs} - R_{free}$ will in the following be referred to as the excess relaxation rate, R_{ex}. Equation (1) is not strictly valid when p_B becomes appreciable – more than 0.1 say – and in many of the applications to be discussed below more complete treatments using the full expressions for chemical exchange in a two site system have actually been used. Nor is equation (1) strictly valid when non-extreme narrowing conditions (i.e. $\omega\tau_c > 1$) apply to the macromolecular binding site. Equation (1) is however adequate to illustrate the limits of applicability of the indirect method.

We may infer from Figure 1A that the method should be applicable to molecules for which the cation binding constants are $k_{Bind}^{Ca} < 10^5$ M^{-1} and $k_{Bind}^{Mg} < 10^2$ M^{-1}. These conditions are not to be regarded as the final word in this matter – future experience will tell.

What can we do when the binding constants of Mg^{2+} and Ca^{2+} considerably exceed the values above and no observable effects on the ^{25}Mg or ^{43}Ca NMR signals are produced by the macromolecule? In such situations at least two alternative approaches are still possible as illustrated by Figure 1B. The use of a monovalent cation like $^{23}Na^+$ to probe the divalent cation binding site is sometimes a very useful techniques. Na^+ has an ionic radius (0.097 nm) similar to that of Ca^{2+} (0.099 nm) but the lower charge-to-radius ratio of the monovalent cation usually gives it a much lower binding constant and a much higher off-rate than for Ca^{2+}. We will not discuss further this technique but refer the reader to the contribution by Prof. Pierre Laszlo in this volume.

The other alternative is to use a spin $I = \frac{1}{2}$ nucleus with suitable properties as a substitute for Mg^{2+} or Ca^{2+}. We will here illustrate the use of NMR of ^{113}Cd (or ^{111}Cd) in the study of calcium binding proteins. The ionic radius of Cd^{2+} (0.097 nm) is very close to that of Ca^{2+} and Cd^{2+} will easily substitute for Ca^{2+} in a number of proteins. The NMR properties of ^{111}Cd and ^{113}Cd are given in Table II. The fact that both cadmium isotopes have spin $I = \frac{1}{2}$ makes direct observation of NMR signals from the cadmium substituted calcium proteins possible(7-9).

3. EXPERIMENTAL ASPECTS

The experiments reported in this article have all been performed on a FT NMR spectrometer built in our laboratory at Lund. The instrument employs a wide-bore 6.0 Tesla superconducting magnet, custom made by Oxford Instruments, Oxford, England. In order to achieve the highest possible sensitivity the probes are of the solenoid type(10) and sample tubes - usually containing about 3 ml of sample - are

Table II. NMR Properties of ^{111}Cd and ^{113}Cd

	^{111}Cd	^{113}Cd
Spin (I)	$\frac{1}{2}$	$\frac{1}{2}$
NMR freq. (MHz) at:		
$B_o = 2.35$ T	21.2	22.2
$B_o = 6.0$ T	54.1	56.7
Natural abundance (%)	12.8	12.3
Relative sensitivity[†] equal no. of nuclei	$0.9 \cdot 10^{-2}$	$1.1 \cdot 10^{-2}$
Effective relative sensitivity "receptivity"	$1.2 \cdot 10^{-3}$	$1.3 \cdot 10^{-3}$

[†]$^1H = 1.000$

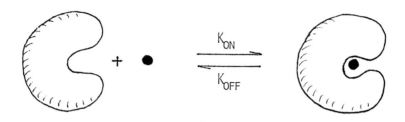

A. CHEMICAL EXCHANGE RATE COMPARABLE TO OR FASTER THAN THE RELAXATION RATE OF BOUND ION ($K_{OFF} \geq R_{iB}$)

RELAXATION RATE OF ^{25}MG AND ^{43}CA NMR SIGNAL IS APPROX. GIVEN BY

$$R_{OBS} = R_{FREE} + \frac{P_B}{R_B^{-1} + K_{OFF}^{-1}}$$

EXPERIMENTAL EXPERIENCE: R_B FOR ^{25}MG AND ^{43}CA IS $10^3 - 10^4$ SEC^{-1}

^{43}CA : $K_{ON} \approx 10^8 - 10^9$ $M^{-1} SEC^{-1}$

THUS $K_{OFF} \geq 10^3$ WHEN $K_{BINDING} \leq 10^5$ M^{-1}

^{25}MG : $K_{ON} \approx 10^5$ $M^{-1} SEC^{-1}$

THUS $K_{OFF} \geq 10^3$ WHEN $K_{BINDING} \leq 10^2$ M^{-1}

B. CHEMICAL EXCHANGE RATE MUCH SLOWER THAN R_{iB}

(1) USE MONOVALENT CATIONS ($^{23}NA^+$, $^7LI^+$,...) AS PROBES OF MG^{2+} OR CA^{2+} BINDING SITES ;

(2) REPLACE MG^{2+} OR CA^{2+} WITH A SUITABLE CATION WITH SPIN $I = 1/2$ --- FOR EXAMPLE $^{113}CD^{2+}$ INSTEAD OF CA^{2+} ;

Fig. 1. Schematic illustration of different NMR approaches in the study of calcium and magnesium binding to biological macromolecules. R_{iB} represents the relaxation rate of the bound ion and p_B is the fraction of ions bonded to the macromolecule. It is assumed that $p_B \ll 1$.

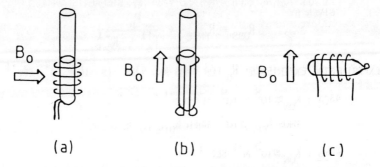

Fig. 2. Different transmitter/receiver coil arrangements in probes of NMR spectrometers: (a) iron core electromagnets; (b) cryomagnets with Helmholtz coils; (c) cryomagnets with solenoids. Arrangement (c) has been used in the studies reported in this chapter.

inserted perpendicular to the cylindrical axis of the proble body (Figure 2). Using isotope enriched material ($\approx$ 90 % ^{25}Mg or 61 % ^{43}Ca) it is possible to obtain a good signal-to-noise ratio after about 100 transients on aqueous solutions with 1 mM ^{43}Ca or 5 mM ^{25}Mg.

The ^{113}Cd NMR studies described below were done on 96 % isotope enriched material.

In the applications of ^{25}Mg and ^{43}Ca NMR to be discussed below we have mostly confined the discussion to the observed line widths assuming Lorenzian or near Lorenzian line shapes. The true relaxation behaviour of I = 5/2 and 7/2 nuclei is however somewhat more complex (11). The conclusions drawn remain unaffected however.

4. APPLICATIONS TO INDIVIDUAL PROTEINS

<u>Troponin C</u>. The contraction of striated muscles is triggered by the release of a pulse of Ca^{2+} ions. On a molecular level the action of Ca^{2+} is generally assumed to be mediated by troponin - a protein complex consisting of three subunits, troponin C, troponin T and troponin I. Of these troponin C is known to bind Ca^{2+} and to undergo conformation changes upon Ca^{2+} binding(12). Troponin C (TnC) is a medium sized protein (MW$\approx$18,000) whose primary structure is known but not its tertiary structure. Equilibrium dialysis studies indicate that TnC has two classes of binding sites, two sites which have a high affinity for Ca^{2+} but also bind Mg^{2+} ("Ca^{2+} - Mg^{2+} sites"; $K^{Ca} = 2.1 \cdot 10^7$ M^{-1} and $K^{Mg} = 5 \cdot 10^3$ M^{-1}(13)) and two sites that bind Ca^{2+} more weakly but not Mg^{2+} ("the Ca^{2+} specific regulatory sites"; $K^{Ca} = 3.2 \cdot 10^5$ M^{-1}(13)). It may be noted that the alleged Ca^{2+} specificity of the latter sites may simply be due to experimental

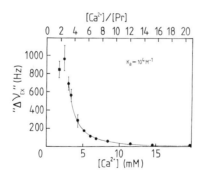

Fig. 3. The ^{43}Ca NMR excess line width (observed line width substracted with the line width of "free", solvated ^{43}Ca^{2+}) as a function of the Ca^{2+} concentration in the presence of: ($\bullet$) 0.94 mM troponin C (TnC); ($\blacksquare$) 1.72 mM TnC $\left[Ca^{2+}\right]$ = 3.16 mM. Temp = 23°C, pH 7.0.

difficulties of measuring Mg^{2+} binding constants of the order of 10^0 or 10^1 M^{-1} (14). In Figure 3 is shown the ^{43}Ca NMR excess line width in the presence of TnC (under the conditions of our experiments the line width of "free" $^{43}Ca^{2+}$ is usually between 5 to 10 Hz - this value is largely due to inhomogeneity broadening since the "natural" line width is less 1 Hz). We clearly see the need to be able to study Ca^{2+} concentrations in the mM range! The ^{43}Ca NMR signal seen at a Ca^{2+} / Tnc ratio of 1.8 is probably due to calcium ions bound to the high affinity Ca^{2+} - Mg^{2+} sites - the line width of 900 Hz is surprisingly small and points to a high symmetry of the binding sites. The ^{43}Ca signals observed for Ca^{2+} / TnC ratios greater than 3 are mainly due to calcium ions exchanging with the two "calcium specific" regulatory sites. The dependence of $\Delta\omega_{ex}$ on Ca^{2+} allows an estimate of the Ca^{2+} binding constant. Assuming for simplicity two independent binding sites with identical binding constants, K^{Ca}, we obtain $K^{Ca} > 10^4$ M^{-1} (the full drawn curve in Figure 3). In the fitting procedure a total band shape analysis was needed to take chemical exchange effects into account. (cfr. Figure 4).

As may be inferred from the simplified expression for R_{obs} in Figure 1A the relaxation rate of the ^{43}Ca signal in the presence of a protein is sensitive to the chemical exchange rate, k_{off}. When exchange rate is very slow there will be no measurable effect on the ^{43}Ca NMR signal for the "free" ions in excess. With increasing exchange rate (increased temperature) the ^{43}Ca NMR signal is expected to broaden. The broadening will reach a maximum that will depend on the line width of the signal due to the bound ions ($\sim R_{2B}$) and at still higher rates (temperatures) the line width will decrease with the line width of the bound ions as the effective correlation time, τ_c, becomes shorter. As is shown in Figure 4a the temperature dependence of the ^{43}Ca NMR line width in the presence of TnC is almost a textbook example of a change from chemical exchange dominated line width at low temperatures to relaxation rate determined line width at higher temperatures. In Figure 4b an example of the observed ^{43}Ca NMR signal is shown. A preliminary analysis of the temperature dependence of the ^{43}Ca signal in Figure 4a gives the following rate parameters for the Ca^{2+} - TnC exchange from the Ca^{2+} specific regulatory sites at 300 K:

$$K_{off} = 10^{-3} \text{ s}^{-1}$$

$$\Delta H^{\ddagger} = 38.5 \text{ kJ mol}^{-1}$$

$$\Delta S^{\ddagger} = 59 \text{ J K}^{-1} \text{ mol}^{-1}$$

The above value for the off-rate of Ca^{2+} from the Ca^{2+} specific sites are the regulatory sites involved in muscle contraction.

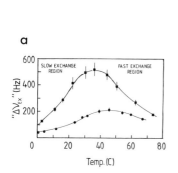
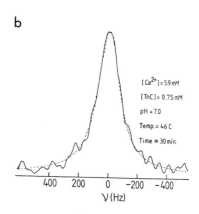

Fig. 4. (a) The temperature dependence of the ^{43}Ca NMR signal in the presence of TnC. (■) 0.86 mM TnC, $[Ca^{2+}]$ = 3.67 mM, pH = 7.0; (●) 0.75 mM, $[Ca^{2+}]$ = 5.92 mM, pH = 7.1.
(b) The experimental ^{43}Ca NMR signal at 46°C obtained under the conditions indicated in the figure. The dotted line is a Lorentzian line shape with the same line width as for the observed signal.

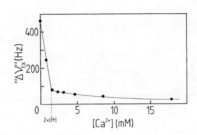

Fig. 5. The effect of added Ca^{2+} on the excess line width of the ^{25}Mg NMR signal in the presence of TnC indicates that a large proportion of the broadening is due to Mg^{2+} binding to the $Ca^{2+} - Mg^{2+}$ sites of TnC. Conditions: $[TnC] = 0.7$ mM; $[Mg^{2+}] = 2.9$ mM, temp. 24°C, pH = 7.1.

In a manner similar to that just discussed for ^{43}Ca it is also possible to use ^{25}Mg NMR to study the interaction of Mg^{2+} with TnC. A study of the dependence on the $^{25}Mg^{2+}$ concentration of the ^{25}Mg line width in the presence of about 1 mM (not shown) has been performed. The data can be analyzed to give a binding constant, K^{Mg}, of about $4 \cdot 10^2$ M^{-1}. This is an average binding constant - no binding constant for individual sites can be obtained from this study alone. The Ca^{2+} competition experiments illustrated in Figure 5 indicate that it is the $Ca^{2+} - Mg^{2+}$ sites that give rise to the major part of the observed broadening. As shown in Figure 5 the ^{25}Mg NMR excess line width in the presence of 0.74 mM TnC is dramatically reduced through the addition of 2 mol of Ca^{2+} per mol of TnC. Our interpretation of this experiment is that $^{25}Mg^{2+}$ is primarily probing the two $Ca^{2+} - Mg^{2+}$ sites although the residual ^{25}Mg excess linewidth after the addition of 2 mol of Ca^{2+} per mol TnC indicates the presence of other protein sites that interact with Mg^{2+}. The temperature dependence of the ^{25}Mg NMR line width in the presence of TnC shows the same general behaviour as that shown for ^{43}Ca in Figure 4. A preliminary calculation of the chemical exchange rate of Mg^{2+} gives $k_{off} = 8 \cdot 10^2$ s^{-1} at 300 K.

The large calcium binding constant to the $Ca^{2+} - Mg^{2+}$ sites ($K^{Ca} = 2.1 \cdot 10^7$ M^{-1}) renders these sites impossible to study with the indirect method - although as we have indicated above the broad ^{43}Ca NMR signal from the bound ions actually can be observed! ^{113}Cd NMR is then perhaps applicable? Indeed it is. Figure 6 shows the ^{113}Cd NMR spectrum of 2.9 mM calcium free TnC to which 2 mol of ^{113}Cd per mol of TnC has been added. No other signals appear to the further addition of $^{113}Cd^{2+}$ nor do the signals undergo a change in chemical shift(8). The two signals, A at -107.5 ppm and B at -111.0 ppm are attributed to the two high affinity sites[†] - structurally similar but not identical as indicated by the presence of two separate ^{113}Cd NMR signals. The chemical shift region in which the ^{113}Cd NMR signals are observed is characteristic of Cd^{2+} coordinate exclusively to oxygens. The ^{113}Cd NMR signals from $(Cd)_2$ - parvalbumin falls in the same region(7). The observed equal intensity - within the experimental error - of the two ^{113}Cd NMR signals already in $(Cd)_1$ - TnC may be due to a fortuitous equality of the binding constant to the two sites but it appears more likely that the two sites display a positive cooperativity.

The $^{113}Cd^{2+}$ ions in $(Cd)_2$ - TnC can be easily displaced by Ca^{2+} but are not noticeably displaced by Mg^{2+} even up to $[Mg^{2+}] \approx 2$ M in agreement with the about 10^4 fold difference between K^{Ca} and K^{Mg}(13).

Calmodulin. As mentioned in the introduction it has recently been demonstrated that calcium regulation of a number of enzyme systems and other cellular events is mediated by a low molecular weight protein termed calmodulin. There is presently good evidence that calmodulin is an ubiquitous calcium regulatory protein whose primary sequence is highly conserved throughout all eucaryotic cells(3).

Calmodulin has a MW $\approx$ 16,700 and its primary sequence is homologous with that of parvalbumins and skeletal muscle TnC. Its amino acid sequence can be divided into four internally homologous domains, each with a potential calcium binding site(16). Most studies have indeed indicated that calmodulin can bind 4 mol of Ca^{2+} per mol of protein but there is some disagreement as regards the relative number of high and low affinity sites.

The biological activity of calmodulin is strongly inhibited by antipsychotic drugs of the phenothiazine type. One of the most effective inhibitors is trifluoperazine (TFP)

TFP

which has been reported to bind to calmodulin with a binding constant of about 10^6 M^{-1}(17).

[†]^{113}Cd NMR chemical shifts are reported relative to 0.1 M $Cd(ClO_4)_2$, shifts to low fields taken as positive.

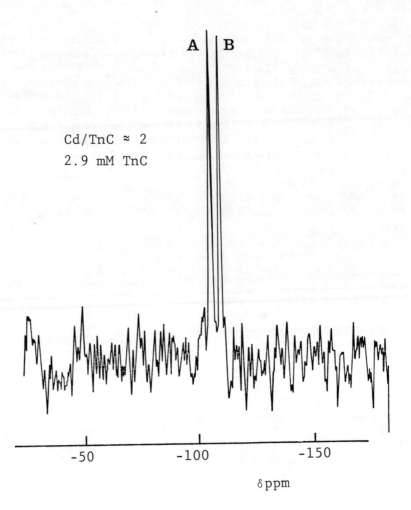

Fig. 6. The ^{113}Cd NMR spectrum at 56.6 MHz of a solution containing 2.9 mM rabbit muscle TnC and 6 mM Cd(ClO$_4$)$_2$ (96.3 % isotope enriched in ^{113}Cd). Temp. = 23°C, pH = 6.4. The shifts are reported relative to 0.1 M Cd(ClO$_4$)$_2$, shifts to low field taken as positive. (From ref. 8).

The homology between TnC and calmodulin makes a comparison of their NMR properties rewarding. When increasing amounts of ^{113}Cd^{2+}

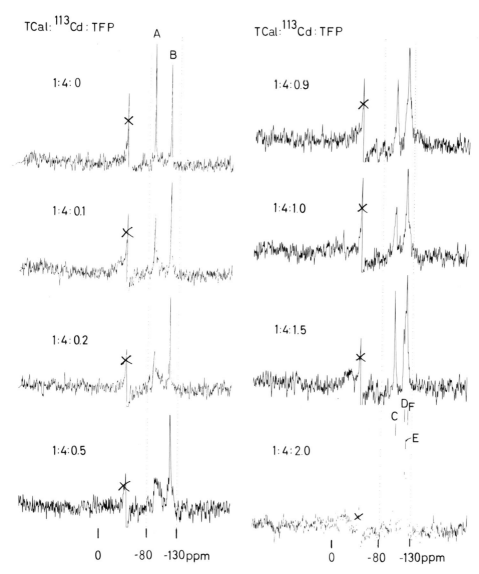

Fig. 7. The effects of trifluoperazine (TFP) on the ^{113}Cd NMR spectrum of a 1.39 mM solution of $(Cd)_4$-calmodulin (bovine testes). In the absence of TFP the two signals occur at $\delta^A = -88.5$ ppm and $\delta^B = -115.0$ ppm. In the presence of 2 mol of TFP per mol calmodulin four signals appear at $\delta^C = -99.7$ ppm, $\delta^D = -113.5$ ppm, $\delta^E = -115.8$ ppm and $\delta^F = -118.7$ ppm. The peak marked with a cross is an instrumental artefact due to an improperly balanced quadrature detector (From ref. 9).

is added to calcium free bovine brain or bovine testes calmodulin (BCal and TCal respectively) two equally intense signals are observed already at a Cd^{2+} to protein ratio below 1:1 This behaviour is strikingly similar to that of TnC and we like to see this as indicative of a strong positive cooperativity in the binding to two high affinity sites. The chemical shifts of the two ^{113}Cd NMR peaks, -88.5 ppm and - 115.0 ppm, are similar but not identical to the shifts observed for $(Cd)_2$ - TnC and $(Cd)_2$-parvalbumin.

^{113}Cd NMR also indicates the existence of two additional binding sites since no signal from "free" $^{113}Cd^{2+}$ appears until more than 4 mol of $^{113}Cd^{2+}$ has been added per mol of calmodulin. The ^{113}Cd NMR signals from the third and fourth site are not observed presumably due to substantial chemical exchange broadening.

The successive addition of the drug TFP to $(Cd)_4$-calmodulin produces dramatic changes in the ^{113}Cd NMR spectrum (Figure 7). A broadening of the ^{113}Cd NMR signals is observed already at a TFP/Calmodulin molar ratio of 0.1:1 and at a ratio of 2:1 a spectrum with four narrow ^{113}Cd NMR signals is observed. The spectral changes observed may be rationalized by assuming the TFP molecules to exchange between the calmodulin molecules with a rate such that the average life-time of a TFP-calmodulin complex is $\approx 10^{-3}$ s. Our interpretation of the results of Figure 7 is that the binding of TFP to calmodulin causes the average conformation of the latter to change in such a way that all four Cd^{2+} binding sites are affected and that the chemical exchange rate of Cd^{2+} to the third and fourth site is considerably reduced as to make detectable also the $^{113}Cd^{2+}$ ions binding to these sites.

^{43}Ca NMR studies of calmodulin further substantiate the above findings. In the presence of 0.4 mM calmodulin the excess line width of a 3 mM $^{43}Ca^{2+}$ solution is close to 80 Hz. When TFP is added to

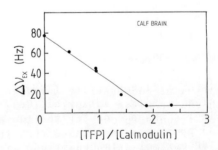

Fig. 8. The addition of TFP to a calmodulin solution containing a slight excess of $^{43}Ca^{2+}$ reduces the ^{43}Ca NMR excess line width considerably and indicates that 2 mol of TFP strongly interact with 1 mol of calmodulin. Calmodulin concentration = 0.41 mM, $[Ca^{2+}]$ = 2.96 mM, temp. = 23°C, pH = 7.1.

this solution the ^{43}Ca excess line width is markedly reduced as is shown in Figure 8. That the reduced line width upon TFP addition is due to a reduction in the chemical exchange rate of Ca^{2+} ions between the free and protein bound states is indicated by the temperature dependence of the ^{43}Ca excess line width in the absence and presence of TFP (not shown). A preliminary analysis of the temperature dependence shows that in the absence of TFP we have for the Ca^{2+} exchange rate - in all likelyhood from the weaker third and fourth binding site - the rate constant $k_{off} = 10^3$ s^{-1}. In the presence of TFP the off rate is reduced by almost an order of magnitude.

A parallel study of the effect of TFP on the Ca^{2+} exchange rate to TnC gives very similar results.

The conformational changes in calmodulin resulting from the binding of TFP as indicated by the present study provides a basis for the study of inhibitory effects of TFP on a variety of calmodulin activated processes(9).

Phospholipase A_2. Both troponin C and calmodulin are calcium binding proteins with a regulatory role. It appears that they have some tertiary structure in the calcium-free form but that they undergo large conformational changes upon binding of Ca^{2+}(18). Phospholipase A_2 by contrast is an enzyme, the physiological function of which is to hydrolize phospholipids in micellar aggregates. The porcine enzyme has a molecular mass of about 14,000. The enzyme occurs also in a proenzyme form and trypsin cleavage of a N-terminal heptapeptide produces the active enzyme. It has been reported to bind 1 mol of Ca^{2+} per mol of enzyme with a binding constant of about 10^3 M^{-1} and an additional mole of calcium with a binding constant of the order of 30 M^{-1} or less depending on pH(19).

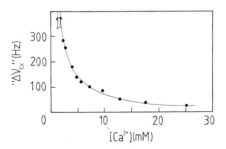

Fig. 9. The ^{43}Ca NMR excess line width as a function of the Ca^{2+} concentration in the presence of 2.0 mM prophospholipase A_2. Temp. = 23°C, pH = 7.5.

The excess line width of the ^{43}Ca NMR signal as a function of the Ca^{2+} concentration in the presence of 2.0 mM of the proenzyme is shown in Figure 9. The pH dependence of the excess line width at a constant Ca^{2+} concentration (Figure 10) shows that the binding is associated with a group or groups with a pK$_a$ value of 5.2.

The data of Figure 9 can be fitted to the simple chemical equilibrium

$$Ca^{2+} + PPLA_2 \underset{}{\overset{K_A}{\rightleftarrows}} (Ca^{2+}) PPLA_2$$

with the binding constant $K_A = 1.1 \cdot 10^3$ M^{-1} (at 277 K). The temperature dependence of the ^{43}Ca line width is again of the type that allows kinetic information to be extracted. A preliminary calculation give $k_{off} = 1.4 \cdot 10^3$ s^{-1} (at 300 K).

A comparison of the relation between the Ca^{2+} binding constant and the off rate obtained for prophospholipase A$_2$ with the corresponding figures for TnC and calmodulin reveals an interesting difference. The off-rates for calcium binding to TnC and calmodulin are of the order of magnitude one would expect if the on-rates were essentially diffusion controlled. By contrast the on-rate of Ca^{2+} to prophospholipase A$_2$ must be several orders of magnitude slower than the on rates to TnC and calmodulin. These differences reflect the different nature of the Ca^{2+} binding sites in prophospholipase A$_2$ on one hand and TnC and calmodulin on the other. Whereas the latter proteins would seem to have flexible amino acid sequences with side chains that easily may wrap around an incoming Ca^{2+} ion, the Ca^{2+} binding site of the enzyme is largely a rigid entity that is unable to bind a Ca^{2+} ion upon each encounter in solution.

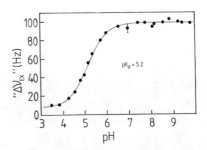

Fig. 10. The pH dependence of the ^{43}Ca NMR signal in a solution containing 1.7 mM prophospholipase A$_2$ and 5.9 mM Ca^{2+}. The changes in the line width can be fitted to a single protonation step with a pK$_a$ of 5.2.

5. SUMMARY

The NMR spectroscopic technique has developed to a state where studies of $^{25}Mg^{2+}$ and $^{43}Ca^{2+}$ in the millimolar concentration range are presently feasible. This has in turn opened up new possibilities in the study of calcium and magnesium binding proteins. The type of information that can be obtained is: binding constants in the range 1 to 10^4 M^{-1}; the chemical exchange rate of the bound ion; the competition of different cations for the calcium or magnesium binding site; the effects of other ligands (drugs etc.) on the protein; pK_a values for the groups responsible for the Mg^{2+} or Ca^{2+} binding; and the dynamics (correlation time) characterizing the binding site. Macromolecular binding sites for Ca^{2+} characterized by binding constants exceeding 10^5 M^{-1} are not generally susceptible to the study by ^{43}Ca NMR. In such situations the use of $^{113}Cd^{2+}$ (or $^{111}Cd^{2+}$) NMR is often a powerful alternative. Since these cadmium isotopes both have spin $I = \frac{1}{2}$ NMR signals from the individual bound ions may be obtained.

The NMR chemical shift of cadmium is very sensitive to the nature of the ligands. Additional information may be obtained from competition experiments, chemical modifications, pH variations etc.

The combined use of both ^{43}Ca and ^{113}Cd NMR appears to be a fruitful tool in the study of calcium binding proteins.

A final word of caution: although the ionic radii of Ca^{2+} and Cd^{2+} are very similar there are after all differences in their solution chemistry, a fact that we must always bear in mind whenever we use Cd^{2+} as a substitute for Ca^{2+} in biological systems.

REFERENCES

1. R. H. Kretsinger, Ann. Rev. Biochem. 239 (1976).
2. M. J. Berridge, Adv. Cyclic Nucleotide Res. 6:1 (1976).
3. W. Y. Cheung, Science, 207:19 (1980).
4. S. Forsén and B. Lindman, Chem. Brit. 14:29 (1978).
5. S. Forsén and B. Lindman, in "Methods of Biochemical Analysis", D. Glick ed. Vol. 27, J. Wiley-Interscience, N.Y. (1980), in press.
6. B. Lindman and S. Forsén, "Chlorine, Bromine and Iodine NMR. Physico Chemical and Biological Applications", Vol. 12 of "NMR Basic Principles and Progress", P. Diehl, E. Fluck, and R. Kosfeld, eds. Springer-Verlag, Berlin (1976).
7. T. Drakenberg, B. Lindman, A. Cavé, and J. Parello, FEBS Lett. 92:346 (1978).
8. S. Forsén, E. Thulin, and H. Lilja, FEBS Lett. 104:123 (1979).
9. S. Forsén, E. Thulin, T. Drakenberg, J. Krebs, and K. Seamon,

FEBS Lett. (1980), in press.
10. D. Hoult, "Progress in NMR Spectroscopy", J. Emsley, J. Freeney, and L. Sutcliffe, eds. Vol. 12, p. 41, Pergamon Press, Oxford (1978).
11. T. E. Bull, S. Forsén, and D. L. Turner, J. Chem. Phys. 70:3106 (1979).
12. S. V. Perry, Biochem. Soc. Transactions. 7:593 (1979).
13. J. D. Potter and J. Gergely, J. Biol. Chem. 250:4628 (1975).
14. R. J. P. Williams, Private communication.
15. J. D. Johnson, S. C. Charlton, and J. D. Potter, J. Biol. Chem. 3497 (1979).
16. T. C. Vanaman, F. Sharief, and D. M. Watterson, in "Calcium Binding Proteins and Calcium Function", R. H. Wasserman, et al., eds. p. 107, Elsevier, N. Y. (1979).
17. R. M. Levine and B. Weiss, Mol. Pharmacol. 13:690 (1979).
18. R. J. P. Williams, Biol. Rev. 54:389 (1979).
19. A. J. Slotboom, E. H. J. M. Jansen, H. Vlijm, F. Pattus, P. Soares de Aranjo, and G. H. deHaas, Biochemistry, 17:4593 (1978).

SYNTHETIC MOLECULAR MEMBRANES AND THEIR FUNCTIONS

Toyoki Kunitake and Yoshio Okahata

Department of Organic Synthesis, Faculty of Engineering
Kyushu University
Fukuoka, 812 Japan

Abstract - Synthetic bilayer membranes can be formed spontaneously from a variety of single-chain and double-chain(dialkyl) amphiphiles. In the case of single-chain amphiphiles, the presence of rigid segments is required in the hydrophobic portion and the aggregate morphology is strongly affected by the structure of the rigid segment. The phase transition and the phase separation are observed in these membranes and the membrane catalysis is affected by these phenomena.

1. INTRODUCTION

The biomembrane is involved in many of the fundamental biological processes. Although its major components are lipid bilayers and proteins, the highly organized nature of the biomembrane is derived from the lipid bilayer. The membrane-forming biolipid is represented by lecithins (1) which possess phosphatidylcholine unit as the hydrophilic moiety and two, higher acyl groups as the hydrophobic moiety.

$$CH_3(CH_2)_{n-1}-CO-CH_2$$
$$CH_3(CH_2)_{n-1}-CO-CHCH_2O-\overset{O}{\underset{O^-}{P}}-OCH_2CH_2-\overset{CH_3}{\underset{CH_3}{N^+}}-CH_3 \qquad (1) \quad \text{lecithin}$$

There have been many attempts to prepare the bilayer structure from amphiphilic compounds other than biolipids; however, the resulting bilayers are either unstable or not well characterized. The major factors which stabilize the lipid bilayer were discussed recently (1,2).

In 1977, we reported for the first time that a totally synthetic bilayer membrane is formed from a simple organic compound (2) (didodecyldimethylammonium bromide)(3). This ammonium bilayer has many physicochemical characteristics that are common to those of the biomembrane. Subsequently, a variety of dialkyl and single-chain amphiphiles with cationic, anionic, nonionic and zwitterionic head groups were shown to form bilayer assemblies in water. These results clearly indicate that the bilayer formation is a general physicochemical phenomenon not limited to the biological system.

The following is a short account of our research on the synthetic molecular membrane since 1977.

2. AMPHIPHILE STRUCTURE AND BILAYER FORMATION

Simple dialkylammonium compounds readily produce the bilayer structure when dispersed in water by sonication. Figure 1 is electron micrographs of didodecyldimethylammonium($2C_{12}N^+2C_1$) aggregates. The micrograph obtained by negative staining clearly shows the formation of multi-walled vesicles. The layer width is about 40 Å, corresponding approximately to two times of the extended molecular length and strongly pointing to the bilayer formation. The freeze-fracture micrograph is evidence of the vesicle formation, since this method is essentially free from deformation during the sample preparation. The molecular weight is about one million, a figure much larger than that of the conventional micelle.

$$CH_3(CH_2)_{n-1} \diagdown \diagup CH_3$$
$$N^+ \quad Br^-$$
$$CH_3(CH_2)_{m-1} \diagup \diagdown R$$

$R = CH_3, CH_2CH_2OH$
CH_2CONH_2, sugar

(2) $C_nC_mN^+2C_1$ (n,m = 8 ~ 22)

The structure requirement for the bilayer formation was subsequently clarified. The higher alkyl chain must be C_{10} or longer, and the head group may be dimethylammonium, modified ammonium or sulfonium(4,5). The aggregate morphology is vesicles or lamellae.

The hydrophilic moiety may be anionic groups such as phosphate (3), sulphonate (4) and carboxylate (5)(6,7).

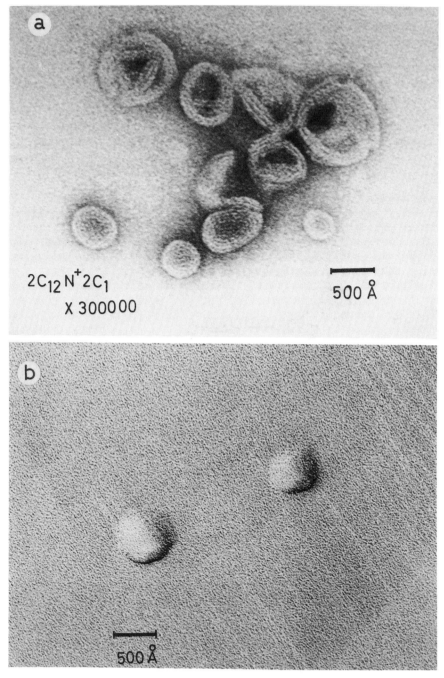

Fig. 1. Electron Micrographs of Didodecyldimethylammonium Vesicles.
 a. Negatively stained sample
 b. Freeze-fracture Replica

$CH_3(CH_2)_{n-1}-O$
$CH_3(CH_2)_{n-1}-O$ P $=O$, O^- (3) $2C_nPO_n^-$ n = 10, 12, 16

$CH_3(CH_2)_{n-1}-O\overset{O}{\overset{\|}{C}}-CH_2$
$CH_3(CH_2)_{n-1}-O\overset{O}{\overset{\|}{C}}-CH-\overset{O}{\overset{\|}{S}}-O^-$ (4) $2C_nSO_3^-$ n = 10, 12, 16

$CH_3(CH_2)_{n-1}-\overset{O}{\overset{\|}{C}}O-CH-\overset{O}{\overset{\|}{C}}-O^-$
$CH_3(CH_2)_{n-1}-\overset{O}{\overset{\|}{C}}O-CH-\overset{O}{\overset{\|}{C}}-O^-$ (5) $2C_n2CO_2^-$ n = 10, 12, 16

We also established the bilayer formation by the following nonionic and zwitterionic dialkyl amphiphiles(8).

$CH_3(CH_2)_{n-1}-OCH_2$
$CH_3(CH_2)_{n-1}-OCH_2$ $CH-O-(CH_2CH_2O)_x-H$ (6) $2C_n$-gl-xG
   (n = 12, 14, 16, 18)
   (x = 6∿30)

$CH_3(CH_2)_{17}-N-(CH_2CH_2O)_{10}-H$
$CH_3(CH_2)_{12}-C=O$ (7) $C_{18}C_{13}$-10G

$CH_3(CH_2)_{11}$ CH_3
$\quad\quad\quad\quad N^+$
$CH_3(CH_2)_{11}$ $(CH_2)_3-\overset{O}{\overset{\|}{S}}-O^-$ (8) $2C_{12}N^+C_1C_3SO_3^-$

$CH_3(CH_2)_{17}$ CH_3
$\quad\quad\quad\quad N^+$
$CH_3(CH_2)_{17}$ $CH_2CH_2O-\overset{O}{\overset{\|}{P}}-O^-$, Na^+ (9) $2C_{18}N^+C_1C_2PO_4^{2-}$

The aggregate morphology includes paired bilayers and fragmentary lamellae in addition to the common vesicle and lamella.

The synthetic bilayer membrane invariably has the liquid crystalline property. We suspected that, if the converse is true, appropriate liquid crystalline amphiphiles could form bilayers. Then, it should be possible to find bilayer-forming amphiphiles whose structures are quite different from those of dialkyl amphiphiles. Diphenylazomethine derivatives belong to a major class of liquid crystalline materials, and, in fact, the following amphiphiles (10) and (11) form stable bilayer membranes(9).

(10) $C_n\text{-BB-N}^+$ n = 0,4,7,12

(11) $C_n\text{-BB-}C_m\text{-N}^+$ n = 0,4,7,12 m = 4, 10

These compounds give clear to slightly turbid solution when dispersed in water by sonication. The flexible alkyl tail must be C_7 or longer for the bilayer formation. The molecular weight of the bilayer aggregate(vesicles and lamellar) is $10^6 - 10^7$. Similar results were obtained from amphiphiles (12) which contained the biphenyl unit as the rigid segment(10).

(12) $C_{12}\text{-BPh-X}$

X: $-(CH_2)_4-\overset{CH_3}{\underset{CH_3}{\overset{|}{N^+}}}-CH_3$, $-\overset{O}{\underset{OH}{\overset{\|}{P}}}-O$, $-(CH_2CH_2-O)_n-H$

An extention of these studies is the formation of monolayer membranes from two-headed ammonium amphiphiles(11). The aggregate structure derived from $N^+C_{10}BBC_{10}N^+$, (13) and $N^+C_{10}BBC_{10}N^+$, (14) is lamellar and tubular, respectively.

(13), $^+\text{N-}C_{10}\text{-BB-}C_{10}\text{-N}^+$

(14), $^+\text{N-}C_{10}\text{-BB-}C_{10}\text{-BB-}C_{10}\text{-N}^+$

A second extention is the use of bent structures as the rigid segment(12). When the rigid segment is composed of diphenyl ether, diphenylketone or meta-substituted diphenylazomethine moiety, rod-like or tubular aggregates are formed. The presence of two ether units in the rigid segment gives disk-shaped aggregates. This morphology is rationalized by assuming a combination of two conformation (zig-zag and bent). The aggregate weight is in the range of 1 - 10 millions. Table I summarizes these results, and the corresponding electron micrographs are shown in Figure 2.

3. FUNCTIONS OF THE BILAYER MEMBRANE

The molecular motion of dialkylammonium bilayers is highly restricted due to their liquid crystalline nature, and the presence of the crystal-liquid crystal phase transition was confirmed by diverse techniques such as differential scanning calorimetry(14), NMR spectroscopy(13), fluorescence polarization(13,15) and positron annihilation(15).

An interesting application of the membrane characteristics is the use of phase separation phenomena. The absorption maximum of the azobenzene unit shifts to shorter wavelengths when stacking of the chromophore occurs. In a mixed bilayer of an azobenzene-containing amphiphile($C_{10}Azo$) and $2C_{16}N^+2C_1$, (15), the phase separation was functions of the phase transition of both of $2C_{10}N^+2C_1$ and $C_{10}Azo$. Two amphiphile phases are completely separated at low temperatures.

$$CH_3-(CH_2)_{11}-O-\phi-N=N-\phi-O-(CH_2)_n-\overset{+}{N}(CH_3)_3 \qquad (15)$$

The bilayer vesicle retains organic and inorganic species much better than the fluid micelle. The vesicle may be cationic(17,18), anionic(7) or nonionic(16).

The tight binding of an ester substrate was used to differentiate the intra-vesicle and inter-vesicle reactions(19). When the catalyst (Cholest-Im), (16), and substrate are solubilized in aqueous solutions of $2C_{12}N^+2C_1$ separately, the inter-vesicle reaction is observed. In contrast, each vesicle can be made to contain both

(16) Cholest-Im

SYNTHETIC MOLECULAR MEMBRANES

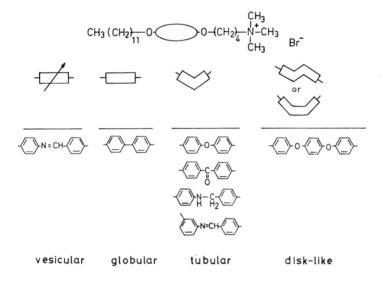

Table I. Relation between the structure of the rigid segment and the aggregate morphology.

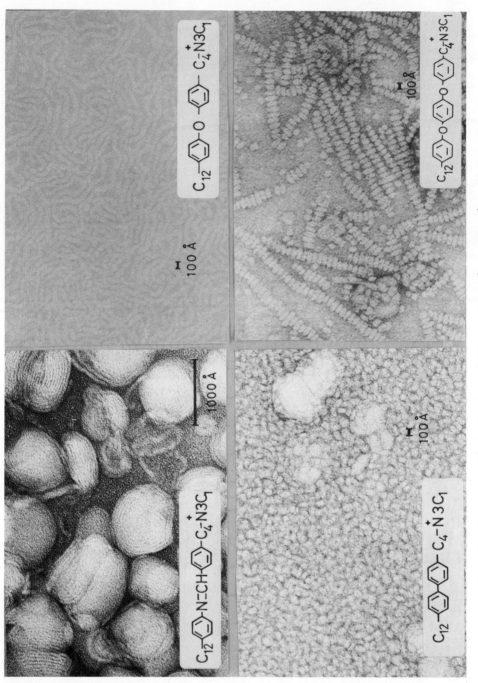

Fig. 2. Electron Micrographs of Representative Ammonium Aggregates.

catalyst and substrate, so that the intra-vesicle reaction is observed. The intra-vesicle reaction was shown to be ca. 200 times faster than the inter-vesicle counterpart in the hydrolysis of PNPP.(17) The rate constant of substrate transfer was estimated to be 10^{-5} s in the subsequent study(20).

$$CH_3(CH_2)_{n-1}\overset{O}{\underset{\|}{C}}O-C_6H_4-NO_2 \quad (17) \quad \begin{array}{l} n = 0; \text{ PNPA} \\ n = 15; \text{ PNPP} \end{array}$$

Cholesterol is found abundantly in animal plasma membranes, and the specific interaction of cholesterol with lipid bilayers in an important biological topic. It appears that cholesterol derivatives interact specifically with ammonium bilayers as well. In the hydrolysis of phenyl esters, cholest-Im showed an especially high reactivity when bound to the $2C_{12}N^+2C_1$ bilayer(18). Cholic acid-derived nucleophiles showed normal reactivity patterns, as may be expected from the fact that cholic acid tends to disintegrate the phospholipid bilayer.

The catalytic action of the dialkylammonium bilayer is affected by the phase transition. The Arrhenius plots of the hydrolysis of PNPP catalyzed by a long-chain imidazole give inflections at temperature close to the phase transition temperature(T_c) of the respective membrane matrix(21).

Similar results were observed in proton abstraction(22):

$$Ph-\overset{O}{\underset{\|}{C}}-CH_2CH_2-O-C_6H_4-NO_2 \xrightarrow[\text{Base} \quad \text{Base·H}^+]{\text{slow}}$$

β-NNP

$$\left[Ph-\overset{O^-}{\underset{|}{C}}-\overset{}{\underset{|}{CH}}-CH_2-O-C_6H_4-NO_2 \right]$$

$$\xrightarrow{\text{fast}} Ph-\overset{O}{\underset{\|}{C}}-CH=CH_2 + {}^-O-C_6H_4-NO_2$$

and in decarboxylation(23):

$$O_2N-\underset{}{\text{benzisoxazole-COO}^-} \longrightarrow O_2N-\underset{}{\text{C}\equiv\text{N, O}^-} + CO_2$$

Figure 3 is the Arrhenius plots of the proton abstraction in the $2C_{16}N^+2C_1$ membrane. The hydroxide-catalyzed reaction gives a straight line but the reaction catalyzed by a hydrophobic base(N-dodecylbenzohydroxamate) gives the Arrhenius plots composed of two lines separated at T_c of the matrix membrane. The activation energy at the temperature range below T_c is much higher than that above T_c, 46 vs. 20 kcal mol^{-1}.

The two kinds of the activation energy are similarly obtained in the decarboxylation reaction(23). Interestingly, the activation energy below T_c increased with increasing chain length of $2C_nN^+2C_1$ (n = 12, 14, 16 and 18), but the reverse was true at temperatures above T_c. These results suggest that the membrane catalysis of the decarboxylation is governed mainly by the hydrophobicity and fluidity at temperatures above and below T_c, respectively.

The influence of T_c on the reaction kinetics was also observed in the cleavage of phosphate esters(24).

At this point it is useful to discuss the general aggregative property of dialkylammonium bilayers in comparison with other ammonium aggregates. As shown in Table II, single-chain ammonium amphiphiles such as CTAB(conventional surfactant) form globular, fluid micelles. The double-chain ammonium amphiphiles can produce bilayers

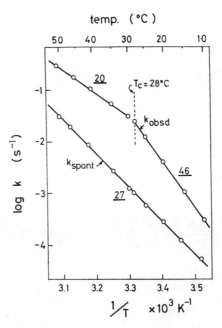

Fig. 3. Arrhenius plots of the proton abstraction in the $2C_{16}N^+2C_1$ membrane.

Table II. Physicochemical properties of ammonium aggregates

Ammonium salts	Total Carbon number	Aggregate morphology	cmc ($\times 10^4$ mol l^{-1})	Aggregate weight ($\times 10^{-4}$)
single-chain CTAB ($C_{16}N^+3C_1 \cdot Br^-$)	19	globular micelle	8	4
double-chain $2C_{12}N^+2C_1 \cdot Br^-$	26	bilayer	0.5	100
$2C_{18}N^+2C_1 \cdot Br^-$	38	bilayer	0.05	1000
triple-chain TMAC ($3C_8N^+C_1 \cdot Cl^-$)	25	small, tight aggregate	ca 0.3	<1

with extraordinarily large aggregate weights($10^6 - 10^7$). The triple-chain amphiphile, typically TMAC, form small, tight aggregates(25). The local hydrophobicity increases in the order of single-chain, double-chain and triple-chain compounds, and this is reflected in the relative rate acceleration of acyl transfer(18), proton abstraction(22) and decarboxylation(23). A wide variety of aqueous aggregates are formed simply by changing the number of the long alkyl chain attached to the ammonium nitrogen.

4. CONCLUSION

The preceding data establish that the bilayer(and monolayer) membrane is formed from a variety of single-chain and double-chain amphiphiles. It is clear that the biolipid bilayer is a special case of the general membrane formation. The peculiar characteristics of these synthetic membranes can be used for constructing new types of the reaction system. The ordered aqueous aggregates will find many applications in the basic and applied research. Some important aspects of the synthetic membrane e.g., chiral synthetic bilayers(26), the membrane-polymer interaction(27), and photochemical functions (28,29) were not mentioned in this paper.

REFERENCES

1. H. Brockerhoff, "Bioorganic Chemistry", Vol. 3, Chapter 1, E. E. vanTamelen, ed., Academic Press, New York (1977).
2. P. L. Yeagle, Accounts Chem. Res. 11:321 (1978).
3. T. Kunitake and Y. Okahata, J. Am. Chem. Soc. 99:3860 (1977).
4. T. Kunitake, Y. Okahata, K. Tamaki, F. Kumamaru, and M Takayanagi, Chem. Lett. 387 (1977).
5. T. Kunitake and Y. Okahata, Chem. Lett. 1337 (1977).

6. T. Kunitake and Y. Okahata, Bull. Chem. Soc. Jpn. 51:1877 (1978).
7. R. A. Mortara, F. H. Quina, and H. Chaimovich, Biochem. Biophys. Res. Comm. 81:1080 (1978).
8. Y. Okahata, S. Tanamachi, N. Nagai, and T. Kunitake, submitted for publication.
9. T. Kunitake and Y. Okahata, J. Am. Chem. Soc. 102:549 (1980).
10. Y. Okahata and T. Kunitake, Ber. Bunsenges. Phys. Chem. in press.
11. Y. Okahata and T. Kunitake, J. Am. Chem. Soc. 101:5231 (1979).
12. T. Kunitake, Y. Okahata, M. Shimomura, and S. Yasunami, Paper presented at the 32nd Colloid and Surface Science Symposium, Kochi, Japan, Oct. 1979.
13. T. Nagamura, S. Mihara, Y. Okahata; T. Kunitake, and T. Matsuo, Ber. Bunsenges. Phys. Chem. 82:1093 (1978).
14. T. Kajiyama, A. Kumano, M. Takayanagi, Y. Okahata, and T. Kunitake, Chem. Lett. 645 (1979).
15. K. Kano, A. Romero, B. Djermouni, H. J. Ache, and J. H. Fendler, J. Am. Chem. Soc. 101:4030 (1979).
16. M. Shimomura, unpublished results in these laboratories.
17. C. D. Tran, P. L. Klahn, A. Romero, and J. H. Fendler, J. Am. Chem. Soc. 100:1622 (1978).
18. Y. Okahata, R. Ando, and T. Kunitake, Bull. Chem. Soc. Jpn. 52:3647 (1979).
19. T. Kunitake and T. Sakamoto, J. Am. Chem. Soc. 100:4615 (1978).
20. T. Sakamoto, unpublished results in these laboratories.
21. T. Kunitake and T. Sakamoto, Chem. Lett. 1059 (1979).
22. Y. Okahata, S. Tanamachi, and T. Kunitake, Nippon Kagaku Kaishi, 442 (1980).
23. T. Kunitake, Y. Okahata, R. Ando, S. Shinkai, and S. Hirakawa, submitted for publication.
24. H. Ihara, unpublished results in these laboratories.
25. Y. Okahata, R. Ando, and T. Kunitake, J. Am. Chem. Soc. 99:3067 (1977).
26. T. Kunitake, N. Nakashima, S. Hayashida, and K. Yonemori, Chem. Lett. 1413 (1979).
27. T. Kunitake and S. Yamada, Polymer Bull. 1:35 (1978).
28. J. R. Escabi-Perez, A. Romero, S. Lukac, and J. H. Fendler, J. Am. Chem. Soc. 101:2231 (1979).
29. M. F. Czarniecki and R. Breslow, J. Am. Chem. Soc. 101:3675 (1979).

NEW INSIGHTS INTO THE HOST-GUEST SOLVENT INTERACTION OF SOME
INCLUSION COMPLEXES. REACTION PATH CONTROL IN CYCLODEXTRIN INCLUSION
AS A LYASE MODEL: SOLVOLYSIS OF β-BROMETHYL-1-NAPHTHALENE

Iwao Tabushi

Department of Synthetic Chemistry
Kyoto University
Sakyo-ku, Kyoto, 606 Japan

Abstract - Since cyclodextrin not only interacts with organic
and/or inorganic compounds through the hydrophobic cavity but
also interacts with water molecules through hydrogen bonding
and/or electronic interactions, several types of host (solute)
- guest (solute) - water (solvent) interactions are investigated
to gain insight into biologically significant recognition
problems. Simultaneous operation of the hydrophobic and other
types of recognition, leading to efficient enzyme model or
receptor model are successfully observed. Detailed discussion
is especially made for the mechanism of "reaction path control"
observed for a cyclodextrin inclusion complex.

1. INTRODUCTION

Among many interesting characteristics of enzymes, the followings are very significant since they are simple catalysts in chemical reactions:

(i)   enormous increasing of rates of chemical reactions,
(ii) strict substrate specificity,
(iii) strict regio and/or stereospecificity for a given specific substrate,
(iv) reaction path control,
(v)   allosteric control.

The first three characteristics are moderately mimicked in the many known examples of homogeneous and heterogeneous nonenzymatic catalysts. Recently the above three enzyme characteristics have been satisfactory mimicked by using specially designed artificial hosts capable of "molecular recognition"(1).

However, almost none of appropriate enzyme models have ever been reported for the reaction path control or the allosteric control in homogeneous aqueous solution(2). In biological systems, both control mechanisms are very frequently and significantly operating as in the case of the reaction path control of the amino acid metabolism by individual B_6 enzyme(3) or the allosteric control of levels of many important compounds by glutamine synthetase.

2. MODIFIED CYCLODEXTRINS

Cyclodextrin (α-, β- or γ-) is known to accomodate appropriate guest molecule(s) in its cavity in aqueous solution, forming an inclusion complex. The major driving force to form this inclusion complex can be derived by thermodynamic consideration(4):

(a) destruction of solvent (water) clusters originally developed along the surface of the guest molecule in water,
(b) Van der Waals interactions between the guest surface and the interior surface of guest molecules,
(c) stabilization of released water molecule(s) which was originally bound in the cavity.

The specificity of the host guest interaction is mainly determined by these factors where the cavity size of a cyclodextrin is very important. Thus, α-cyclodextrin (cyclic hexamer) is specific for benzene derivatives, while β-(heptamer) or γ-(octamer)cyclodextrin is specific for naphthalene or anthracene derivatives, respectively.

Since destruction of water network and host-guest Van der Waals stabilizations are significant to force a hydrophobic guest to come into the host cavity, some artificial situation for enlargement of hydrophobic area should enhance the host-guest binding. Capped cyclodextrin (I) thus shows the remarkably enhanced Van der Waals stabilization, together with effective destruction of water network around a guest molecule, leading to very strong bindings as shown by the following equilibrium constants:

$$I + II \underset{}{\overset{H_2O}{\rightleftharpoons}} I \cdot II \qquad K = 1,300 \text{ M}^{-1}$$

$$\beta\text{-CD} + II \underset{}{\overset{H_2O}{\rightleftharpoons}} \beta\text{-CD} \cdot II \qquad K = 53 \text{ M}^{-1}$$

It is therefore evident the strong hydrophobic nature of the capped cavity(5,6).

This type of "rigid capping" enhances binding as well as more important chemical interactions between the host and the guest: a typical example of the latter is shown in the unusually effective host guest triplet energy transfer from benzophenone capped β-CD(III) to a naphthalene derivative(6).

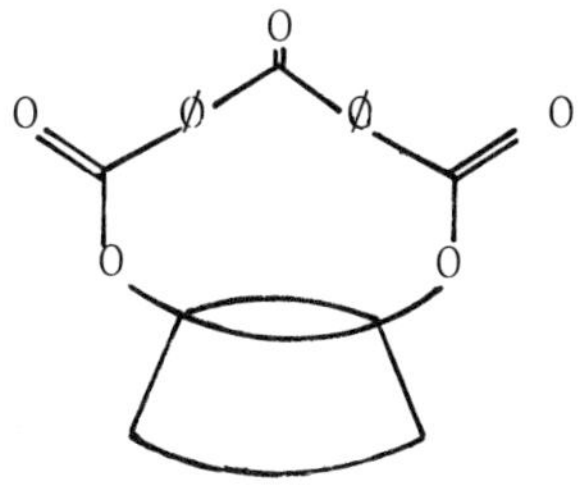

(III)

The hydrophobic host-guest interaction mentioned above is called "hydrophobic recognition" by the author and may be part of a more sophisticated "multiple recognition" of a guest by a host in water; this is exemplified by a metal capped cyclodextrin (IV) where the central metal ion is involved in the coordination recognition site, and the β-cyclodextrin cavity acts as a hydrophobic recognition site. Toward this special host, adamantane-2-keto-1-carboxylate (V) behaves as a doubly recognized specific guest molecule being 300 times stronger with respect to the parent β-CD.

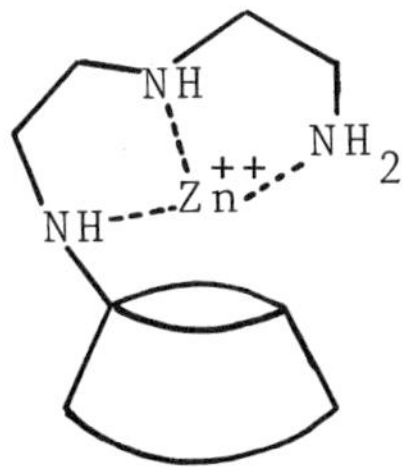

(IV)

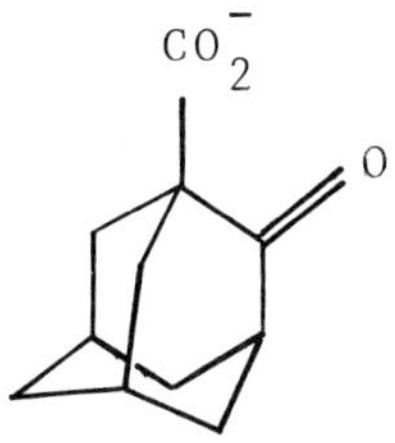

(V)

Capped cyclodextrin, I, is also very helpful for preparation of regionspecifically bifunctionalized cyclodextrin (VI) via successive S_N2 reaction with X^- and Y^-(7). Compounds of this type are important to mimic a variety of enzymes since they have a hydrophobic recognition site as well as a bifunctional catalytic site (X...Y) which is very often encountered in native enzyme mechanisms.

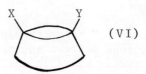

(VI)

A typical example is shown below (Scheme 1). Carbonic anhydrase activity is said to require CO_2 and H_2O in hydrophobic environment around $(imidazole)_3Zn^{2+}$. This situation is reasonably modeled by (VIIa) or (VIIb),

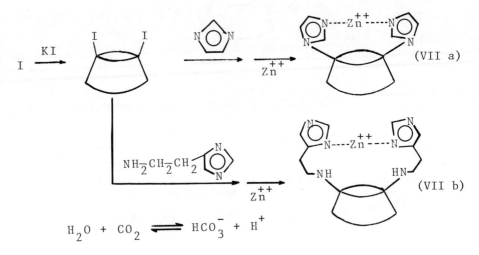

Scheme 1

where H_2O solute seems to be strongly activated by coordination at the metal ion and by the hydrophobic environment as indicated by the catalytic constants (Table I).

Table I. Mimicked Carbonic Anhydrases

	k_{cat} ($M^{-1} s^{-1}$)
imidazole	negligible
$(imidazole)_2 Zn$	2.0
VIIa	16.2
$(histamine)_2 Zn$	57.9
VIIb	166

Triple recognition model "duplex cyclodextrin", (VIII), is also successfully constructed from capped cyclodextrin (I)(8), which may accomodate two different guest molecules to interact with each other in the central catalytic site.

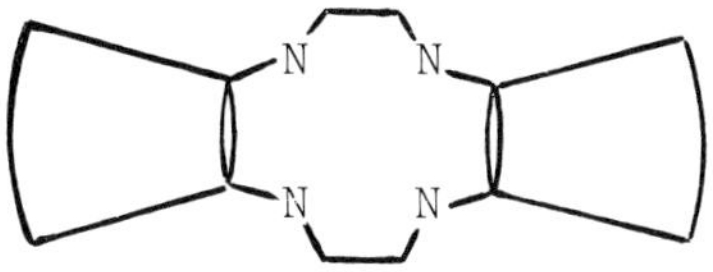

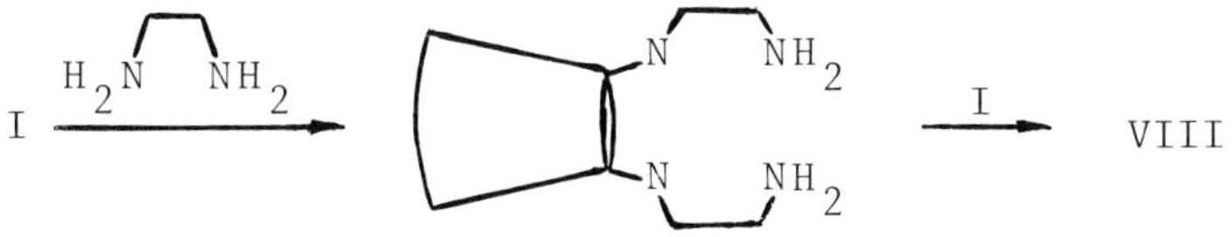

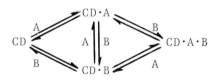

As already cited, bifunctionalized or multifunctionalized cyclodextrins have versatile applications to biomimetic chemistry as it is seen in a retinal pigment model(9).

Great attention should be paid for the basic investigation and useful application of parent cyclodextrins themselves. Cyclodextrin can interact with two kinds of solutes A and B in water to form a a ternary inclusion complex, CD·A·B· leading to specific condensation of two solutes to form

Scheme 2

A-B under certain appropriate conditions (Scheme 2). A typical example of this type of catalysis is seen for the synthesis of vitamine K(10,11).

3. SOLVOLYSIS OF β-BROMOETHYLNAPHTHALENE

β-Bromoethyl-1-naphthalene was used as a specific guest for the present lyase model (reaction path control) by β-cyclodextrin, since the naphthalene moiety is known to be appropriately included by the β-cyclodextrin cavity(5) and β-Bromoethyl-1-naphthalene (IX) is well assumed to be a good substrate capable of solvolysis via S_N2, S_N1, or E_2 pathway depending on the conditions(12). Solvolysis of (IX) was carried out in 10% ethanol at pH 13.1, 12.1 and 11.1 in the presence or absence of 10^{-2} molar β-cyclodextrin.

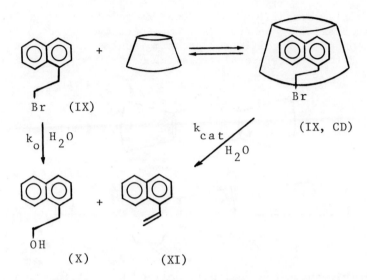

Exclusive formation of β-hydroxyethyl-1-naphthalene (X) and vinyl-1-naphthalene (XI) as products, together with recovery of the starting material (IX) was observed; the product composition was dependent on the conditions as shown in Table II.

The rate of the product formation gave satisfactory pseudo first order kinetics. Based on the glc and the spectroscopic determination, the ratio of (X)/(XI) was found to be independent of time (13).

In the absence of β-CD, the observed product distribution was that expected from the competition among S_{N1}, S_{N2}, E_1 and E_2 mechanisms. Under strongly alkaline conditions, predominant formation

Table II. Composition of Products obtained from β-Bromoethylnaphthalene at various pH in presence or absence of β·CD.

pH	with β-CD (10^{-2} M)		without β-CD	
	(X)	(XI)	(X)	(XI)
11.1[a]	9	91	55	45
12.1[b]	1	99	17	83
13.1[c]	1	99	10	90

[a] 0.05 M Na_2CO_3 + 0.05 M KCl
[b] 0.02 M Na_2HPO_4 + 0.02 M NaOH + 0.06 M KCl
[c] 0.1 M KCl + 0.1 M NaOH

of the elimination product, (X), was observed, indicating the E_2 mechanism is most important under these conditions. The rate constant of each mechanism was estimated by the following equations (1-5).

$$v = k_o [(IX)] \tag{1}$$

$$k_p = k_{SN}^o + k_E^o \tag{2}$$

$$k_{SN}^o / k_E^o = d[(X)] / d[(XI)] \tag{3}$$

$$k_{SN}^o = k_{SN1}^o + k_{SN2}^o [OH^-] \tag{4}$$

$$k_E^o = k_{E1}^o + k_{E1}^o + k_{E2}^o [OH^-] \tag{5}$$

The rate constants thus obtained are listed in Table III: in the absence of β-CD the following characteristics were observed:

(a) a remarkable rate enhancement,
(b) saturation kinetics (with increase in the cyclodextrin concentration),
(c) almost exclusive formation of the elimination product, (XI).

Since the effective catalysis by β-cyclodextrin is evident from these observations, the observed rates were analyzed by the Lineweaver-Burk treatment to give satisfactory straight line from which K_m and K_2 were estimated, where k_{obs} is the observed pseudo first order rate constant and each k^c is the rate constant of the cyclodextrin catalysis. The rate constants thus obtained are listed in Table III.

The most significant contribution of cyclodextrin in the present solvolysis is concluded to be the enormous acceleration of E reaction (expecially E_2) together with remarkable deceleration of S_N2 reaction. The rate constant of each elementary reaction was estimated by the equations (11) and (12).

Table III. Rate Constants of Elementary Processes

noncatalytic		CD catalytic	
k_{E1}^o	2.0×10^{-7} s^{-1}	k_{E1}^c	180×10^{-7}
k_{E2}^o	2.1×10^{-4} s^{-1}M^{-1}	k_{E2}^c †	320×10^{-4} s^{-1}M^{-1}
k_{SN1}^o	2.2×10^{-6} s^{-1}	k_{SN1}^c	1.5×10^{-6} s^{-1}
k_{SN2}^o	2.0×10^{-6} s^{-1}M^{-1}	k_{SN2}^c	v. small

†Dependent on pH if OH⁻ is taken as the base

$$k^c \cdot (f_E^c / f_{SN}^c + f_E^c) = k_{E1}^c + k_{E2}^c [B] \quad (11)$$

$$k^c \cdot (f_{SN}^c / f_{SN}^c + f_E^c) = k_{SN1}^c + k_{SN2}^c [B] \quad (12)$$

where k^c is the combined catalytic constant, f_E^c and f_{SN}^c are the fraction of E and SN reactions of the cyclodextrin catalysis, respectively, and B is the base participating in E or S_N reaction. Based on the rate and the product analyses, contribution of $S_N 2$ was very small under these conditions. Based on the reasonable assumption that the E_1 reaction is not appreciably affected by pH(14), the pH dependence of the E_2 term was studied. As seen from the E_2 term in Table IV, the pKa value of B was very close to 12, which strongly indicates B in the monoanion of cyclodextrin. By taking 12.1 as a pKa value of β-cyclodextrin(15), k_{E2}^c can be calculated to be 2.4 x 10^{-2} $s^{-1}M^{-1}$ at pH 13 and 12 and 2.5 x 10^{-2} $s^{-1}M^{-1}$ at pH 12 and 11. Therefore, a conclusion can be drawn that a base partecipating in the E_2 mechanism is the monoanion of β-cyclodextrin. Careful examination of the recovered cyclodextrin did not show the formation of any naphthylethyl derivative of cyclodextrin, suggesting that the $S_N 2$ reaction of the cyclodextrin monoanion does not contribute appreciably. The present reaction may be conveniently depicted in the Scheme 3. The fact that the E_2 reaction is specifically accelerated in the cyclodextrin catalysis is in accord with the observation that the solvolysis of the present substrate in 50% EtOH was considerable decelerated, even though the predominant elimination reaction took place.

$$k_{E1}^c / k_{E1}^o = 120 \quad \text{in CD cavity}$$

$$0.30 \quad \text{in 50 \% EtOH}$$

Mechanism of Catalytic Solvolysis

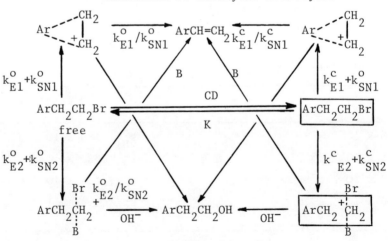

Scheme 3

Table IV. E_2-term of Cyclodextrin Catalysis at different pH.

pH	$k^c_{E2}[B]$ (s^{-1})
11.1	0.68×10^{-5}
12.1	4.2×10^{-5}
13.1	7.4×10^{-5}

These facts strongly indicate that the present remarkable reaction path control is due to the enormous acceleration of the E_2 reaction, most probably via a very appropriately aligned acid base catalytic process as shown in Scheme 4.

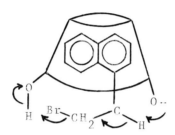

REFERENCES.

1. (a) Crown ethers and cryptands. R. C. Helgeson, K. Koga, J. M. Timko, and D. J. Cram, J. Am. Chem. Soc. 95:3021 (1973); E. Graf and J. M. Lehn, ibid. 98:6403 (1976). (b) Modified cyclodextrins. J. Emert and R. Breslow, J. Am. Chem. Soc. 97:670 (1975); I. Tabushi, N. Shimizu, T. Sugimoto, M. Shiozuka, and K. Yamamura, ibid. 99:7100 (1977). (c) Cyclophanes. I. Tabushi, Y. Kimura, and K. Yamamura, ibid. 100:1304 (1978).
2. It should be pointed out that the reaction path is readily controlled by changing the reaction media, but if every event is restricted to proceed in aqueous solution, the reaction path control is not easily done.
3. E. E. Snell, A. E. Braunstein, E. S. Severin, and Yu. M. Torchinsky, eds. "Pyridoxal Catalysis", Interscience Publishers, New York (1968).
4. I. Tabushi, Y. Kiyosuke, T. Sugimoto, and K. Yamamura, J. Am. Chem. Soc. 100:916 (1978).
5. I. Tabushi, K. Shimokawa, N. Shimizu, H. Shirakata, and K. Fujita, J. Am. Chem. Soc. 98:7855 (1976).

6. I. Tabushi, K. Fujita, and L. C. Yuan, <u>Tetrahedron Lett.</u> 2503 (1977)
7. I. Tabushi, K. Shimokawa, and K. Fujita, <u>Tetrahedron Lett.</u> 1527 (1977).
8. I. Tabushi, Y. Kuroda, and K. Shimokawa, <u>J. Am. Chem. Soc.</u> 101:1614 (1979).
9. I. Tabushi, Y. Kuroda, and K. Shimokawa, <u>J. Am. Chem. Soc.</u> 101:4759 (1979).
10. I. Tabushi, K. Fujita, and H. Kawakubo, <u>J. Am. Chem. Soc.</u> 99:6456 (1977).
11. I. Tabushi, Y. Kuroda, K. Fujita, and H. Kawakubo, <u>Tetrahedron Lett.</u> 2083 (1978).
12. C. Lapinte and P. Viout, <u>Tetrahedron Lett.</u> 1113 (1973).
13. Since the electronic spectrum of (X) resembles to that of (IX), the glc determination was essential.
14. Minor change may be brought by the change of microenvironment due to either the dissociation of cyclodextrin or the hydroxide anion binding by cyclodextrin.
15. T. F. Chin, P. H. Chung, and J. L. Lach, <u>J. Pharm. Sci.</u> 57:44 (1968).

THE COMPOSITE PHYSICAL AND CHEMICAL APPROACH TO THE SOLUTION

SPATIAL STRUCTURE OF POLYPEPTIDE NEUROTOXINS

V. F. Bystrov, V. T. Ivanov, V. V. Okanov,
A. I. Miroshnikov, A. S. Arseniev, V. I. Tsetlin and
V. S. Pashkov

Shemyakin Institute of Bioorganic Chemistry, USSR Academy of Sciences, GSP-1, Moscow 117988, USSR

E. Karlsson

Institute of Biochemistry, S-75123 Uppsala, Sweden

Abstract - The crucial details for the interpretation of spatial structures and intermolecular interactions of peptides and proteins could be revealed by proper combination of physical and chemical techniques. The paper presents the results of the combined approach for the evaluation of the conformation in solution of honey-bee venom component apamin (18 membered polypeptide), of three dimensional structure of Central Asian cobra neurotoxin II (61 amino acid residues), and of the topography of its binding site with acetylcholine receptor Torpedo Marmorata.

1. INTRODUCTION

It is fully recognized that a crucial factor in the functioning of physiologically active polypeptide molecule is its conformation. To this fact one must add that it is becoming more and more recognized that of equal importance is also the dynamic conformation of the molecule in solution. It should be stressed that the conformational analysis in solution can be best accomplished by combining the information from several physical techniques (experimental and theoretical), supported by extensive chemical studies - primary structure elucidation, selective modification, label incorporation, synthesis of analogs and model compounds (Figure 1). In the composite approach the results obtained by each technique are used to answer the questions they are best fitted to deal with, and in this way the over-all problem can be solved in a most efficient manner.

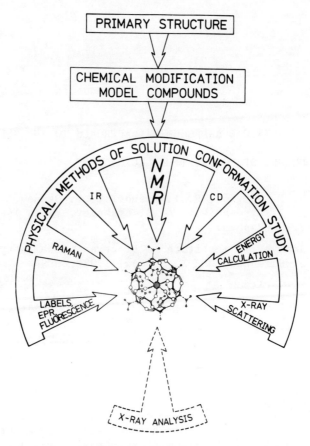

Fig. 1. The composite physical and chemical approach to the solution spatial structure of polypeptides.

Although the X-ray analysis gives unsurpassed accurate and lucid molecular three-dimensional structures in solid state, the combined use of other physical and chemical methods affords additional information in that one can:

(i) disclose dynamic conformations and intermolecular actions under various environment, resembling the physiological conditions,

(ii) shed light on the specific intramolecular interactions leading to predominance of a given conformation,

(iii) identify the particular groups participating in interactions with biological targets, for instance with ligands and receptors,

(iv) reveal the role of the spatial structure in given biological processes performed by the molecule, that is to answer the question of just why has Nature constructed such a molecule for a given task and not for another.

The nuclear magnetic resonance is the most powerful technique in the composite approach, as proved by the fact that application of NMR spectroscopy has had many important results in the study of detailed spatial structures and conformational dynamics of peptides in solution. These results have had a significant impact on the understanding of physiological mechanism of action of such substances as gramicidins, oxytocin, valinomycin, etc. [for a review, see(1)].

This paper presents the results of the composite physical and chemical approach, with the main stress on NMR spectroscopy data, in an analysis of the spatial structure of some polypeptide neurotoxin components of bee and snake venoms - apamin and short neurotoxins, respectively.

In the NMR conformational study of folded polypeptides, in addition to the well established NMR techniques for small peptides (2-4) one may also rely on the NMR parameters of the polypeptide side chain signals for detection of local conformational transition, intramolecular interactions and chemical modification effects, resulting not only from neighbouring amino acid residues, but also from residues which, while remote in the primary structure, become close to one another by the folding of the polypeptide backbone(5).

The interchain nonbonded interactions (Figure 2) amenable to the NMR technique, includes

  (a) electrostatic dipole-dipole interactions,
  (b) hydrogen bonding,
  (c) π-electron anisotropy effect of aromatic rings,

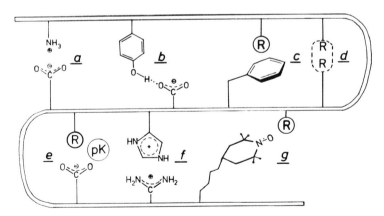

Fig. 2. The interchain non-covalent interactions ("contacts") in polypeptides, detectable by NMR.

(d) van der Waals and hydrophobic interactions,
(e) response of R-group chemical shift to the ionization of closely lying ionogenic groups,
(f) effect of a neighbouring charge on the dissociation constant of a particular ionogenic group,
(g) paramagnetic effect of spin labels on relaxation and line width.

Through detection of such direct "contacts" and of indirect interactions due to conformational changes, arising from electronic and structural perturbation of particular ionogenic group, one could obtain information on the general folding of the polypeptide molecule.

2. APAMIN

Apamin, an active principle of the honey-bee venom, is an octadecapeptide with two disulphide bonds (Figure 3). It is unique in being the smallest known peptide neurotoxin with an excitatory effect on the central nervous system and the only known polypeptide that passes the blood-brain barrier(6).

Signals in the proton and carbon-13 NMR spectra of apamin have been attributed(7) by the two levels of assignment. On their first level the proton signals are interconnected by homonuclear proton-proton spin decoupling and assigned to a particular spin multiplet system according to the specific NMR types of amino acid residues(8). On the second level of assignment the individual spin systems are attributed to a specific position of an amino acid residue in the apamin primary structure by heteronuclear C-13 {H-1} selective decoupling of carbonyl signals(9). The results of assignment are

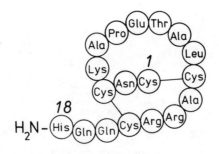

Fig. 3. The primary structure of honey bee apamin.

shown in Table I where the proton signal parameters are attributed to the particular amino acid residues. Due to signal overlap we were not able to directly differentiate signals on the second level between Arg^{13} and Arg^{14} residues, and Gln^{16} and Gln^{17} residues. The assignment of Cys^3 and Cys^{15} shown in Table I is tentative as based on the influence of N-terminal amino group deprotonation on the signal position of the former residue and of the His^{18} deprotonation of the last residue.

Table I presents the NMR parameters most important for the apamin conformational study. Backbone $H-NC^\alpha-H$ and side chain $H-C^\alpha$ $C^\beta-H$ vicinal proton-proton coupling constants provide information on the torsional angles ϕ and χ^1 of $N-C^\alpha$ and $C^\alpha-C^\beta$ bonds, respectively(4). Deuterium exchange half-time $t_{1/2}$ discriminates intramolecular hydrogen bonded backbone NH groups(2). Chemical shift pH-dependence indicates the proximity of corresponding ionogenic groups (5): N-terminal α-amino, ε-amino of Lys^4, γ-carboxyl of Glu^7 and imidazole ring of His^{18}. The pK_a values of α-amino group and His^{18} are too similar to be differentiated for some weakly pH dependence chemical shifts. Therefore the chemically modified apamin with blocked α-amino group ($[Ac-Cys^1]$-apamin) was studied(7).

The conformational stability of apamin follows from CD spectra (10), which demonstrates persistence against pH change in wide range, presence of organic solvents and 6 M guanidinium-HCl, and chemical modifications.

When analyzing the NMR parameters (Table I) attention is captured by the three, or even four, residues in succession (Leu^{10}, Cys^{11}, Ala^{12} and possibly Arg^{13}) with both very slow deuterium exchange ($t_{1/2}$ > 14 hours) and low values of $H-NC^\alpha-H$ proton coupling (3.0 - 5.9 Hz). Both of these features are inherent to the polypeptide right-handed helical structure (α-, 3_{10}- or π-helix). Thus, it was assumed that the 10-13 fragment is incorporated in a helix, and its backbone NH groups are involved in hydrogen bonding with the carbonyls of n-i residues (i = 3,4,5 for 3_{10}-, α- and π-helix). It is essential that the preceeding to this fragment Ala^9 residue has also low $^3J(H-NC^\alpha-H)$ value.

By model building it was found that only α-helix could be constructed, due to steric hindrances introduced by 1-11 and 3-15 disulphide links. The presence of α-helical region in apamin is in accord with CD(10) and laser Raman spectra(11), and also with the secondary structure prediction(12) based on the Chou and Fasman method.

The model building around the α-helical fragment 6-13 permits the formation of β-turns for other very slow exchanging NH groups of Ala^5, Cys^{15} and Gln^{16} or Gln^{17} residues. The most plausible

Table I. Apamin proton NMR signal parameters.

Apamin Residues	$^3J_{HNC^\alpha H}$ Hz	$^3J_{HC^\alpha C^\beta H}$ Hz	$t_{1/2}$[a] hours	Backbone NH (in H_2O)	$C^\alpha H$ (in D_2O)	Side Chain CH (in D_2O)
Cys[1]	-	7.8, 7.8	-	-	0.70(αN)	C^β:0.04(Glu), 0.29(αN)
Asn[2]	9.5	11.2, 4.2	<0.1	-1.17(Glu), 0.77(αN)	0.03(Glu), 0.04(αN)	
Cys[3](a)	5.2	11.9, 2.8	0.2	0.02(Glu), 0.12(αN)	-0.02(Glu), 0.02(αN)	
Lys[4]	7.6		4	0.04(Glu), -0.03(αN)		C^δ:0.23(Lys); C^ε:0.43(Lys)
Ala[5]	6.8	6.1	11	0.07(Glu), -0.09(αN)	0.05(Glu), -0.01(αN)	C^β:0.03(Glu), -0.01(αN)
Pro[6]	-	6.2, 6.2	-	-	0.05(Glu)	
Glu[7]	6.9	9.6, 4.4	0.1	-1.64(Glu), 0.61(αN)	0.22(Glu)	C^γ:0.14(Glu), 0.06(αN)
Thr[8]	8.5	1.9	4	0.22(Glu), -0.06(αN)	no dependence	C^γ:0.03(Glu)
Ala[9]	2.7	6.1	<0.1	-0.04(Glu), 0.07(αN)	-0.05(Glu)	C^β, no dependence
Leu[10]	5.5		20	0.05(αN)		C^δ: no dependence
Cys[11]	5.9	7.6, 7.6	14	-0.17(Glu), -0.04(αN)	0.06(Glu)	C^β:-0.03(Glu), 0.02(αN)
Ala[12]	3.0	6.1	26	0.07(Glu), 0.02(αN)	0.02(Glu)	C^δ: no dependence
Arg[13,14](c) 4.2 (c') 3.5		>100	-0.04(Glu), 0.05(αN)		C^δ: no dependence
Cys[15](b)	5.4	9.6, 4.2	2.7	0.08(Glu)		
			26	-0.06(Glu)		
Gln[16,17](d) 7.1 (d') 6.8		0.5	no dependence	0.03(Glu), -0.01(His)	C^δ:0.08(His)
			11	no dependence		C^γ: no dependence
His[18]	7.9	8.6, 6.1	<0.1	0.04(Glu), 0.20(His)	0.16(His)	C_2:0.97(His); C_4:0.35(His)

a) Backbone NH deuterium exchange half-time at pH 2.9 and 14°C.
b) Chemical shift changes in ppm; positive values correspond to high field shift with increasing pH. The symbols in parentheses mean that the chemical shift pH-dependence has an inflection point at the pH value, which corresponds to the pK_a value of N-terminal α-amino group (pK_a 6.7), Glu[7] carboxyl group (pK_a 3.6), His[18] imidazole ring (pK_a 6.6) and Lys[4] ε-amino group (pK_a 10.8).

intramolecular hydrogen bond system is schematically depicted in Figure 4. The Raman-laser study(11) indicates the presence of at least two β-turns.

In construction of the apamin spatial structure (Figure 5) also the following main features of the NMR study(5,7) were taken into account:

(a) Carboxyl group of Glu^7 is ~ 5 Å apart from the N-terminal α-amino group.
(b) Carboxyl group of Glu^7 is also close to the methyl groups of Ala^5 and Thr^8.
(c) The side chain of at least one arginine residue is in the vicinity of Leu^{10} methyl groups.
(d) The ε-amino group of Lys^4 is distant with respect to other residues.
(e) The side chains of His^{18} and of both arginine residues are distant to other ionogenic groups of the molecule.
(f) The backbone NH groups of Lys^4, Thr^8 and Arg^{14} are capable of forming weaker intramolecular hydrogen bonds, presumably with the side chain functions.
(g) The torsional angles ϕ must be consistent with the measured vicinal coupling constants $^3J(H-NC^\alpha-H)$ and $^3J(^{13}C'-NC^\alpha-H)$.

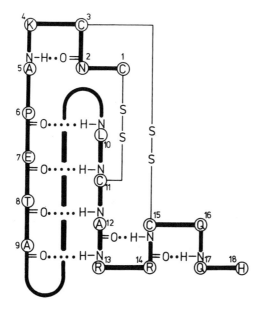

Fig. 4. Diagram of intramolecular hydrogen bonds in apamin. The standard one-letter code(25) is used.

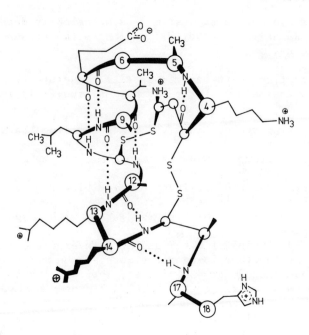

Fig. 5. Schematic representation of experimental spatial structure of apamin in solution.

The laser Raman spectra study(11) demonstrates the similarity of the conformational arrangement of both disulphide bridges 1-11 and 3-15. On the basis of correlation(13) the trans-gauche-gauche-gauche-trans orientation for the $H-C^{\alpha}-C^{\beta}-S-S-C^{\beta}-C^{\alpha}-H$ fragments is predicted.

The spatial structure of apamin presented in this paper explains the spectroscopic results(5,7,10,11) much better than structures proposed earlier on the basis of conformational energy calculations using the atomic(14) and residual representation(15) as well as secondary structure prediction(12). The rigidity of the structure shown in Figure 5 is achieved by relative orientation of the α-helical unit 6-13 and two β-turns 2-5 and 12-15 about the disulphide bridges Cys^1-Cys^{11} and Cys^3-Cys^{15}. Additional, but weaker hydrogen bonds could be formed by the backbone NH of Lys^4 with hydroxyl group of Thr^8, by the backbone NH of Arg^{14} with the side chain carbonyl groups of Gln^{17} and by the backbone NH of Thr^8 with the side chain carbonyl of Asn^2, or carboxyl group of Glu^7, or the backbone carbonyl of Ala^5. The C-terminal His^{18} is the only residue which is not involved in the apamin tertiary structure.

The spatial structure derived for apamin in solution could be considered to be similar to the conformation on a receptor owing to the persistence of the conformation against changes of the medium

SPATIAL STRUCTURE OF POLYPEPTIDE NEUROTOXINS

and chemical modification, as well as because the synthetic apamin possesses biological activity(16-19) and the same CD and NMR spectra (19): a fact which is considered(12) an evidence that the information on the correct folding of the peptide chain is contained in the chain and not in an apamin precursor.

However, the overall rigid spatial structure is not the only feature important for a biological action. The chemical structure of arginine side chain is also crucial, as evinced from a dramatic loss of neurotoxic activity when both Arg^{14} are substituted either by lysine(20) or ornithine(18) residues.

Further progress in the study of structure-activity relationship of apamin depends on localization and isolation of a receptor system and on the knowledge of its interaction with apamin and its derivatives.

3. SNAKE VENOM SHORT NEUROTOXINS

The direct evaluation of the conformation in solution of even a small protein poses a challenge to the techniques now available. One of the most favourable group of proteins for such study is the so-called short neurotoxins isolated from elapidae and hydrophidae snake venom(21). They consist of only 60-62 amino acid residues with four disulphide bridges, they possess higly stable conformations in solution(22) and have an abundant set of homologs(23)

			1	5	10	15	20	25
1. Neurotoxin II	(N. n. oxiana)		Ⓓ E C Ⓗ N	Q Q S S	Q P P T T	Ⓧ T C S	- G E T	N C Ⓨ
2. α-Toxin	(N. nigricollis)						P -	
3. Neurotoxin III	(N. mossambica m.)		N		Ⓜ Ⓐ		T R	R Ⓦ
4. Neurotoxin I	(N. mossambica m.)				E		T R	G
5. Cobrotoxin	(N. n. atra)				T		T G	G
6. Erabutoxin a	(L. semifasciata)		R I F	Ⓗ	Q		P S	S S
7. Erabutoxin b	(L. semifasciata)		R I F	H	Q		P S	S S
8. Erabutoxin c	(L. semifasciata)		R I F	H	Q		P S	S S

		26	30	35	40	45	50	55	60 62
1. Nt-II		Ⓚ Ⓚ Ⓦ Ⓜ Ⓢ D Ⓗ R	G T Ⓓ Ⓘ E	R G C G C P Ⓚ Ⓥ Ⓚ P	G Ⓥ N Ⓛ N C C R T D R C N N
2. α-T			V R		T		I K		T K
3. Nt-III			R R	Ⓥ R T	T	K	Ⓘ Q Ⓗ	T S	N
4. Nt-I			R R	Y R T	T	K	I E		T
5. Ct			R R	Ⓥ R T	S	N	Ⓘ E Ⓘ		T
6. Er-a	N	Q	F		T		I K	S	E S E V
7. Er-b	Ⓗ	Q	F		T		I K	S	E S E V
8. Er-c	H	Q	F		T		I N	S	E S E V

Fig. 6. Amino acid sequences of short neurotoxins(23) studied by NMR: 1 - in(26-35), 2 - in(27), 3 - in(36), 4 - in(37), 5 - in(38,39), and 6-8 - in(40). The standard one-letter code(25) is used. The signals of circuled residues were used as reporter resonance; underlined residues were revealed by effect of their charges; double underlined residues were used as substituents for signal assignment and "contact" identification.

The mechanism of neurotoxin action on the nerve systems involves blocking of the nerve impulse transmission at the site of the neuromuscular junction by strong specific binding to the acetylcholine receptor in the postsynaptic membrane, thus interrupting the pathway whereby the neurotransmitter released in the synaptic cleft could affect the excitability of the postsynaptic neuron. The availability of highly purified acethylcholine receptor protein(24) allows a direct study of the structure-activity relationship of neurotoxins and their bindings with the receptor.

This study is concentrated on the neurotoxin II isolated from the Central Asian cobra Naja naja oxiana (see Figure 6 for the primary structure). The signal assignment and microenvironment of aromatic (His^4, His^{32}, Trp^{28}, Trp^{29}, Tyr^{25}) and aliphatic side chains (Figure 7) were delineated and their intramolecular interactions were revealed in(5,26,27) and in(28,29) respectively. The pK values of ionogenic groups were measured by pH dependence of these signals in D_2O solution as well as of NH signals in H_2O solution(28,29). To obtain information on the overall folding of the molecule in solution, an extensive use was made of specific NMR, EPR and fluorescence reported groups selectively incorporated in neurotoxin II. Retaining of intact conformation by these chemically modified derivatives was carefully controlled by CD spectra. Fluorine label, with its ^{19}F NMR

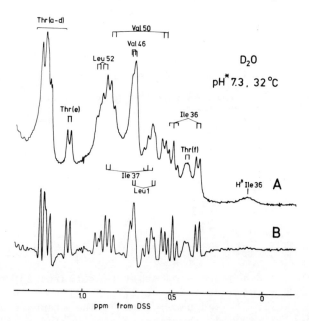

Fig. 7. Signal assignment in high field region of 300-MHz proton NMR spectrum of neurotoxin II in D_2O solution at pH 7.3 and 32°C. A - Normal spectrum. B - resolution enhanced spectrum by means of a triangular function(41).

chemical shifts very sensitive to the environment, was incorporated by selective trifluoro-acetylation of the ε-amino groups of the five lysine residues and of the N-terminal α-amino group(5,30). Covalently bound spin labels (six- or five-membered ring nitroxyl stable radicals) were incorporated (31-35,42) in the identified positions of neurotoxin II molecules (ε-amino groups of lysines, N-terminal α-amino group, imidazole ring of His^{32}) by using spin labeling reagents (2,2,6,6-tetramethyl-4-carboxymethyl-piperidyl and 2,2,5,5-tetramethyl-3-carboxypyrrolidyl N-hydroxysuccinimide esters) and a wide range of conditions (various reagent/protein ratios and pH values, introduction of denaturants). The obtained di-spin labeled derivatives were used for interspin distances evaluation by EPR detection of dipole-dipole interaction(43) at 77°K(31,33). Tryptophan fluorescence of mono-spin labeled and of dansyl-Lys^{27} derivatives provide additional interresidue distances (44) collected in Table II.

The mono-spin labeled derivatives (Lys^{15}, Lys^{26}, Lys^{27}, Lys^{45} and Lys^{47}) of neurotoxin II were used to estimate apparent distances between the spin label and the protons whose signals in the NMR spectrum were assigned. The unpaired spin paramagnetic contributions were measured and the Solomon-Bloembergen equation(45) was applied with isotropic correlation times τ_c estimated from the EPR spectra of corresponding derivatives. In particular, the results shown in Figure 8 for the SL-Lys^{27} and the SL-Lys^{26} neurotoxin II derivatives substantiate, along with other interresidue nonbonded contacts, the β-structure of the central loop of the molecule(28,29,33-35). Broadening of the His^{32} imidazole CH protons is pH-dependent,

Table II. Interresidue distances (in Å) in neurotoxin II (pH 5-6) as evaluated by (a) EPR on di-spin labeled derivatives (31-33) and by (b) fluorescence study on spin- and dansyl-labeled derivatives(44).

Residue	Lys^{15}	Lys^{26}	Lys^{27}	Trp^{28}	Trp^{29}	Lys^{45}	Lys^{47}
Lys^{15}	18a		14a			17a	16a
Lys^{26}			15a	4-6b		17a	18a
Lys^{27}				13-14b	4-6b	15a	12a†
Trp^{28}					10b		
Trp^{29}							
Lys^{45}							17a
Lys^{47}							

†Less than 8.5 Å, as follows from the ^{19}F NMR spectrum of (SL-Lys^{27})(CF$_3$CO)$_5$-neurotoxin II at pH 6.1. For distances between Lys^{27} and other lysine residues the results are coincident.

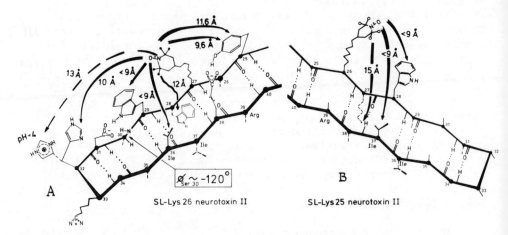

Fig. 8. Conformation of the central loop of the neurotoxin II molecule. A - Upper view. Proton - spin label distances are obtained for SL-Lys^{27} derivative(29,33-35). B - Bottom view. Distances are obtained for SL-Lys^{26} derivative. The Lys^{26}-Lys^{27} distance is for di-spin labeled neurotoxin II (Table II).

i.e. protonation of the imidazole ring is accompanied by a change of its orientation as shown in Figure 8A.

The backbone folding of the whole protein molecule can be unequivocally derived, as shown in Figure 9, by collecting all the data on nonbonded contacts obtained for neurotoxin II in this study (5,22,26-36,44) by pH dependence of chemical shifts, influence on pK values of nearby charges, nuclear Overhauser effect and chemical shift differences arising from selective chemical modification, as well as by the apparent distances estimated for spin- and dansyl-labeled derivatives. Valuable additional information have been obtained from proton NMR studies of homologues neurotoxins (Figure 6): toxin α from Naja nigricollis(27), neurotoxin III(36) and neurotoxin I(37) Naja mossambica mossambica, erabutoxins a, b and c Laticauda semifasciata(40) and cobrotoxin Naja naja atra(38,39). The pK values for these compounds are collected in Table III.

The derived solution conformation of short neurotoxins (Figures 8 and 9) is in general agreement with the X-ray crystal structure of erabutoxins(46,48), although the amino acid composition of these sea snake toxins differs by ca 30% from that of the terrestrial snake neurotoxins mostly studied by NMR. The conformational reorientation of the fragment 30-34 (β-turn in the central loop, Figure 9)

SPATIAL STRUCTURE OF POLYPEPTIDE NEUROTOXINS   243

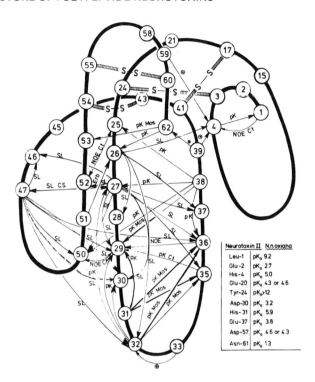

Fig. 9. Backbone folding of short neurotoxin molecule as derived
from Naja naja neurotoxin II study. Additional contacts
from the studies of other neurotoxin are indicated as
follows: Mos - neurotoxin III Naja mossambica mossambica
(36), Erb - erabutoxin b Laticauda semifasciata(40) and
Ct - cobrotoxin Naja Naja atra(39). Abbreviation: pK -
deprotonation of ionogenic group influence on chemical
shifts, + - charge effect on pK values, NOE - nuclear
Overhauser effect, CS - chemical shift changes upon selective modification, SL - spin label broadening effects.

with respect to the crystallographic structure proposed by Drake,
Dufton and Hider(49,50) has been advanced on the basis of NMR data
(5,26) and by comparison with an antagonist (C-alkaloid E) spatial
structure: this fact could be responsible for His^{32} pH-dependent
reorientation (Figure 8). Another confirmation of the transition
at pH $\sim$ 6 comes from the distance between $SL-Lys^{47}$ and His^{32} protons
estimated in this study which seems to be too short ($\sim$ 8 Å at pH > 6)
to be properly explained by the X-ray structure. At pH < 6 the
distance is longer ($\sim$9.5 Å).

Table III. pK Values of short neurotoxins revealed by NMR studies. For abbreviations see Table II.

Residue	NT-II	α-T (27)	Nt-III (36)	Nt-I (37)	Ct (39)	Er-b (40)
N-terminal	9.2		9.1	9.3	9.35	
His^4	5.02	5.16	5.99	5.06	5.35	
His^7	-	-	-	-	-	<3
Tyr^{25}	>12		>12	11.9	11.9	12.0
His^{26}	-	-	-	-	-	5.8
Asp^{31}	3.2	3.2	2.6	2.6	3.0	
His^{32}	5.93	5.91	5.76	5.83	5.93	-
Tyr^{35}	-	-	9.75	9.8	9.75	-
Glu^{38}	3.8	3.8	3.6	3.4	3.4	
His^{53}	-	-	5.61	-	-	-
C-terminal	1.3		1.3	1.3		

4. NEUROTOXIN BINDING TO ACETYLCHOLINE RECEPTOR

Despite extensive studies on acetylcholine receptor (AChR) and snake venom neurotoxins(23,24), direct evidence is lacking as to what region of the neurotoxin molecule the receptor bind to. The availability of series of neurotoxin II derivatives with spin or fluorescence labels in known positions provides the means for outlining the topography of a neurotoxin binding site.

The interaction of selectively spin and dansyl mono-labeled neurotoxin II derivatives with solubilized nicotinic AChR protein isolated from Torpedo marmorata electric organ, was monitored by EPR and fluorescence spectroscopy(42). All the compounds specifically bind to the AChR, with dissociation constants ranging from 3 to 80 nM. The stoichiometry of the toxin-AChR complex is 2:1, assuming a molecular weight of 250 000 daltons for AChR(51). The two binding sites were found to be independent and indiscernible by dissociation constants and neurotoxin II spin label microenvironment, however they were not identical in that the nontoxic hexatrifluoroacetyl neurotoxin II interacts only with one AChR binding site. This parallels the kinetic study conclusions(52) that the two sites are initially kinetically identical with respect to toxin binding, and yet differ in reactivity toward the affinity alkylating agent, 4-(N-maleimido)-benzyltrimethylammonium iodide. The 2:1 stoichiometry and structural similarity of the toxin combining sites of AChR have been recently confirmed also by EPR studies of interaction of nonselectively spin labeled Naja naja siamensis "long-type" α-neuro toxin with Torpedo californica AChR solubilized protein and AChR-rich membranes(53).

The mobility of spin labels grafted on neurotoxin II and the degree of their exposure towards solvent environment containing non-specific paramagnetic probes $K_3Fe(CN)_6$(42) or $Ni(Ac)_2$, were examined both for AChR bonded AchR and non-bonded states (Table IV). Isotropic correlation times for spin labeled compounds (τ_{obs}) within the 50 ps - 1 ns interval were determined from the EPR spectra as described in(54). For slower rates the τ_{obs} values were evaluated by comparison with calculated EPR spectra(55), assuming isotropic rotation. The correlation time of spin label tumbling relative to the protein (τ_{SL}) was evaluated in terms of simple isotropic rotation model by the equation

$$1/\tau_{SL} = 1/\tau_{obs} - 1/\tau_p$$

where τ_p is the overall rotation time of the protein. For neurotoxin II derivatives τ_p = 4.3 ns was found from polarisation and excited state life-time measurements with dansyl-Lys^{27} derivative and for AChR τ_p = 0.4 ms was calculated from(56). The accessibility of spin labels on neurotoxin molecule towards paramagnetic probes was estimated relatively to a spin label in aqueous solution, as a dimensionless parameter(43,57)

$$k_w/2\, k_p = \frac{k_w \cdot C}{2\, \Delta H_o} \cdot 6.5\, 10^{-8}$$

where ΔH_o is broadening of the central component in the EPR spectrum (in Gauss), C is concentration of the paramagnetic probe (in M), k_p is the rate constant of spin exchange between the neurotoxin spin label and probe, and k_w is the exchange rate between a free spin label and the probe in aqueous solution (k_w = 12·10^8 M^{-1} s^{-1} for $K_3Fe(CN)_6$ and 11·10^8 M^{-1} s^{-1} for $Ni(Ac)_2$).

No difference was found for (SL-Lys^{45})- neurotoxin II between non-bonded and AChR-bound states, indicating that this fragment does not penetrate into the receptor matrix. The Lys^{15} residue retains mobility in the AChR-neurotoxin complex, but becomes somewhat shielded from the medium. The tumbling of Lys^{26} and His^{32} spin labels is considerably restricted in the complex although they are still accessible to the solvent. The N-terminal and Lys^{27} spin labels, also hindered in the complex, are almost inaccessible to the solvent, whereas the highly shielded SL-Lys^{47} possesses the similar degree of mobility as in the free neurotoxin.

These results demonstrate the orientation of neurotoxin molecule in the complex with the receptor (Figure 10). The N-terminal and Lys^{27} residues are tightly bound in the receptor cavity, whereas the Lys^{47} is situated in a more spacious "pocket". This conclusion is confirmed by a fluorescence study of neurotoxin II derivatives with dansyl or dansyl-Gly-grouping grafted on Lys^{27} or Lys^{47} residues.

Table IV. EPR data on spin labeled neurotoxin II derivatives and their interaction with AChR at 21°C and pH 7.5 (0.05 M Tris buffer, 0.1 M NaCl, 0.025 % Triton X-100, 0.02% NaN$_3$).

Derivative	LD_{50} μg/kg	Correlation time, ns					Broadening by non-specific paramagnetic probe, $k_w/2k_p$			
		Non-bonded		AChR-bound			$K_3Fe(CN)_6$		$Ni(Ac)_2$	
		τ_{obs}	τ_{SL}	τ_{obs}	τ_{SL}		Non-bonded	AChR-bound	Non-bonded	AChR-bound
SL-N-terminal		1.1	1.5	>10	>10				1.17	7.4
SL-Lys^{15}	110	0.8	1.0	2	2		1.29	4.37	0.94	
SL-Lys^{26}	85	2.0	3.7	20	20		2.29	4.23	1.47	
SL-Lys^{27}	300	2.0	3.7	20	20		2.93	>10	1.51	6.20
SL-Hys^{32}		0.6	0.8	>20	>20				1.30	1.61
SL-Lys^{45}	152	0.5	0.6	0.9	0.9		1.02	1.25	1.18	0.83
SL-Lys^{47}	240	0.9	1.1	2	2		1.13	10.1	0.91	

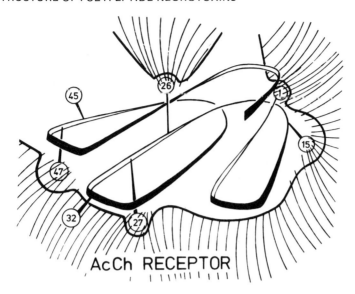

Fig. 10. A scheme for neurotoxin binding to AChR as inferred from the EPR and fluorescence studies.

Thus, we provided direct spectroscopic evidences that the neurotoxin binding site is quite extended and that the neurotoxin-receptor interaction is of multipoint nature, as was assumed from the effect of chemical modification on biological activity of short neurotoxins(23) and from consideration of erabutoxin X-ray structure(58).

5. MATERIALS AND METHODS

The NMR spectra were recorded on a Varian SC-300 spectrometer in Fourier transform mode at 300 MHz. When measured in H_2O solution, the intensity of the solvent line was reduced by irradiation of water signal for 2s prior to the observation pulse. The pH measurements were carried out directly in the 5 mm o.d. NMR tubes using a 180-mm Ingold 405 M3 combination microelectrode. For D_2O solutions the values used are the direct reading of digital pH meter (Orion Research Model 601). For analysis of the experimental pH dependence of the chemical shifts an iterative curve-fitting software was used.

Apamin was isolated from the venom of honey bee Apis mellifera and chemically modified derivatives of apamin were obtained as described(10).

The EPR spectra were obtained on a Varian E-109 (X-band) spectrometer equipped with the Varian E-900 Data Acquisition System. Fluorescence measurements were made using an Aminco SPF-1000 instrument. CD spectra were taken on a Jobin-Yvon III Dichrograph. Raman-laser spectra were obtained on a Ramanor HG-2S spectrometer.

Neurotoxin II Naja naja oxiana was isolated by the procedure of reference(26). Selectively trifluoroacetylated derivatives, spin labeled and dansyl labeled neurotoxin II were obtained as described in(30-33,42).

Acetylcholine receptor protein from the electric organ of Torpedo marmorata was isolated by the procedure of reference(59). The preparations of the receptor solubilized in Triton X-100 showed one band on polyacrylamide gel electrophoresis and had an activity of 10 nM α-toxin Naja naja siamensis binding sites per 1 mg receptor protein.

Acknowledgment. This work is a cooperative effort. Drs. V. A. Afanasiev, E. S. Efremov, E. G. Elyakova, A. Z. Gurevich, A. B. Kudelin, I. R. Nabiev, A. K. Nurkhametov and L. B. Senyavina participated in the apamin study. Drs. E. S. Efremov, A. Z. Gurevich, K. A. Pluzhnikov, L. B. Senyavina, A. M. Surin and Yu. N. Utkin collaborated in the neurotoxin and AChR study. Many fruitful discussion with Professor Yu. A. Ovchinnikov are gratefully acknowledged.

REFERENCES

1. Yu. A. Ovchinnikov and V. T. Ivanov, Tetrahedron, 31:2177 (1975).
2. H. R. Wyssbrod and W. A. Gibbons, Surv. Progr. Chem. 6:209 (1973).
3. L. C. Craig, D. Cowburn, and H. Bleich, Ann. Rev. Biochem. 44:477 (1975).
4. V. F. Bystrov, Progr. NMR Spectr. 10:41 (1976).
5. V. F. Bystrov, A. S. Arseniev, and Yu. D. Gavrilov, J. Magn. Res. 30:151 (1978).
6. E. Habermann, Science, 177:314 (1972).
7. V. V. Okhanov, V. A. Afanasiev, A. Z. Gurevich, E. G. Elyakova, A. I. Miroshnikov, V. F. Bystrov, and Yu. A. Ovchinnikov, Bioorgan. Khim. (Soviet J. Bioorgan. Chem.), 6:840 (1980).
8. V. F. Bystrov, V. V. Okhanov, A. S. Arseniev, V. A. Afanasiev, and A. Z. Gurevich, Bioorgan. Khim. (Soviet J. Bioorgan. Chem.), 6:386 (1980).
9. V. V. Okhanov, V. A. Afanasiev, and V. F. Bystrov, J. Magn. Res. 39:000 (1980).
10. A. I. Miroshnikov, E. G. Elyakova, A. B. Kudelin, and L. B. Senyavina, Bioorgan. Khim. (Soviet J. Bioorgan. Chem.), 4:1022 (1978).
11. E. S. Efremov, I. R. Nabiev, and A. Kh. Nurkhametov, IX National Conference on Molecular Spectroscopy, Albena, Bulgaria, September 29 - October 3 (1980).

12. R. C. Hider and U. Ragnarsson, FEBS Letters, 111:189 (1980).
13. H. E. Van Wart and H. A. Scheraga, J. Phys. Chem. 80:1823 (1976).
14. E. M. Popov and P. N. Melnikov, Bioorgan Khim. (Soviet J. Bioorgan. Chem.), 5:828, 1011, 1471 (1979); 6:21 (1980).
15. B. Busetta, FEBS Letters, 112:138 (1980).
16. J. van Rietschoten, C. Granier, H. Rochat, S. Lissitzky, and F. Miranda, Eur. J. Biochem. 56:35 (1975).
17. W. L. Cosand and R. B. Merrifield, Proc. Nat. Acad. Sci. USA, 74:2771 (1977).
18. B. E. B. Sandberg and U. Ragnarsson, Int. J. Peptide Prot. Res. 7:503 (1977); 11:238 (1978).
19. A. R. Nuriddinov, G. F. Zhukova, V. I. Tsetlin, and V. T. Ivanov, Bioorgan. Khim. (Soviet J. Bioorgan. Chem.), 4:1322 (1978).
20. C. Granier, E. Podroso Muller, and J. van Rietschoten, Eur. J. Biochem. 82:293 (1978).
21. A. T. Tu, "Venoms: Chemistry and Molecular Biology", Wiley Interscience, New York (1977).
22. V. I. Tsetlin, I. I. Mikhaleva, M. A. Myagkova, L. B. Senyavina, A. S. Arseniev, V. T. Ivanov, and Yu. A. Ovchinnikov, in "Peptides: Chemistry, Structure and Biology", R. Walter and J. Meienhofer, eds. p. 935, Ann Arbor Science, Ann Arbor, Michigan (1975).
23. E. Karlsson, in "Snake Venoms (Handbook of Experimental Pharmacology)", Vol. 52, C. -Y. Lee, ed. p. 159, Springer Verlag, West Berlin (1978).
24. T. Heidmann and J. -P. Changeux, Ann. Rev. Biochem. 47:317 (1978).
25. International Union of Biochemistry, "Biochemical Nomenclature and Related Documents", The Biochemical Society, London (1978).
26. A. S. Arseniev, T. A. Balashova, Yu. N. Utkin, V. I. Tsetlin, V. F. Bystrov, V. T. Ivanov, and Yu. A. Ovchinnikov, Eur. J. Biochem. 71:595 (1976).
27. A. S. Arseniev, A. M. Surin, Yu. N. Utkin, V. I. Tsetlin, V. F. Bystrov, V. T. Ivanov, and Yu. A. Ovchinnikov, Bioorg. Khim. (Soviet J. Bioorgan. Chem.), 4:197 (1978).
28. V. S. Pashkov, A. S. Arseniev, and V. I. Kondakov, "Magnetic Resonance and Related Phenomena", E. Kundla, E. Lippmaa, T. Saluvere, eds. p. 544, Springer Verlag, West Berlin (1979).
29. A. S. Arseniev and V. F. Bystrov, 4th Specialized Colloque AMPERE on Dynamic Processes in Molecular Systems Studied by RF-Spectroscopy, Abstracts, p. 95-99, Leipzig (1979).
30. V. I. Tsetlin, A. S. Arseniev, Yu. N. Utkin, A. Z. Gurevich, L. B. Senyavina, V. F. Bystrov, V. T. Ivanov, and Yu. A; Ovchinnikov, Eur. J. Biochem. 94:337 (1979).
31. V. T. Ivanov, V. I. Tsetlin, E. Karlsson, A. S. Arseniev, Yu. N. Utkin, V. S. Pashkov, A. M. Surin, K. A. Pluzhnikov

and V. F. Bystrov, Toxicon, Suppl.1:77 (1979).
32. Yu. N. Utkin, V. S. Pashkov, A. M. Surin, V. I. Tsetlin, andV. T. Ivanov, in "Peptides 1978", G. Kupryszewski, I. Z. Siemion, eds. p. 397, Wroclaw University Press, Poland (1979).
33. V. T. Ivanov, V. I. Tsetlin, E. Karlsson, A. S. Arseniev, Yu. N. Utkin; V. S. Pashkov, A. M. Surin, K. A. Pluzhnikov, and V. F. Bystrov, to be published in Proceedings of the 6th International Symposium on Animal, Plant and Microbial Toxins, Uppsala (1979).
34. V. F. Bystrov, in "Biomolecular Structure, Conformation, Functions and Evolution", R. Srinivasan, E. Subramanian, N. Yathindra, eds. Vol. 2, p. 3, Pergamon Press, Oxford (1980).
35. V. T. Ivanov, V. I. Tsetlin, I. I. Mikhaleva, O. M. Volpina, A. R. Nuriddinov, Yu. N. Utkin, A. S. Arseniev, V. S. Pashkov, E. Karlsson, A. M. Surin, and V. F. Bystrov, in "Peptides 1978", G. Kupryszewski, I. Z. Siemion eds. p. 41, Wroclaw University Press, Poland (1979).
36. A. S. Arseniev, H. Rochat, and V. F. Bystrov, in preparation.
37. J. Lauterwein, M. Lazdunski, and K. Wüthrich, Eur. J. Biochem. 92:361 (1978).
38. C. H. Fung, C. C. Chang, and R. K. Gupta, Biochemistry, 18:457 (1979).
39. T. Endo, F. Inagaki, K. Hayashi, and T. Miyazawa, Eur. J. Biochem. 102:417 (1979).
40. F. Inagaki, T. Miyazawa, H. Hori, and N. Tamiya, Eur. J. Biochem. 89:433 (1978).
41. A. Z. Gurevich, paper in preparation.
42. V. I. Tsetlin, E. Karlsson, A. S. Arseniev, Yu. N. Utkin, A. M. Surin, V. S. Pashkov, K. A. Pluzhnikov, V. T. Ivanov, V. F. Bystrov, and Yu. A. Ovchinnikov, FEBS Letters, 106:47 (1979).
43. G. I. Likhtenstein, "Spin Labeling Methods in Molecular Biology", Wiley, New York (1976).
44. A. M. Surin, K. A. Pluzhnikov, and Yu. N. Utkin, IX National Conference on Molecular Spectroscopy, Albena, Bulgaria, September 29 - October 3 (1980).
45. R. W. Wien, J. D. Morrisett, and H. McConnell, Biochemistry, 11:3707 (1972).
46. D. Tsernoglou and G. A. Petsko, FEBS Letters, 68:1 (1976).
47. B. W. Low, H. S. Preston, A. Sato, L. S. Rosen, J. E. Searl, A. D. Budko, and J. S. Richardson, Proc. Nat. Acad. Sci. USA, 73:2991 (1976).
48. M. R. Kimball, A. Sato, J. S. Richardson, L. S. Rosen, and B. W. Low, Biochem. Biophys. Res. Commun. 88:950 (1979).
49. A. F. Drake, M. J. Dufton, and R. C. Hider, FEBS Letters, 83:202 (1977).

50. M. J. Dufton and R. C. Hider, Trends in Biochem. Sci. 53 (1980); to be published in Proceedings of the 6th International Symposium on Animal, Plant and Microbial Toxins, Uppsala (1979).
51. J. A. Reinolds and A. Karlin, Biochemistry, 17:2035 (1978).
52. V. N. Damle and A. Karlin, Biochemistry, 17:2039 (1978).
53. J. F. Ellena and M. G. McNamee, FEBS Letters, 110:200 (1980).
54. A. N. Kuznetsov, A. M. Wasserman, A. I. Volkov, and N. N. Kost, Chem. Phys. Letters, 12:103 (1971).
55. L. I. Antsyferova, A. M. Wasserman, A. N. Ivanova, V. A. Lifshits, and N. S. Nazemets, "Atlas of EPR Spectra of Spin Labels and Probes", Nauka Publ. Moscow (1977).
56. V. Sator, M. A. Raftery, and M. Martinez-Carrion, Arch. Biochem. Biophys. 190:57 (1978).
57. K. M. Salikhov, A. B. Doctorov, and Yu. N. Molin, J. Magn. Res. 5:189 (1971).
58. B. L. Low, Adv. Cytopharmacology, 3:141 (1979).
59. E. Karlsson, J. Fohlman, and M. Groth, Bull. Inst. Pasteur, 74:11 (1976).

METALLOENZYMES AND MODEL SYSTEMS. CARBONIC ANHYDRASE: SOLVENT AND BUFFER PARTICIPATION, ISOTOPE EFFECTS, ACTIVATION PARAMETERS AND ANIONIC INHIBITION [†]

Y. Pocker, Thomas L. Deits, and Nobuo Tanaka

Department of Chemistry, University of Washington
Seattle, Washington 98195, U.S.A.

Abstract - The interconversion between CO_2 and HCO_3^- in the presence of bovine carbonic anhydrase (BCA) was studied by initial rate measurements using a stopped-flow indicator method. The results analyzed in terms of the Michaelis-Menten Scheme exhibited a sigmoidal variation of the turnover number (k_{cat}) against pH with a pK_a value of about 6.8 at 25.0°. For CO_2 hydration large k_{cat} values were observed in the high pH region, while for HCO_3^- dehydration large k_{cat} values were observed in the low pH region. For both substrates K_m values did not show any significant variation with pH in the region studied. Solvent deuterium isotope effects were found to be between 2.5 and 3.0 at 25.0° for both k_{cat} and K_m, and did not vary significantly with change of pH or substrate. Activation parameters for CO_2 hydration catalyzed by H_2O, OH^-, and BCA were obtained in the temperature range 7.0° - 35.0°C. For BCA catalysis, temperature effects were confined largely to k_{cat}. Inhibition of BCA by monoanions was studied over a pH range of 6.6 to 9.0 for CO_2 hydration, and 6.6 to 7.0 for HCO_3^- dehydration. Anions were found to exhibit linear mixed inhibition of CO_2 hydration at low pH, linear uncompetitive inhibition of CO_2 hydration at high pH, and linear competitive inhibition of HCO_3^- dehydration at all pH values studied. The implications of these results are discussed, a formal kinetic scheme is proposed, and a mechanism presented to account for these observations.

[†] This investigation was supported by grants from the National Institutes of Health, National Science Foundation and Muscular Dystrophy Association.

1. INTRODUCTION

Bovine carbonic anhydrase (BCA) is a highly efficient catalyst for the interconversion between CO_2 and HCO_3^-. In addition to studies of these physiological substrates(1-3) much work has been done on the hydrase and esterase activity of BCA(4-17). The structure and activity of carbonic anhydrase has been the subject of a recent extensive review(18).

It is generally accepted that a zinc(II) ion ligated to three imidazoles of histidines and a water molecule are essential for catalytic activity. An ionization with a pK_a of approximately 7.0 controls the activity of BCA. The basic form of the enzyme is active towards CO_2 hydration, but the acidic form is active towards HCO_3^- dehydration. This ionization has been attributed by some authors to an equilibrium between a zinc-aquo and a zinc-hydroxo form of BCA. Others have assigned the activity linked ionization to a non-ligated active site histidine conserved in all carbonic anhydrase isozymes for which structures have been determined.

In all instances, the activity of BCA towards its physiological substrates has been found to fully conform to the Michaelis-Menten formalism. Under such analysis, BCA exhibits an extremely high turnover number, the magnitude of which is sufficient to render diffusion limitations on protons and substrate a significant constraint on any proposed mechanism.

The inhibition of carbonic anhydrase by anions has long been recognized. The inhibitory effect of Cl^- was noted by early investigators(19). A later study of anion inhibition by stopped-flow techniques was compromised by the presence of 80mM Cl^- in buffers used (20). The inhibition of the hydrase activity of the Cobalt(II) substituted enzyme has been investigated over the full pH range(21). Anionic inhibition of esterase activity has been studied by initial rate techniques(11-13) and by complexometric titration(22). None of the work thus far published has included full scale Michaelis-Menten analysis of the inhibition of the native Zinc(II) enzyme towards its natural substrate over an extended pH range.

The present study represents a full characterization of the effects of pH, buffer concentration, and monoanions on the hydration and dehydration activity of BCA, studied by the method of initial rates using stopped-flow techniques. The results of this study impose further constraints on acceptable mechanisms for carbonic anhydrase activity. We propose a kinetic rationalization and a mechanistic scheme consistent with the data presented.

2. EXPERIMENTAL

Enzyme: Bovine carbonic anhydrase was purchased from Schwarz/Mann as a highly active lyophilized form and was used without further purification.

Reagents: 3-picoline, 1-methylimidazole, and 1,2-dimethylimidazole were distilled under reduced pressure and stored at 5°C under a N_2 atmosphere. Imidazole, diethylmalonic acid, p-nitrophenol (λ_{max} = 400 nm) were purified by recrystallization. Reagent grade sodium dihydrogen phosphate, disodium hydrogen phosphate, sodium sulfate, sodium salts of anions, bromocresol purple (λ_{max} = 558 nm) and metacresol purple (λ_{max} = 578 nm) were used without further purification. Indicators were paired with buffers of similar pK_a values. D_2O used had a deuterium content greater than 99%.

Substrates: Saturated CO_2 solutions were made by bubbling CO_2 gas (Airco, Research Grade, 99.99%) into deionized distilled water at 25°C. The concentration of CO_2 in the saturated solution was taken as 0.034 M in H_2O and 0.038 M in D_2O(3) and then diluted with degassed deionized distilled water to the desired concentration. Reagent grade sodium bicarbonate was used in the dehydration reaction. The ionic strength of sodium bicarbonate solutions was kept at 0.1 with sodium sulfate. Substrate concentrations used to determine kinetic parameters ranged from 0.017 to 0.0034 M for CO_2, and from 0.05 to 0.005 M for HCO_3^-.

Buffer Solutions: For studies of buffer strength, isotope effects, and activation parameters, ionic strength was made constant at 0.1 by addition of sodium sulfate. In order to accommodate inhibitors with a wide range of K_1 values ionic strength was maintained at 0.2 with sodium sulfate in anion inhibition studies.

Kinetic Procedures: A changing pH-indicator method(23) was employed to follow reaction rates. Initial rates were determined using a Durrum-Gibson stopped-flow spectrophotometer interfaced to a PDP 8L computer. Transmittance data was converted to absorbance and plotted versus time using unweighted least square analysis. 10 to 15 runs were analyzed for each kinetic determination. The standard deviation of a replicate set was approximately independent of the magnitude of the velocity of the set and was in no case greater than 5% of the mean value. Replicate determination of Michaelis-Menten parameters agreed to within 10-15%.

Inhibition Constants: In inhibition experiments, data were collected for 4 to 5 different substrate concentrations at each of 4 to 6 different inhibitor concentrations (ranging from zero added inhibitor, to sufficient inhibitor to reduce the observed activity by approximately 80%). Lineweaver-Burk plots of the resulting data were analyzed by unweighted linear least squares analysis, yielding a slope and ordinate intercept at each inhibitor concentration. (Due to its prevalence in the literature, Lineweaver-Burk analysis was the analytical method of choice. Because of the magnitude of error in our data was independent of velocity, and because alternate plotting methods ($V/[S]$ versus V, etc.) yielded no significantly different values, this choice does not compromise our findings.)

Replots of slopes and/or ordinate intercepts versus inhibitor concentration were analyzed by unweighted linear least squares analysis. All replots were found to be linear, with correlation coefficients greater than 0.99+. The inhibitor constants derived from slope replots are designated K_I^{Sl}, and those from ordinate intercept replots are designated K_I^{int}. The values so obtained were inserted into the velocity equation describing the observed mechanism of inhibition and were used to generate the lines shown in the final figures.

3. RESULTS

Carbon dioxide hydration in the absence of enzyme at pH below 7.5 exhibited a first-order rate constant of 0.037 s^{-1}, which is in good agreement with the values reported earlier(3,23). Enzyme catalyzed rates were obtained by subtracting solvent catalyzed rate from total reaction rates. The solvent catalyzed rate was generally kept at less than 15% of the total reaction rate. The buffer concentration did not affect appreciably the solvent catalyzed rate of CO_2 hydration or HCO_3^- dehydration. A typical plot of V against $V/[S]$ is shown in Figure 1 at various buffer concentrations.

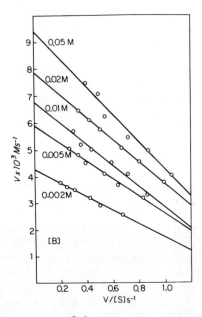

Fig. 1. Plot of V against $V/[S]$ for CO_2 hydration in D_2O at pD 7.5, 25°C in 1-methylimidazole buffer. Buffer concentrations are indicated $[BCA] = 3.9 \times 10^{-8}$M.

For CO_2 hydration, K_m values were found to be essentially independent of pH except at low pH where a small increase was observed. The minor variations of K_m values with buffer composition for each buffer system are very similar to Khalifah's earlier observations(23) as seen in Figure 2a. The turnover number k_{cat} was a sigmoidal function of pH and reached a maximum of 10^6 at pH above 8. Thus, k_{cat}/K_m also showed a sigmoidal variation against pH with an inflection at pH 6.8. In HCO_3^- dehydration K_m was independent of pH except in the high pH region. Both k_{cat} and k_{cat}/K_m were sigmoidal functions of pH with maxima at low pH. These results are in good agreement with previous reports(23-26). The variation of log K_m and log (k_{cat}/K_m) with respect to pH were not appreciably affected by a moderate change in buffer concentration, as shown in Figures 2 and 3.

Table I shows the solvent deuterium isotope effect on k_{cat} and K_m in this system. In the plateau region of the plot of k_{cat}/K_m against pH, both k_{cat} and K_m exhibited solvent deuterium isotope effects of about 2.5 in hydration. The magnitude of these isotope effects did not vary appreciably with the change of buffer concentration. There seems to be no significant isotope effect on k_{cat}/K_m regardless of buffer concentration in hydration. The small isotope effect found for k_{cat}/K_m in dehydration may be due to the fact that

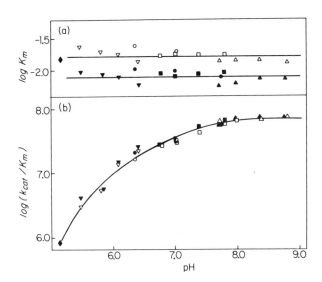

Fig. 2. Plot of log K_m (a) and log (k_{cat}/K_m) (b) against pH in CO_2 hydration. Buffer species are pyridine (◇), 3-picoline (▽), phosphate (○), 1-methylimidazole (□), and 1,2-dimethylimidazole (△). Open symbols are for the measurement at 0.05M and filled symbols at 0.005M buffer concentration.

Table I. Effect of Buffer Concentration on k_{cat}, K_m and k_{cat}/K_m in CO_2–HCO_3^- Interconversion Catalyzed by BCA in H_2O and D_2O.[a]

Substrate	[B] M	pH	pD	$K_m \times 10^3$ M			$k_{cat} \times 10^{-5}$ s^{-1}			$(k_{cat}/K_m) \times 10^{-7}$ M^{-1}s^{-1}		
				H_2O	D_2O	$(K_m^{H_2O}/K_m^{D_2O})$	H_2O	D_2O	$(k_{cat}^{H_2O}/k_{cat}^{D_2O})$	H_2O	D_2O	$(k_{cat}^{H_2O}/K_m^{H_2O})/(k_{cat}^{D_2O}/K_m^{D_2O})$
CO_2[b]	0.002	6.98	7.50	6.2	2.3	(2.7)	2.4	1.0	(2.4)	3.8	4.4	(0.86)
	0.005	7.01	7.52	8.6	3.1	(2.8)	3.5	1.5	(2.4)	4.1	4.8	(0.85)
	0.01	7.01	7.53	11	3.6	(3.0)	4.3	1.7	(2.5)	4.0	4.8	(0.82)
	0.02	7.02	7.53	12	4.2	(2.8)	4.8	2.0	(2.4)	4.1	4.9	(0.83)
	0.05	7.02	7.53	18	5.2	(3.4)	6.7	2.4	(2.8)	3.8	4.6	(0.83)
CO_2	0.002	8.75	9.26	5.1	2.1	(2.5)	3.8	1.6	(2.4)	7.4	7.8	(0.96)
	0.005	8.77	9.32	7.4	3.1	(2.4)	5.4	2.3	(2.4)	7.2	7.3	(0.99)
	0.01	8.80	9.34	9.0	3.9	(2.3)	6.7	2.8	(2.4)	7.4	7.2	(1.02)
	0.05	8.79	9.34	13	6.0	(2.2)	10	5.1	(2.0)	7.6	8.5	(0.90)
HCO_3^-[b]	0.005	5.95	6.57	18	7.1	(2.5)	4.0	1.5	(2.7)	2.3	2.1	(1.09)
	0.01	5.95	6.52	24	8.4	(2.9)	5.5	1.7	(3.3)	2.3	2.0	(1.16)
	0.02	5.95	6.52	28	10	(2.7)	6.3	1.9	(3.3)	2.2	1.9	(1.29)
	0.05	5.95	6.52	33	11	(3.0)	7.0	1.9	(3.7)	2.2	1.8	(1.20)

[a]Buffer species used in this study are, 1-methylimidazole (pH 7.0), 1,2-dimethylimidazole (pH 8.8), and 3-picoline (pH 5.9).

[b]These experiments were not done in the plateau region of the plot of k_{cat}/K_m versus pH and hence the solvent deuterium associated with k_{cat} or k_{cat}/K_m may include a small effect due to the pH (pD) difference. In the plateau region no such ambiguity exists.

measurements were not carried out in the plateau region of the plot of k_{cat}/K_m against pH. A similar isotope effect was reported for the dehydration of HCO_3^- in the presence of HCA-C(24).

Values of k_{cat} and K_m were obtained at different temperatures varying from 7.0° to 35.0°C in 1,2-dimethylimidazole buffers for hydration and in 3-picoline buffers for dehydration.

As shown in Figure 4, K_m exhibited very small variation with the temperature change from 7.0° to 35°C in CO_2 hydration and in HCO_3^- dehydration. Thus, k_{cat} represents the major variation of the rate of enzymatic reaction with temperature. The temperature dependence of these parameters was not affected appreciably by changes in buffer concentration.

Activation parameters calculated for CO_2 hydration catalyzed by BCA, by the solvent, and by OH^-, are listed in Table II. Although temperature effects were examined in the plateau region of the plot k_{cat}/K_m against pH for CO_2 hydration, similar measurements of temperature coefficients were found to be impractical for HCO_3^- dehydration. Since pK_a values at different temperatures are not yet known for BCA, the activation parameters for HCO_3^- dehydration are not available at this time.

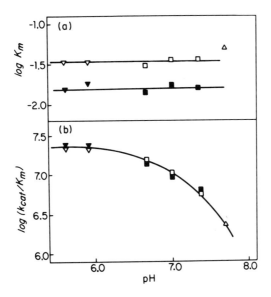

Fig. 3. Plot of log K_M (a) and log (k_{cat}/K_M) (b) against pH in HCO_3^- dehydration. The same symbols are used for buffer species and concentration as in Figure 2.

Table II. Activation Parameters for CO_2 Hydration.

Catalyst		E_a (kcal mol^{-1})	$\Delta S^{\ddagger}$ (cal K^{-1} mol^{-1})	$\Delta G^{\ddagger}$ (kcal mol^{-1})
BCA[a]	k_{cat}	5.0	-16	9.3
	(k_{cat}/K_m)	5.0	-7	6.6
OH^-		7.2	-18	11.9
H_2O[b]		14.3	-19	19.5

[a] Small slopes found in the plot of log K_M against T^{-1} in Figure 4 were neglected.
[b] The activation parameters were derived from first order rate constants.
1 cal = 4.187 J

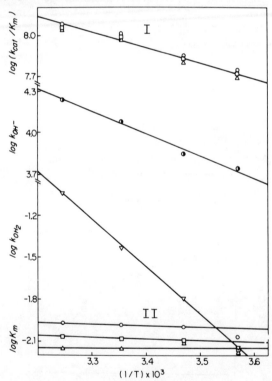

Fig. 4. Arrhenius plot for CO_2 hydration catalyzed by OH^- (●), H_2O (▽), and BCA (I: k_{cat}/K_m, II: K_m) at pH above 8.5, with buffer concentration 0.02M (○), 0.01M (□), and 0.005M (△).

The monoanionic inhibition of BCA undergoes dramatic changes in mechanism with changes in substrate and pH. Figure 5 shows the Lineweaver-Burk plot for inhibition by SCN^- of BCA catalyzed CO_2 hydration at pH 6.6. The mechanism observed is best described as linear mixed inhibition, with $K_I^{sl} = 1.6 \times 10^{-3} M$ and $K_I^{int} = 3.2 \times 10^{-3} M$ which closely resembles noncompetitive inhibition. At the same pH, SCN^- inhibition of BCA catalyzed HCO_3^- dehydration (Figure 6) is linear competitive in mechanism, with $K_I^{sl} = 1.7 \times 10^{-3} M$.

The inhibitory power of anions towards BCA catalyzed HCO_3^- dehydration is reduced at higher pH, while the formal mechanism remains competitive, as reported earlier(3). The inhibition of BCA catalyzed CO_2 hydration by SCN^- at pH 9.0 (Figure 7) represents linear uncompetitive inhibition, with $K_I^{int} = 4.2 \times 10^{-3} M$. This represents the first observation of uncompetitive inhibition of carbonic anhydrase activity, and has significant implications for any proposed mechanism.

The inhibition patterns reported above are largely independent of the structure or inhibitory power of the monoanion tested with one important exception, noted below. Observed inhibition constants for a number of anions of diverse structure with K_I values ranging over 3 orders of magnitude are listed in Table III.

The observed patterns are established extremely rapidly. Parallel experiments were performed comparing preincubation of monoanions

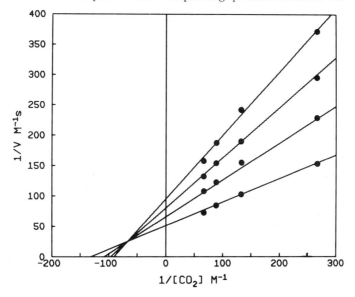

Fig. 5. Lineweaver-Burk plot for SCN^- inhibition at BCA catalyzed CO_2 hydration, pH 6.6 [BCA] = $2.71 \times 10^{-7} M$ [SCN^-] = 0, 9.1 $\times 10^{-4} M$, $1.82 \times 10^{-3} M$, and $2.73 \times 10^{-3} M$.

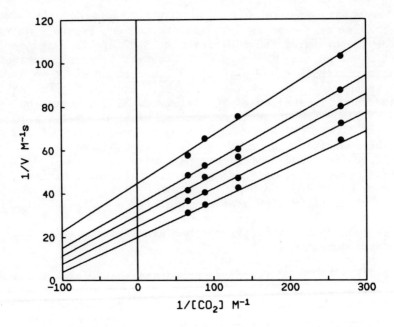

Fig. 6. Lineweaver-Burk plot for SCN^- inhibition of BCA catalyzed HCO_3^- dehydration, pH 6.6 [BCA] = 2.54×10^{-7}M [SCN^-] = 0, 9.1×10^{-4}M, 1.82×10^{-3}M, and 2.73×10^{-3}M.

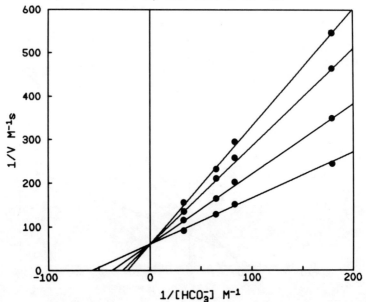

Fig. 7. Lineweaver-Burk plot for SCN^- inhibition of BCA catalyzed CO_2 hydration, pH 9.0 [BCA] = 3.29×10^{-7} [SCN^-] = 0, 1.03×10^{-3}M, 2.06×10^{-3}M, 3.09×10^{-3}M, 5.15×10^{-3}M.

Table III. Anionic Inhibition of Carbonic Anhydrase.

Anion	vs. CO_2 pH 6.6 K_I^{sl}	K_I^{int}	vs. CO_2 pH 9.0 K_I^{int}	vs. HCO_3^-, pH 6.6 K_I^{sl}
Cl^-	7.8×10^{-2}	1.3×10^{-1}	2.2×10^{-1}	7.3×10^{-2}
OAc^-	3.8×10^{-2}	5.5×10^{-2}	1.1×10^{-1}	2.5×10^{-2}
Br^-	1.6×10^{-2}	3.8×10^{-2}	1.3×10^{-1}	3.0×10^{-2}
SCN^-	1.6×10^{-3}	3.2×10^{-3}	4.2×10^{-3}	1.7×10^{-3}
N_3^-	3.8×10^{-4}	5.6×10^{-4}	1.4×10^{-3}	2.2×10^{-4}

with BCA with experiments in which enzyme and inhibitor were allowed to interact only 10 ms before initial rates were determined. No qualitative or quantitative difference in inhibition was observed.

Experiments at intermediate pH values showed patterns of inhibition of BCA catalyzed CO_2 hydration undergoing a smooth transition between the mechanisms observed at the two pH extrema, with the change in mechanism attributable to an ionization with a pK_a of approximately 6.8.

The possibility that changing concentrations of SO_4^{2-} (or of trace equilibrium concentrations of HSO_4^-) used to maintain ionic strength could be affecting the observed pattern was eliminated by control experiments. Anions of low inhibitory power were tested in two sets of experiments. In one the ionic strength was kept constant with sodium sulfate; in the other the concentration of sodium sulfate was constant and the ionic strength allowed to vary with inhibitor concentration. No significant difference was observed between the two experiments. whose contents are mixed in a reaction cell within 10 ms of initia-

The design of the stopped-flow instrument, two separate syringes whose contents are mixed in a reaction cell within 10 ms, of initiation of a kinetic run, permits the analysis of the inhibitory power of the HCO_3^- anion on BCA catalyzed CO_2 hydration at pH 9.0. Solutions containing buffer, indicator, and BCA were brought to a desired concentration of HCO_3^- with solid $NaHCO_3$. (Ionic strength was kept constant with added solid sodium sulfate.) Since at pH 9, an equilibrium solution of carbonate species, $CO_2/HCO_3^-/CO_3^{2-}$, is more than 95% in the form of the anion HCO_3^-, this equilibrium mixture can be treated as essentially another anionic inhibitor. The BCA/HCO_3^- equilibrium mixture was then tested against CO_2 solutions in the other drive syringe, and the results plotted as for other anionic inhibitors. The result (Figure 8) is unique among anions tested. The mechanism of inhibition is mixed, with $K_I^{int} = 0.1M$, $K_I^{sl} = 0.04M$.

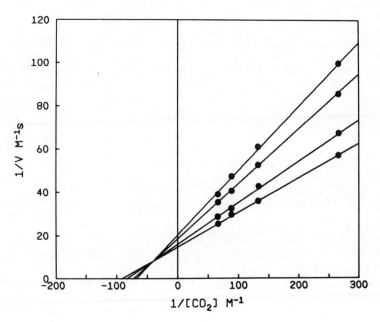

Fig. 8. Lineweaver-Burk plot of HCO_3^- inhibition of BCA catalyzed CO_2 hydration, pH 9.0 [BCA] - 3.29×10^{-7}M [HCO_3^-] = 0, 7.56×10^{-3}M, 2.24×10^{-2}M, 3.25×10^{-2}M.

4. DISCUSSION

Plots of V versus V/[S] were sensibly linear for both substrates as shown in Figure 1, indicating that for all practical purposes this system can be treated by the Michaelis-Menten formalism. For both substrates, k_{cat} and K_m increase with an increase in buffer concentration, but the ratio k_{cat}/K_m remains virtually unchanged (Figures 2 and 3). This observation is very similar to that described for CO_2 hydration catalyzed by BCA(26) and by human carbonic anhydrase, isozyme C (HCAC)(2,27). The Haldane relation is obeyed for our data at every buffer concentration and pH, showing that the kinetic parameters determined under initial rate conditions describe a system consistent with the attainment of chemical equilibrium.

The observation at high buffer concentration of K_m values independent of pH and k_{cat} values varying in sigmoidal fashion for both CO_2 and HCO_3^- agree well with earlier reports(24,26). This serves to confirm the kinetic similarity of HCAC and BCA.

The kinetic solvent deuterium isotope effects on the Michaelis-Menten parameters, $k_{cat}^{H_2O}/k_{cat}^{D_2O}$, are essentially identical for both CO_2 hydration and HCO_3^- dehydration activity. Changes in buffer concentration sufficient to significantly affect the values of both k_{cat} and K_m resulted in no significant change in solvent deuterium isotope effect. The two most conservative interpretations to be drawn from these observations are either that even at reduced buffer concentration no change in rate determining step occurs in BCA catalysis or that if such a change occurs the respective proton inventories are similar. These observations are somewhat at variance with experiments performed earlier on the buffer concentration and solvent isotope effects on HCAC catalysis(28).

In all respects, CO_2-HCO_3^- interconversion catalyzed by BCA exhibits a high degree of symmetry. Both substrates can be described as exhibiting pH independent K_m values, a sigmoidal variation of k_{cat} with pH, and a proportionate change in both k_{cat} and K_m with reduced buffer concentration and with a change in solvent from H_2O to D_2O. This striking symmetry relation must be given due weight in any consideration of a catalytic mechanism.

In studies of HCAC catalysis at equilibrium, measured by isotopic double labeling experiments, both k_{cat} and K_m were found to remain constant with changes in buffer concentration, provided that the enzyme could alternate hydration and dehydration steps(29).

This difference in the magnitude of the buffer effect in equilibrium studies versus initial rate studies could be accounted for by at least three possibilities. The first is that under equilibrium and initial rate conditions, BCA exhibits different mechanisms. This is an unattractive hypothesis, and is inconsistent with the well characterized adherence of BCA catalysis to Michaelis-Menten kinetics and to the Haldane relation under initial rate conditions. The possibility that the concentration of HCO_3^- present in equilibrium studies acts as a buffer in the absence of added buffer species is made untenable by the present observation of a similar dependence of k_{cat} and K_m upon buffer concentration for both CO_2 and HCO_3^-. Significant HCO_3^- buffering activity would be expected to ameliorate this effect in the dehydration reaction.

The third, and, we feel most reasonable explanation is the presence in the active site of a bound buffer species. At equilibrium, this proton transfer group would alternately protonate and deprotonate, as the enzyme correspondingly dehydrated and hydrated CO_2 species, free of proton diffusion limitations. The most likely candidate for this internal buffer is the imidazole of the conserved non-ligand histidine in the active site of carbonic anhydrase (His - 64 in HCAC), as proposed earlier(3,24,26,29).

The above evidence appears to suggest a close kinetic similarity between elementary kinetic steps affecting the Michaelis-Menten constants k_{cat} and K_m. This close relationship breaks down, however, when the temperature dependence of the respective values are considered. As is evident from Figure 4, the variation of activity of BCA catalyzed CO_2 hydration resides entirely in the

k_{cat} term. K_m is virtually unaffected over this substantial temperature range. These values are determined at pH values above 8.5, sufficiently far above the pK_a value of the ionizing group so that changes in pK_a associated with temperature changes will have a negligible effect. The activation parameters for HCO_3^- dehydration activity of BCA in the corresponding 'plateau' region are not yet available, but near the plateau the variation of activity again resides almost entirely in the k_{cat} term.

It is clear, moreover, that the Arrhenius plots for K_m and k_{cat} are quite linear over the temperature range considered, furnishing presumptive evidence that no mechanistic change occurs in the processes described by either parameter. A further implication of this result is that if an ice-like "structured water" network is an important component of the active site of BCA, then little change in this structure occurs over the temperature range studied. Such a pronounced temperature stability of an extended hydrogen bonded network seems unlikely.

We were able, under the experimental conditions chosen, to perturb the value of k_{cat}/K_m by a factor of approximately 2 to 3 by changes in buffer concentration, solvent changes from H_2O and D_2O, and by temperature changes from 7° to 35°. It seems reasonable that a change in rate determining step induced by one of these perturbations would be likely to have a measureable effect in the other experiments. In this light, the presumption that k_{cat} and K_m are governed by the same elementary step, based on solvent isotope and buffer concentration effects, is not supported by the separate temperature effects for these two parameters. Further, the interpretation, based on the observed buffer concentration effects, that a change in the rate determining step occurs is not supported by the observed linearity of Arrhenius plots for k_{cat} and K_m, while the solvent isotope effect is interpretable under either hypothesis. The only interpretation consistent with all data presented, then, appears to be that k_{cat} and K_m do not share a common rate determining step, and are therefore kinetically distinct, and that no change in the rate determining step occurs for either parameter under the experimental perturbations induced to date.

The above discussion bears heavily on the analysis of the patterns of anionic inhibition of BCA catalysis. Anionic inhibitors, depending on the choice of pH and substrate, are able to inhibit solely through the apparent K_m of HCO_3^- dehydration activity; largely through only the apparent $V_{max}(=k_{cat}/[E])$ of CO_2 hydration activity at low pH; and to inhibit equally through the apparent V_{max} and K_m of CO_2 hydration at high pH. Combined with the fact that anionic inhibition is rapidly established (see Section 3), and that under all conditions anionic inhibition is explicable in terms of simple linear inhibition mechanisms, it seems appropriate to propose a scheme

METALLOENZYMES AND MODEL SYSTEMS

based on elementary rapid equilibrium kinetics, at least as a first order analysis.

We wish to propose that K_m does in fact represent a substrate dissociation constant, and that the observed K_I values represent actual inhibitor dissociation constants. This leads to the proposition that the dissociation of substrate is in actual fact independent of the state of ionization of the enzyme, and that the sigmoidal variation in k_{cat} is a reflection of the changes in concentration of the appropriate state of ionization of the enzyme-substrate complex.

Scheme 1 below, then, is proposed to account for the anionic inhibition of CO_2 hydration by all monoanions other than HCO_3^- (HCO_3^- inhibition will be discussed below). In this and in subsequent schemes, S refers to CO_2 and P refers to HCO_3^-; E and EH^+ refer to the two states of ionization of the enzyme; I refers to any monoanionic inhibitor.

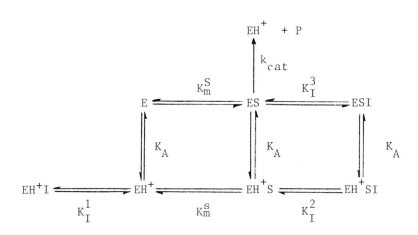

Scheme 1

Rapid equilibrium analysis of Scheme 1, followed by casting the resulting velocity equation in Lineweaver-Burk form yields equation (1)

$$\frac{1}{V} = \frac{1}{[S]} \frac{K_m^S}{V_{max}} \left(1 + \frac{[H^+]}{K_A} \left(1 + \frac{[I]}{K_I^1} \right) \right)$$
$$+ \frac{1}{V_{max}} \left(1 + \frac{[I]}{K_I^3} + \frac{[H^+]}{K_A} \left(1 + \frac{[I]}{K_I^2} \right) \right) \quad (1)$$

The individual K_I values in Scheme 1 can be determined as follows K_I^1 is the K_I^{sl} term for anionic inhibition of CO_2 hydration activity at limiting low pH. K_I^2 is the K_I^{int} term under the same conditions. K_I^3 is the K_I^{int} term for inhibition of CO_2 hydration activity at limiting high pH values.

This equation completely accounts for the changes in mechanism observed for the anionic inhibition of BCA catalyzed CO_2 hydration, and, when observed values for K_m, V_{max}, K_A, and the various K_I values are inserted, quantitatively accounts for the mechanism observed at the pH extrema and at all intermediate pH values.

Anionic inhibition of BCA catalyzed HCO_3^- dehydration is accounted for by Scheme 2.

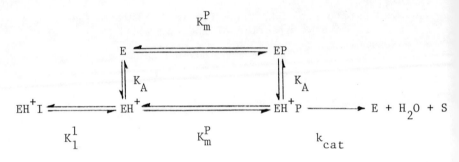

Scheme 2

The Lineweaver-Burk velocity equation for this scheme is presented in equation (2).

$$\frac{1}{V} = \frac{1}{[S]} \frac{K_M^P}{V_{max}} \left(1 + \frac{[H^+]}{K_A} \left(1 + \frac{[I]}{K_I^1} \right) \right) + \frac{1}{V_{max}} \quad (2)$$

Equation (2) is consistent with the observed diminution of inhibitor power of anions towards HCO_3^- dehydration as the pH is increased. An important prediction evident from a comparison of Scheme 1 and 2 is that the term referred to in both schemes as K_I^1 represents exactly the same dissociation equilibrium. If these schemes do in fact hold, then the K_I^{sl} values determined at low pH for the inhibition of both CO_2 hydration and HCO_3^- dehydration should be identical. Inspection of Table III bears this out completely

The third case to be dealt with is kinetically the most involved; the inhibition of CO_2 hydration activity by HCO_3^-. For the following Scheme 3, Schemes 1 and 2 are combined and references to I are replaced by P.

We propose that HCO_3^- can occupy the anionic binding sites common to all anionic inhibitors and that HCO_3^- can also inhibit CO_2

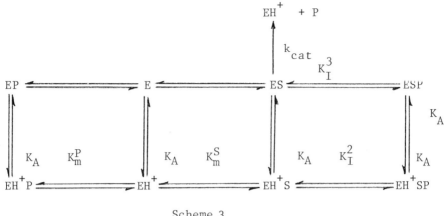

Scheme 3

hydration by binding directly to the basic form of the enzyme, a mode of binding denied to other anions lacking exchangeable protons. Quantitatively, the inhibition by HCO_3^- of CO_2 hydration at limiting high pH predicted by this scheme would be linear mixed in mechanism, with a K_I^{sl} term equal to the K_M^P value determined in experiments on HCO_3^- dehydration catalysis. The K_I^{int} term should be comparable in magnitude to that of anions of similar electronic structure, as the binding site for HCO_3^- and anions in this mode is expected to be identical. We observe the K_I^{sl} term to be 0.04M, in excellent agreement with the K_m^P value of 0.032 from HCO_3^- dehydration catalysis experiments (Figure 3). The K_I^{int} value for HCO_3^- inhibition is 0.1, quite compatible with the K_I^{int} value of 0.11 for the electronically similar acetate anion under the same conditions (Table III).

These K_I values for HCO_3^- inhibition are in agreement with those reported earlier for HCAC(28) except that the earlier authors considered the mechanism of inhibition to be non-competitive (K_I^{sl} = K_I^{int}) and apparently attributed the deviations from this mechanism evident in their data (Figure 3 of reference 28) to experimental error. They report a value corresponding to K_I^{int} of 0.031, an excellent quantitative agreement for this particular parameter. We attribute the deviations in their data to the mixed mode of inhibition discussed above.

The schemes presented above offer a consistent rationalization of the binding of substrate and inhibitors to carbonic anhydrase under a broad range of conditions, and are generally consistent with earlier, more limited studies of the anionic inhibition of the Zinc(II) carbonic anhydrase catalyzed hydration of CO_2 and dehydration of HCO_3^-(3,20). Carbonic anhydrase catalyzed esterase activity does not conform to this scheme, unless the binding represented by K_I^3 is eliminated or greatly reduced. Esterase activity inhibition studies have shown that even the most potent anionic inhibitors of CO_2 hydration activity are dramatically less effective in inhibiting esterase activity(11-13) at pH values greater than 8.

A study of the anion inhibition of Cobalt(II) HCAC CO_2 hydration activity reported a loss of inhibitory power of anions at high pH(21). Further scrutiny of the high pH anion inhibition of metal substituted carbonic anhydrase is certainly in order. If, as it appears, high pH anion activity inhibition is a property confined to the natural substrate of the Zinc(II) enzyme, the implications for conclusions drawn from studies of model substrates and metal substituted enzymes need to be carefully scrutinized.

The electronic structure of the bound species and transition states of carbonic anhydrase have remained elusive. We wish to outline structures of the bound states of carbonic anhydrase which rationalize the data presented above and which, we believe, constitut a catalytic mechanism not previously considered.

We propose that the catalysis of Zinc(II) carbonic anhydrase proceeds through a variety of four- and five-coordinate Zinc(II) species in rapid equilibrium. The extensive and stimulating series of spectral studies recently undertaken by Bertini and colleagues (30-32) has strongly suggested to us that the expansion of the coordination sphere of metal-substituted carbonic anhydrase is an important mode of accommodation of anions in the enzyme active site. X-ray evidence(23) and ^{35}Cl NMR data(34) as well as other spectral studies, have localized the binding of simple anionic inhibitors to the inner coordination sphere of the zinc atom in the acidic form of carbonic anhydrase.

Given this placement of anionic inhibitors in the inhibited acidic form of carbonic anhydrase, and based on the strict retention of order of binding affinity of anions to the low pH form of BCA versus the binding to the high pH ES complex (Table III) we propose that this mode of binding characterizes anionic inhibition wherever it is observed. Since this inner sphere binding is so dramatically influenced by pH, we further propose that another inner sphere effect, the equilibrium conversion of a Zn-OH, is responsible for the observed pH effects in anion binding and in catalysis. We represent the binding of anions to the acidic form of carbonic anhydrase as

$$Zn-OH_2 + I \rightleftharpoons Zn\begin{matrix} I \\ OH_2 \end{matrix}$$

Any similar equilibrium at high pH, when Zn-OH is the predominant species, then, is negligible. Evidently, addition of an anionic ligand to Zn-OH increases the electron density at zinc beyond its capacity to function as a Lewis acid.

It has been noted previously(22) that HCN and other ligands supplying an exchangeable proton bind to carbonic anhydrase by donation of a proton and association as an anion. We attribute the ability of HCO_3^- to bind to the basic form of carbonic anhydrase to this effect. The relevant equilibrium is

$$Zn-OH + HCO_3^- \rightleftharpoons Zn\begin{pmatrix} O\cdots H\cdots O \\ O-C \end{pmatrix}=O$$

The reduction in electron density accompanying this proton donation permits expansion of the coordination sphere of zinc.

An important consideration in any proposed scheme is the rationalization of the uncompetitive inhibition of carbonic anhydrase at high pH. We suggest that CO_2 itself acts as an electron sink for Zn-OH, and thus its presence facilitates anion binding, acting in a Lewis acid sense much as the HCO_3^- proton participates as a Brønsted acid;

It is intriguing to consider this equilibrium when I is HCO_3^-. The five coordinate species formed closely resembles (but is not identical to) two bicarbonate molecules bound to zinc. This structure may have a considerable bearing on experimental data previously interpreted as representing two bicarbonate binding sites on carbonic anhydrase(35). This mode of binding accounts for the mixed mode of inhibition observed in HCO_3^- inhibition of BCA catalyzed CO_2 hydration.

The catalytic steps of carbonic anhydrase can, we feel, be analyzed along closely related lines. A consequence of the increasing electron density on zinc during binding of a fifth ligand is an increase in electron density on the hydroxyl ligand, as noted above. This process is equally well represented as an increase in nucleophilicity of a hydroxyl with a pK_B near neutrality when a substrate CO_2 is present. In the catalytic steps, we suggest that a water molecule, entering the fifth coordination site, (possibly with general base assistance from the conserved active site histidine or a buffer molecule, with one or more intervening solvent molecules) both initiates the nucleophilic attack on CO_2 and stabilizes the developing carboxylate anion;

$$\begin{array}{c}\text{B}\quad\text{H}\\\quad\searrow\\\text{O—H}\\\text{Zn—O}\searrow\\\quad\text{H}\\\quad\text{O—H}\\\quad\diagup\\\text{H}'\end{array}\quad\overset{\text{B----H}}{\underset{\begin{array}{c}\text{O}\\\parallel\\\text{C}\\\parallel\\\text{O}\end{array}}{}}\rightleftharpoons\begin{array}{c}\text{O—H}\\\text{Zn}\\\text{O----C}\\\text{H}'\quad\text{O}\\\quad\diagdown\\\text{O----H}'\\\text{H}'\end{array}\rightleftharpoons\begin{array}{c}\text{BH}^+\\\text{O—H}\\\text{Zn}\searrow\qquad\text{O}\\\qquad\text{O—C}\\\qquad\diagdown\text{O}\\\text{H}\qquad\text{O}\quad\text{H}'\\\quad\diagdown\quad\diagup\\\quad\text{H}'\end{array}$$

The productive binding of HCO_3^- to $Zn-OH_2$ is then established as the reverse of the above scheme. We defer consideration of the exact order of covalent bond formation and proton transfer, and present the above as a working hypothesis. Clearly, however, if one or more of the steps involved in the above scheme represents the rate determining step of BCA catalyzed CO_2 catalysis, as seems likely, then the observed primary isotope effect is readily explicable. Earlier steps, also involving proton transfer and solvent reorganization would account similarly for the isotope effect in K_M. The identity of the group B and the proton transfer group represented by the lower water molecule is a matter for further consideration.

The constraints on solvent orientation and the solvent reorganization accompanying the making and breaking of the C-O bond in the above scheme is entirely consistent with the significant temperature dependence of k_{cat}. K_M, by this criterion, would represent binding prior to the most significant alterations in active site solvent structure.

The high pH anionic inhibition of BCA catalyzed CO_2 hydration is, we believe, the first well characterized example of uncompetitive inhibition in the class of single substrate enzymes. It is worth noting that carbonic anhydrase is in fact not a single substrate enzyme. Its second substrate is solvent water, a substrate of high and essentially invariant concentration. The postulated simultaneous operation of solvent water as ligand, nucleophile, and proton transfer agent outlined above is eloquent testimony to the importance of solvent in enzyme catalysis.

Acknowledgments. The authors thank Mr. Donald B. Moore for his diligent assistance in all aspects of computer interfacing and programming and Drs. Conrad T.O. Fong, Simo Sarkanen and John E. Stein for helpful discussions.

REFERENCES

1. H. Devoe and G. B. Kistiakowsky, J. Am. Chem. Soc. 83:274

(1961).
2. B. -H. Jonsson, H. Steiner, and S. Lindskog, FEBS Lett. 64:310 (1976).
3. Y. Pocker and D. W. Bjorkquist, Biochemistry, 16:5698 (1977); J. Am. Chem. Soc. 99:6537 (1977).
4. Y. Pocker and J. E. Meany, J. Am. Chem. Soc. 87:1809 (1965)
5. Y. Pocker and J. E. Meany, Biochemistry, 4:2535 (1965).
6. Y. Pocker and J. E. Meany, Biochemistry, 6:239 (1967).
7. Y. Pocker and J. E. Meany, J. Am. Chem. Soc. 89:631 (1967).
8. Y. Pocker and J. E. Meany, J. Phys Chem. 71:3113 (1967).
9. Y. Pocker and J. T. Stone, J. Am. Chem. Soc. 87:5497 (1965).
10. Y. Pocker and J. T. Stone, Biochemistry, 6:668 (1967).
11. Y. Pocker and J. T. Stone, Biochemistry, 7:2936 (1968).
12. Y. Pocker and J. T. Stone, Biochemistry, 7:3021 (1968).
13. Y. Pocker and J. T. Stone, Biochemistry, 7:4139 (1968).
14. Y. Pocker and L. J. Guilbert, Biochemistry, 11:180 (1972).
15. Y. Pocker and L. J. Guilbert, Biochemistry, 13:70 (1974)
16. Y. Pocker and S. Sarkanen, Abstract 10th Meeting of Federation of European Biochemical Societies, Abstr. 782 (1975).
17. Y. Pocker and S. Sarkanen, Biochemistry, 17:1110 (1978).
18. Y. Pocker and S. Sarkanen, Adv. Enzymol. Relat. Areas. Mol. Biol. 47:149 (1978)
19. F. J. W. Roughton, Harvey Lectures, 39:96 (1943).
20. J. C. Kernohan, Biochim. Biophys. Acta, 96:304 (1965).
21. S. Lindskog, Biochemistry, 5:2641 (1966).
22. J. E. Coleman, J. Biol. Chem. 242 (1967).
23. R. G. Khalifah, J. Biol. Chem. 246:2561 (1971).
24. H. Steiner, B. -H. Jonsson and S. Lindskog, Eur. J. Biochem. 59:253 (1975).
25. D. W. Bjorkquist, Ph. D. Dissertation, University of Washington (1975).
26. Y. Pocker and D. W. Bjorkquist, Abstract 10th Meeting of Federation of European Biochemical Societies, Abstr. 783 (1975).
27. B. -H. Jonsson, H. Steiner, and S. Lindskog, Abstract 10th Meeting of Federation of European Biochemical Societies, Abstr. 784 (1975).
28. H. Steiner, B. -H Jonsson, and S. Lindskog, FEBS Lett. 62:16 (1976).
29. C. K. Tu and D. N. Silverman, J. Am. Chem. Soc. 97:5935 (1975).
30. I. Bertini, C. Luchinat and A. Scozzafava, J. Chem. Soc. Dalt. Trans. 20:1962 (1977).
31. I. Bertini, G. Canti, C. Luchinat, and A. Scozzafava, Inorg. Chim. Acta, 36:9 (1979).
32. I. Bertini, E. Borghi, and C. Luchinat, J. Am. Chem. Soc. 101:7071 (1980).
33. I. Vaara, "The Molecular Structures of Human Carbonic Anhydrase C and Inhibitor Complexes, Inaugural Disser-

tation", UVIC-B22-2-Uppsala University (1974).
34. R. L. Ward, Biochemistry, 9:2447 (1970).
35. P. L. Yeagle, C. H. Lochmuller and R. W. Henkens, Proc. Natl. Acad. Sci. U.S.A., 72:454 (1975).

PYRIDINIUM-N-PHENOXIDE BETAINE DYES AS SOLVENT POLARITY INDICATORS.
SOME NEW FINDINGS

C. Reichardt, E. Harbusch, and R. Müller

Department of Chemistry, University of Marburg
D-3550 Marburg, Fed. Rep. Germany

Abstract - Empirical parameters of solvent polarity seem nowdays as useful for the prediction of solvent effects as Hammett's substituent constants for the estimation of substituent effects on chemical reactions. The extreme sensitivity of the position of the long-wavelength UV/VIS absorption band of pyridinium-N-phenoxide betaine dyes not only changes in temperature (thermo-solvatochromism), external pressure (piezo-solvatochromism), and the introduction of substituents, but also to small changes in solvent polarity (negative solvatochromism) has been used to establish a comprehensive empirical scale of solvent polarity, called the E_T-scale. In addition, the influence of solvents and substituents on the chemical reactivity of the pyridinium-N-phenoxides has now also been studied.

1. INTRODUCTION

Whenever a chemist wishes to carry out a certain chemical reaction, he not only has to look at the right reaction partners, the proper reaction vessels and the appropriate reaction temperature, but also one of the most important factors is the selection of a suitable solvent for the planned reaction. This often determines the success or failure of the reaction under study. Since there are now about 300 different organic solvents available (apart from the infinite number of possible solvent mixture), the chemist needs, in addition to his experience and intuition, some general rules for this often difficult choice.

One of the important criteria in this connection is a solvent property generally known as "solvent polarity". As every chemist knows, solvents may have a strong influence on reaction rates and on the position of chemical equilibria as well as on the position of spectral absorption data. Solvent effects on organic reactivity have been studied for more than a century(1).

As early as 1862, Berthelot at the Collège de France in Paris found, together with his coworker Saint-Gilles, that the rate of the esterification of acetic acid with ethanol is strongly influenced by the solvent(2). He noticed, that the esterification is disturbed or retarded if the reaction is carried out in various neutral solvents, which do not participate in the reaction.

This work was continued later on by Menshutkin in Petersburg in 1890, who studied the quaternization of tertiary amines with alkyl halides - a reaction which now is commonly known as the 'Menshutkin reaction'. Menshutkin found, for example, that the reaction rate between triethylamine and ethyl iodide increases with solvent polarity, up to a factor of 742 in benzyl alcohol relative to n-hexane as solvent(3).

However, the property "solvent polarity" is difficult to define precisely and to express quantitatively.

The simplicity of idealized electrostatic solvation models has led to the use of dielectric constant (ε) and of the permanent dipole moment (μ) as parameters of the so-called solvent polarity. However, the dielectric constant describes only the change in the electric field intensity that occurs between the plates of a condenser, when the latter is removed from vacuum and placed into a solvent. This induces a dipole moment in nonpolar solvent molecules and dipolar molecules are aligned. Hence, the dielectric constant describes only the ability of a solvent to separate electrical charges and orient its dipolar molecules. The intermolecular forces between solute and solvent molecules are, however, much more complicated: in addition to the non-specific coulombic, directional, inductive and dispersion interactions, can also be present specific hydrogen bond, electron-pair donor (EPD)/electron-pair acceptor (EPA), and solvophobic interactions in solutions.

Hence, from a more practical point of view, it seems more advantageous to include in the term "solvent polarity" the <u>overall solvation ability</u> of a solvent for reactants, activated complexes, and products, as well as for molecules in the electronic ground and excited states. This in turn depends on the action of <u>all</u> possible, specific and non-specific intermolecular forces between solute and solvent molecules(1).

It stands to reason that solvent polarity, as defined, cannot be described quantitatively by a single physical parameter such as the dielectric constant or the dipole moment. This has stimulated attempts to introduce pure empirical scales of solvent polarity(4,5). The procedure is to start with well-known, easily measurable, and strongly solvent-dependent processes, which are investigated in the greatest number of commonly used solvents. For example, one can investigate the rate or the position of equilibrium of a carefully selected reference reaction, or the spectral absorption of a suitable solvatochromic standard dye. From the solvent-dependent rate or equilibrium constants, or from the absorption maxima it is then possible to derive empirical parameters of solvent polarity, which are believed to provide a more comprehensive measure of the overall solvation ability of the solvents than do their individual physical data.

This kind of procedure is very common in organic chemistry. For example, the well-known Hammett equation, used for the estimation of substituent effects on rates and equilibria, is likewise based on an empirical reference process - the dissociation of substituted benzoic acid in water at 25°C(6).

The relationships between these empirical substituent or solvent parameters and the substituent- or solvent-dependent processes under study take usually the form of a linear free-energy (LFE) relationship(5,7,8).

2. SOLVATOCHROMISM OF PYRIDINIUM-N-PHENOXIDE BETAINES-E_T-VALUES

One of the attempts to introduce empirical parameters of solvent polarity uses the light absorption of solvatochromic dyes such as shown in Figure 1.

The pyridinium-N-phenoxide betaine dye shown in Figure 1 exhibits a negative solvatochromic $\pi-\pi^*$ absorption with intramolecular charge-transfer character. The outstanding sensitivity of this dye to small changes in the surrounding medium has been used by Dimroth, Reichardt et al. to establish an empirical scale of solvent polarity, called the E_T-scale†. The E_T-values are simply defined as the

---

† Since the solvatochromic standard betaine dye (see Figure 1) was numbered 30 in the first publication(9), its molar transition energies were also designated at $E_T(30)$-values. This transition energy E_T or $E_T(30)$ can be used either directly or in the form of a relative measure RPM (i.e. Relative Polarity Measure defined as follows(5):

$$RPM = \frac{E_T(n-hexane) - E_T}{E_T(n-hexane)}$$

The dimensionless RPM-values are then equal to zero for n-hexane, the least polar solvent. RPM-values are negative for negative solvatochromic dyes; however, for correlation analysis it is sufficient to use absolute |RPM| - values(5).

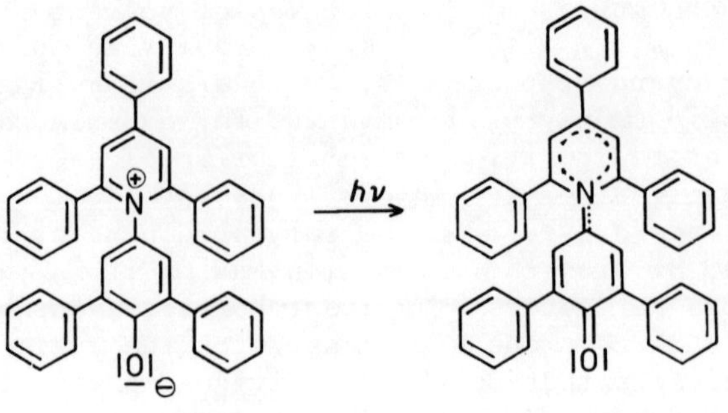

Fig. 1. Solvent-dependent intramolecular charge-transfer absorption of 2,4,6-triphenyl-N-(2,6-diphenyl-4-phenoxide)-pyridinium betaine, proposed as solvent polarity indicator by Dimroth, Reichardt et al.(9).

μ_G and μ_E are the permanent dipole moments in the electronic ground and first excited state of a betaine dye with two t-butyl groups instead of the two 2,6-diphenyl groups next to the oxygen atom(17,18).

$\mu_G \approx 14.8$ D

$\mu_G \approx 6.1$ D → $\Delta\mu \approx 9$ D

Solvent	$C_6H_5OCH_3$	CH_3COCH_3	$i\text{-}C_5H_{11}OH$	C_2H_5OH	CH_3OH
λ [nm]	769	677	608	550	515
E_T [kcal/mol]	37.2	42.2	47.0	51.9	55.5
Solution colour	yellow	green	blue	violet	red

$$E_T [\text{kcal/mol}] = h \cdot c \cdot \tilde{v} \cdot N_L = 2.859 \cdot 10^{-3} \cdot \tilde{v} \, [\text{cm}^{-1}]$$

transition energy for the long-wavelength solvatochromic absorption band of the dissolved betaine dye, measured in kcal mol^{-1} or kJ mol^{-1}. A high E_T-value corresponds to high solvent polarity or, in other words, to high overall solvation capability of the solvent. At present E_T-values are known for more than 160 organic solvents(9-12) and for many binary solvent/water mixture(13,14), and they have found wide application in single (5,7) and multiparameter(15,16) correlation equations.

These E_T-values provide a very useful empirical characterization of the solvent polarity owing to the exceptionally large displacement of the long-wavelenght absorption band by solvents (see Figure 2).

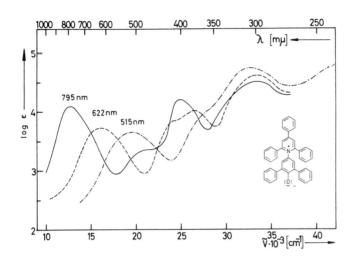

Fig. 2. UV/Vis absorption spectra of 2,4,6-triphenyl-N-(2,6-diphenyl-4-phenoxide)-pyridinium betaine in 1,4-dioxane (———), acetonitrile (-------), and methanol (-.-.-.-) at 25°C(9).

For example, the solvatochromic band is situated at 795 nm in 1,4-dioxane, at 622 nm in acetonitrile, and at 515 nm in methanol. With a hypsochromic shift of more than 350 nm in the case of a solvent change from diphenyl ether (810 nm) to water (453 nm), this betaine dye holds to date the world record in solvatochromism (ΔE_T = 28 kcal mol^{-1} = 117 kJ mol^{-1}).

Since the greater part of this solvatochromic range lies within the visible region, it is even possible to make a visual estimation of the solvent polarity. The solution colour is red in methanol, violet in ethanol, green in acetone, blue in isoamyl alcohol, and greenish-yellow in anisole. By applying suitable binary solvent mixtures, nearly every colour of the visible spectrum can be obtained. This has been used, for example, for the spectro-photometric determination of water in organic solvents(13,14).

In order to overcome solubility problems in nonpolar organic solvents, lipophilic alkyl-substituted betaines as secondary standard dyes have been synthesized, some of which are shown below(10,11).

The synthetic route leading to these betaine dyes allows one to modify these solvatochromic dyes over a wide range, tailored for our special purpose. Just out of interest, a dimeric betaine dye with a zero dipole moment is also shown, which again exhibits a strong negative solvatochromism(10).

It is believed that the use of the pyridinium-N-phenoxide betaine dyes as solvent polarity indicators follows from some peculiar properties of these dyes, summarized in Figure 3.

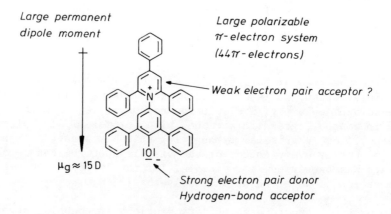

Fig. 3. Ground state properties of the 2,4,6-triphenyl-N-(2,6-diphenyl-4-phenoxide)-pyridinium betaine.

Our standard betaine dye exhibits

- a large permanent dipole moment of about 15 D(17,18), suitable for the registration of dipole-dipole and dipole-induced dipole interactions;
- a large polarizable π-electron system, consisting of 44 π-electrons, suitable for the registration of dispersion interactions;
- and with the phenolic oxygen atom it exhibits a highly basic electron-pair donor (EPD) centre, suitable for hydrogen-bond interactions with protic solvents.

The positive charge of the cationic solvation centre is partly delocalized over the pyridinium ring and shielded by the phenyl substituents. Therefore, the E_T-values depend strongly on the electrophilic solvation power of the solvents, i.e. on their hydrogen-bond donor (HBD) ability or Lewis acidity, rather than on their nucleophilic solvation capability, for which the Donor Numbers of Gutmann (19) are the better empirical solvent parameters.

Direct experimental proof for the existence of hydrogen-bonding between the phenolic oxygen atom and protic solvents comes - at least for the crystalline state - from an X-ray analysis of a p-bromo-substituted betaine dye, carried out by R. Allmann(20). Some results of this X-ray analysis are given in Figure 4.

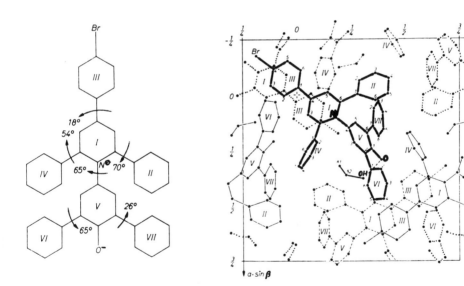

Dihedral angles between the aryl groups

Projection of an unit cell onto the x,y-plane

Fig. 4. Results of an X-ray analysis of 2,6-diphenyl-4-(4-bromophenyl)-N-(2,6-diphenyl-4-phenoxide)-pyridinium-betaine-monoethanolate(20).

The betaine dye was recrystallized from an ethanol/water mixture (2:1 by volume) and, therefore, the crystal lattice includes ethanol molecules which are hydrogen-bonded to the phenolic oxygen atom of the betaine molecule. The small O-O distance of only 271 pm is indicative of the presence of a hydrogen-bond.

In addition it should be mentioned, that pyridinium-N-phenoxide betaines are not only sensitive to changes in solvent polarity, but also their long-wavelenght absorption band depends on changes in temperature and pressure, as well as on the introduction of substituents in the peripheric phenyl group.

3. THERMO-SOLVATOCHROMISM

Figure 5 demonstrates the thermochromism of our betaine dye, dissolved in ethanol(21). A temperature rise of only 50°C leads to a bathochromic shift of the long-wavelength absorption band of 18 nm. This thermochromism can be easily visualized by means of an ethanolic betaine solution in a test tube. At -78°C the betaine solution is red-coloured, and at +78°C the solution is blue-violet, which corresponds to a wavelength shift of 55 nm.

This new kind of thermochromism(22), which has been called thermo-solvatochromism(1), is obviously caused by increased stabilization of the dipolar electronic ground state of the betaine dye relative to its excited state with decreasing temperature.

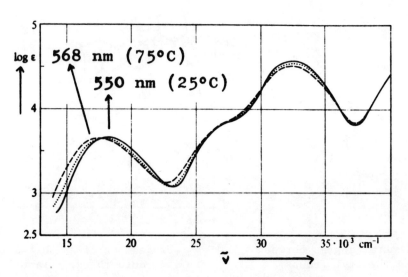

Fig. 5. UV/Vis absorption spectra of 2,4,6-triphenyl-N-(2,6-diphenyl-4-phenoxide)-pyridinium betaine in ethanol at 25°C (─────), 45°C (······), and 75°C (-------)(21).

This is due to the stronger solute-solvent interactions at low temperatures. In other words, the lower the temperature, the higher the corresponding E_T-values, and the better the solvation capability of the solvent. That means, solvent polarity is temperature-dependent!

4. PIEZO-SOLVATOCHROMISM

Recently, Tamura and Imoto(23) as well as Kelm and coworkers (24) have observed pressure effects on the solution spectra of our pyridinium-N-phenoxide betaines, which are illustrated in Figure 6.

In all solvents used, they found a hypsochromic shift of the long-wavelength absorption band with increasing pressure. The observed shift runs up to 10 nm by raising the pressure from 1 to about 2000 atmospheres, applied to a betaine solution in acetone(23).

On the supposition that this "piezo-solvatochromism" results from a better ground state stabilization of the betaine dye with increasing pressure, it can be stated that, the higher the pressure, the higher the E_T-value, and the more polar the solvent. In other words, solvent polarity is also pressure-dependent!

However, this conclusion must be considered very carefully, because the observed effects are small and other factors such as conformational changes of the betaine molecule may also contribute to the observed spectral changes.

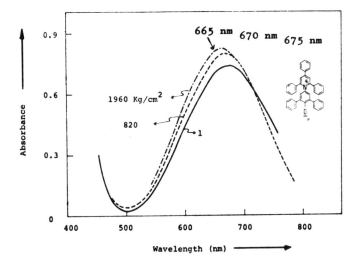

Fig. 6. UV/Vis absorption spectra of 2,4,6-triphenyl-N-(2,6-diphenyl-4-phenoxide)-pyridinium betaine at 25°C in acetone at 1 kg cm² (————), 820 kg cm² (--------), and 1960 kg cm² (-·-··-··-)(23).

5. INFLUENCE OF SUBSTITUENTS

Not only changes of solvent, temperature, and pressure can affect the betaine absorption spectrum, but also the introduction of substituents in the peripheric phenyl groups of the betaine dye (see Figure 7).

Introduction of a nitro group in the p-position of the p-phenyl groups leads, as expected, to a bathochromic shift of the long-wavelength absorption band, whereas electron-donating substituents such as the p-methoxy group lead to a hypsochromic shift.

This substituent effect on the betaine spectrum can be quantitatively described by means of a Hammett correlation, as shown in Figure 8.

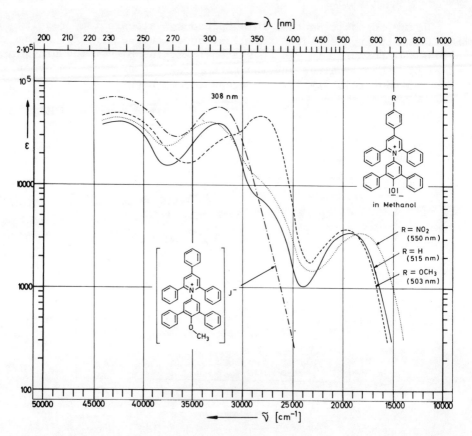

Fig. 7. UV/Vis absorption spectra of 2,4,6-triphenyl-N-(2,6-diphenyl-4-phenoxide)-pyridinium betaine without a p-substituent (———), with a p-nitro group (·······), and with a p-methoxy group (-----), measured in methanol(25).

An interesting feature is the observation, that the sensitivity of the betaine absorption to changes of substituents is solvent-dependent too. For each solvent used, a separate correlation line with a slightly different slope has been found. The gradient of the slope is largest in the non polar solvent chloroform ($\rho = -3.64$) and lowest in the more polar solvent acetone ($\rho = -2.72$)(25).

This results seems at least qualitatively understandable. Introduction of a substituent in the betaine dye dissolved in a polar solvent with strong solute-solvent interactions, should lead to a

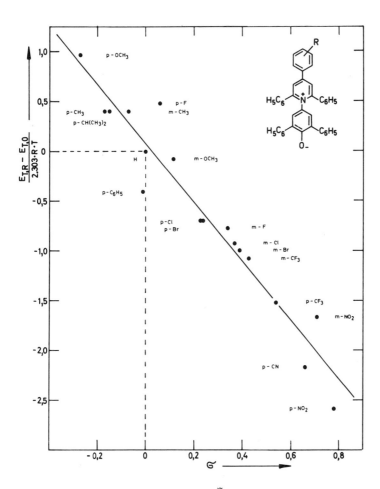

Fig. 8. Hammett correlation between σ^*-values and the modified transition energies of the longest wavelength $\pi-\pi^*$-absorption band of the substituted pyridinium-N-phenoxides in methanol at 25°C: $(E_{T,R} - E_{T,0})/2.3 \cdot RT = -2.97 \cdot \sigma + 0.08$ (n = 18; correlation coefficient r = -0.962)(25).

weak perturbation only, in contrast to the behaviour in a nonpolar solvent.

Substituents can be introduced not only into the betaine solute, but also into the surrounding solvent molecules. Let us consider a betaine dye solution in various alcohols R-OH with different alkyl groups R. An enhanced hypsochromic shift of the solvatochromic absorption band with increasing hydrogen-bond donor ability of the alcohols should be obtained. This is indeed the case(25). As Figure 9 shows, even exists a linear correlation between the modified E_T-values, measured in nine different alcohols, and Taft's polar substituents constants for the alkyl groups R of these alcohols(25).

This behaviour again demonstrates how sensitive this betaine dye is to small changes in the electrophilicity (or Lewis acidity) of the surrounding medium. By the way, this linear correlation can be used for the calculation of new σ^*-values(25).

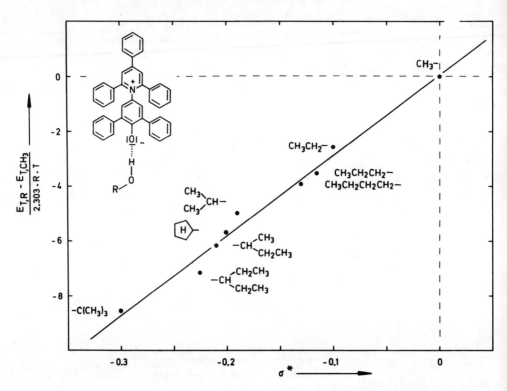

Fig. 9. Taft correlation between σ^*-values and the modified transition energies of the longest wavelength $\pi-\pi^*$-absorption band of the pyridinium-N-phenoxide betaine, measured at 25°C in different aliphatic alcohols R-OH: $(E_{T,R} - E_{T,CH_3})/2.3 \cdot RT = 29.2 \cdot \sigma^* + 0.03$ (n = 9; r = 0.992)(25).

6. CHEMICAL REACTIVITY

Finally, the influence of solvents and substituents on the chemical reactivity of the standard betaine dye should be mentioned.

The negatively charged oxygen atom of the betaine dye can act as a nucleophilic centre: on the one hand, it can act as a hydrogen-bond acceptor in protic solvents, on the other hand, it can react with methyl iodide in a typical S_N2 reaction to give the corresponding colourless pyridinium salt (see following Scheme).

$$R^1\text{-}\underset{R^2}{\bigcirc}\text{-}\underset{C_6H_5}{\overset{C_6H_5}{\bigcirc}}\text{-}{}^+N\text{-}\underset{C_6H_5}{\overset{C_6H_5}{\bigcirc}}\text{-}\bar{O} + CH_3I \xrightarrow{k_2 \atop 25°C}$$

$$\left[R^1\text{-}\underset{R^2}{\bigcirc}\text{-}\underset{C_6H_5}{\overset{C_6H_5}{\bigcirc}}\text{-}{}^+N\text{-}\underset{C_6H_5}{\overset{C_6H_5}{\bigcirc}}\text{-}\bar{O}CH_3\right] I^- \qquad \frac{k_2(R_1 = OCH_3\,;\,R_2 = H)}{k_2(R = NO_2\,;\,R_2 = H)} \approx 3$$

$$(\text{in } CHCl_3)$$

This alkylation reaction has been studied as a function of 17 substituents and 15 solvents(11,26).

Although the distance between substituents and nucleophilic oxygen atom consists of no less than 12 bonds, a rate factor of 3 was nevertheless obtained for the alkylation of m- and p-substituted betaine dyes with methyl iodide in chloroform as solvent.

As shown in Figure 10, a very good correlation exists between the relative rate constants and Hammett's σ-values.

Here again the sensitivity of the betaine dye to changes of substituents – as given by its different chemical reactivity – is solvent-dependent. This sensitivity is lower in the more polar solvents, and vice versa. For example, the Hammett reaction constant ρ for this alkylation reaction, measured in polar acetonitrile as solvent, is with –0.15 considerably smaller than the corresponding ρ-value measured in the less polar solvent chloroform, which amounts to –0.40(26).

In addition, the alkylation rate of the unsubstituted betaine dye has been measured in 15 solvents of different polarity (see the Scheme in the following page and Table I)(11,26). We have done this in the hope of getting a complementary solvent scale by means of kinetic instead of spectroscopic measurements with the same standard betaine dye. Indeed, the rate of alkylation of the

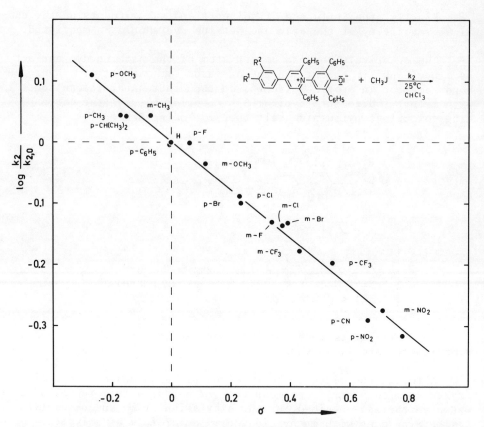

Fig. 10. Hammett correlation between σ-values and the logarithms of the relative rate constants of the S_{N2} alkylation reaction of substituted pyridinium-N-phenoxides with methyl iodide in chloroform at 25°C: lg $(k_2/k_{2,0})$ = $-0.40 \cdot \sigma + 0.0003$ (n = 18; r = -0.993)(26)

$$\frac{k_2(\text{DMSO})}{k_2(\text{EtOH})} = 2530$$

betaine dye with methyl iodide is solvent-dependent, being faster in dipolar aprotic solvents (e.g. dimethylsulfoxide) by a factor of about 2500 compared with protic solvents (e.g. ethanol)

To our disappointment, but not unexpected, there is no correlation at all between the solvent-dependent alkylation rate constants and the spectroscopic E_T-scale of solvent polarity (see Figure 11), in spite of the fact that both solvent scales have been derived from the same betaine dye. There is also no correlation of the alkylation rate constants with other single parameters of solvent polarity.

This failure can be explained as follows: application of the Hughes-Ingold rules (1,27) for solvent effects on S_N2 reactions does not lead to a clear statement in this case (see Scheme in the next page).

The dispersal of the negative charge between the oxygen and the iodine atom during the activation process should lead to a

Table I. Rate constants of the S_N2 alkylation reaction between 2,4,6-triphenyl-N-(2,6-diphenyl-4-phenoxide)-pyridinium betaine with methyl iodide, measured in 15 solvents at 25°C(11,26).

Solvent	E_T $\lvert\text{kcal mol}^{-1}\rvert$	$k_2 \cdot 10^4$ $\lvert\text{mol}^{-1}\text{ s}\rvert$	lg k_2
Formamide	56.6	2.55	-3.59
Methanol	55.5	1.30	-3.89
Ethanol	51.9	0.77	-4.11
1-Butanol	50.2	1.68	-3.77
Acetonitrile	46.0	271	-1.57
Dimethylsulfoxide	45.0	1948	-0.710
2-Methyl-2-propanol	43.9	14.1	-2.85
N,N-Dimethylformamide	43.8	1205	-0.919
Acetone	42.2	1140	-0.943
Dichloromethane	41.1	79.1	-2.10
Chloroform	39.1	52.4	-2.28
Ethyl acetate	38.1	807	-1.09
Chlorobenzene	37.5	118	-1.93
Tetrahydrofuran	37.4	926	-1.03
1,4-Dioxane	36.0	104	-1.98

[a] 1kcal = 4.187 kJ

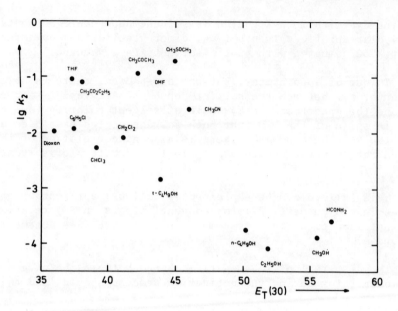

Fig. 11. Non-correlation between the E_T-values and the logarithms of the rate constants of the S_N2 alkylation reaction of 2,4,6-triphenyl-N-(2,6-diphenyl-4-phenoxide)-pyridinium betaine with methyl iodide, measured in 15 solvents at 25°C (26).

decrease of the reaction rate with increasing solvent polarity. But the simultaneous separation of unlike charges between the positive nitrogen atom and the negative iodine atom should result in a rate increase with increasing solvent polarity. The contribution of these two counteracting effects to the net solvent effect is not clear and may be even different from solvent to solvent.

The rate deceleration in protic solvents by a factor of about 2500 compared with the alkylation rate in dipolar aprotic solvents demonstrates that hydrogen-bonding to the phenolic oxygen atom not only raises the spectroscopic E_T-values, but also reduces considerably the nucleophilicity of the phenoxide part against methyl iodide as alkylating agent.

7. CONCLUSION

In conclusion, it can be said that the outstanding sensitivity of the spectral absorption and chemical reactivity of our pyridinium-N-phenoxide betaine dyes to small changes in solvent, temperature, pressure, and substituents makes these dyes a very useful class of compounds. They are not only solvatochromic compounds, but exhibit also the phenomena of thermo and piezo-solvatochromism. Their use for setting up different reaction series in order to get Linear Free-Energy Relationships has been demonstrated by the fact that the same betaine dye can be used not only for the introduction of a spectroscopic solvent polarity scale, the so-called E_T-scale, but also for the establishment of kinetic and spectroscopic scales of substituents.

It seems that empirical parameters of solvent polarity such as the E_T-scale are nowadays as useful for the prediction of solvent effects as Hammett's substituent constants for the prediction of substituent effects with one and the same standard molecule.

The behaviour of these betaine dyes may be compared to that of the Princess in Hans Christian Andersen's fairy-tale "The Princess and the Pea"(28). As one perhaps remember, that Princess was so sensitive to her surroundings, that she was able to feel a pea through 20 mattresses and 20 eider-down quilts in her bed.

Aknowledgment. Part of the work reported was done in cooperation with Professor K. Dimroth at Marburg. I should like to thank Professor K. Dimroth as well as my coworkers Reinhard Müller and Erwin Harbusch.

Financial support of the Fonds der Chemischen Industrie, Frankfurt, is gratefully acknowledged.

REFERENCES

1. C. Reichardt, "Solvent Effects in Organic Chemistry", Verlag Chemie, Wienheim, New York (1979).
2. M. Berthelot and C. Péan de Saint-Gilles, Ann. Chim. Phys. 3. Sér., 65:385 (1862); 66:5 (1862); 68:225 (1863).

3. N. Menshutkin, Z. Phys. Chem. 1:611 (1887); 5:589 (1890); 34:157 (1900).
4. C. Reichardt, Angew. Chem. 77:30 (1965); Angew. Chem. Internat. Edit. Engl. 4:29 (1965).
5. C. Reichardt, Angew. Chem. 91:119 (1979); Angew. Chem. Internat. Edit. Engl. 18:98 (1979).
6. L. P. Hammett, J. Am. Chem. Soc. 59:96 (1937); Trans. Farad. Soc. 34:156 (1938); L. P. Hammett, "Physical Organic Chemistry", McGraw-Hill, New York (1970).
7. C. Reichardt and K. Dimroth, Fortschr. Chem. Forsch. 11:1 (1968).
8. N. B. Chapman and J. Shorter, eds., "Advances in Linear Free Energy Relationships", Plenum Press, London, New York (1972). N. B. Chapman and J. Shorter, eds., "Correlation Analysis in Chemistry - Recent Advances", Plenum Press, London, New York (1978).
9. K. Dimroth, C. Reichardt, T. Siepmann, and F. Bohlmann, Liebigs Ann. Chem. 661:1 (1963).
10. K. Dimroth and C. Reichardt, Liebigs Ann. Chem. 727:93 (1969); C. Reichardt, Liebigs Ann. Chem., 752:64 (1971).
11. C. Reichardt and E. Harbusch, unpublished results.
12. J. Hormadaly and Y. Marcus, J. Phys. Chem. 83:2843 (1979).
13. Z. B. Maksimovic, C. Reichardt, and A. Spiric, Z. Anal. Chem. 270:100 (1974).
14. K. Dimroth and C. Reichardt, Z. Anal. Chem. 215:344 (1966).
15. T. M. Krygowski and W. R. Fawcett, J. Am. Chem. Soc. 97: 2143 (1975); Austr. J. Chem. 28:2115 (1975); Canad. J. Chem. 54:3283 (1976).
16. I. A. Koppel and V. A. Pal'm, Reakts. Sposobnost. Org. Soedin. 8:291 (1971); Organic Reactivity, 8:296 (1971); I. A. Koppel and V. A. Pal'm, "The Influence of the Solvent on Organic Reactivity" in N. B. Chapmann and J. Shorter, eds., "Advances in Linear Free Energy Relationships", Plenum Press, London, New York, Chapter 5, p. 203 ff. (1972).
17. C. Reichardt and A. Schweig, Z. Naturforsch. 21a:1373 (1966).
18. W. Liptay, H.-J. Schlosser, B. Dumbacher, and S. Hünig, Z. Naturforsch. 23a:1613 (1968).
19. V. Gutmann and A. Scherhaufer, Monatsh. Chem. 99:335 (1968); V. Gutmann, "The Donor-Acceptor Approach to Molecular Interactions", Plenum Press, New York (1978).
20. R. Allmann, Z. Kristallogr. 128:115 (1969).
21. K. Dimroth, C. Reichardt, and A. Schweig, Liebigs Ann. Chem. 669:95 (1963).
22. For a review on thermochromism see J. H. Day, Chem. Rev. 63:65 (1963).
23. K. Tamura, Y. Ogo, and T. Imoto, Chem. Lett.(Tokyo) 1973: 625; K. Tamura and T. Imoto, Bull. Chem. Soc. Japan,

       48:369 (1975).
24. J. v. Jouanne, D. A. Palmer, and H. Kelm, Bull. Chem. Soc. Japan, 51:463 (1978).
25. C. Reichardt and R. Müller, Liebigs Ann. Chem. 1976:1937.
26. C. Reichardt and R. Müller, Liebigs Ann. Chem. 1976:1953.
27. E. D. Huges and C. K. Ingold, J. Chem. Soc.(London) 1935: 244; Trans. Farad. Soc. 37:603, 657 (1941).
28. H. C. Andersen, "Complete Fairy-Tales and Stories" (translated by E. Haugard), Gollantz, London (1974).

SOME APPLICATIONS OF LIQUID CRYSTALS IN ORGANIC CHEMISTRY

G. Gottarelli and B. Samorì

Istituto di Chimica degli Intermedi
Viale Risorgimento 4, Università di Bologna, Italy

Abstract - Some applications of thermotropic liquid crystals (LC) as anisotropic solvents for stereochemical, spectrochemical and reactivity studies are presented. The study of cholesteric mesophases induced in a nematic LC by traces of optically active substances having a single asymmetric centre gives information about the configuration and/or the relative size of the groups connected to the chiral carbon or sulphur atom. Liquid crystalline solvents, transparent in the UV-Visible region, can be used as orientating matrices in order to obtain linear dichroism spectra of "guest molecules". Finally the problem of a possible influence of LC used as a solvent for chemical reactions will be discussed.

1. INTRODUCTION

The term "Liquid Crystal" refers to a state of matter that is intermediate between the solid crystalline phase and the isotropic liquid phase. Liquid crystals flow like liquids, but they are anisotropic like crystals. They possess many of the mechanical properties of ordinary fluids, e.g. they adopt the shape of their container. At the same time they exhibit anisotropy in their optical, electrical and magnetic properties as solid crystals do.

The characteristic properties of liquid crystals arise from the orientational order. The long axes of the molecules tend to align along a preferred direction called director.

In the present applications we have used only thermotropic liquid crystals of the nematic and cholesteric type.

In the nematic phase, long range order in the position of the centres of mass of the molecules is absent. The difference between a nematic and an ordinary liquid is solely the presence of the long range orientation. The cholesteric mesophase is similar to the nematic one but is characterized by the fact that the director rotates continuously to form a helical structure.

A very interesting property of liquid crystals is their ability to orientate molecules of solute. The anisotropic solute-solvent interactions depend critically on the geometry of the guest molecule and the applications reported are based on this property.

2. LIQUID CRYSTALLINE SOLVENTS IN LINEAR DICHROISM (L.D.) SPECTROSCOPY

The anisotropy of the absorption of linearly polarized light by an oriented substance gives information about the electronic states involved, which is not present in the spectroscopic studies carried out in isotropic solvents.

Liquid crystalline solvents, transparent in the UV-Visible region, can be orientating matrices of guest molecules; if the mean orientation of the guest molecule is known, information about the electronic states involved become available or, viceversa, if the polarizations of the transitions are known, the orientation is obtained.

A very simple relation links the L.D. $\Delta\varepsilon = (\varepsilon_\parallel - \varepsilon_\perp)$ of an absorption band to the mean absorption ($\bar{\varepsilon}$) and to the order parameter (S_{ii}) when no mixed polarizations are present

$$\Delta \varepsilon_i = 3\,\bar{\varepsilon}\,S_{ii} \quad (i = x,y,z) \tag{1}$$

Provided the symmetry of the oriented molecule is at least C_{2v}, the determination of the polarization of the transition can be obtained very easily and directly from the sign of the LD by following a "shape model" approach(1).

The majority of relatively symmetric organic molecules are ellipsoidally shaped, which is an intermediate case between two limiting shape models, the rod (Figure 1a) and the disc (Figure 1b).

The factor $|S'x| = |S_{yy} - S_{xx}/3S_{zz}|$ can define the orientational anisotropy in the xy plane and if it is $> 1/3$ or $< 1/3$ we may consider the molecule as rod-like or disc-like respectively.

The important turning point is at $|S'_x| = 1/3$.

For rod-like molecules (Figure 1c), S_{xx} only is greater than 0 and for disc-like molecules, S_{zz} only is lower than 0 and, since the

APPLICATIONS OF LIQUID CRYSTALS IN ORGANIC CHEMISTRY

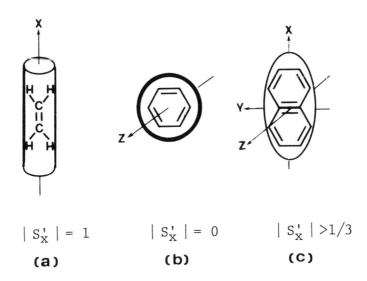

Fig. 1. Models of rod (a), disc (b) and rod-like (c) shaped molecules.

order parameter and the corresponding L.D. are equally signed (see eq. 1), it results that:

i) in all cases S_{xx} exceeds 0, therefore long-axis in plane polarized transitions must have a positive L.D.
ii) in all cases S_{zz} is lower than 0, therefore out-of-plane polarized transitions must have a negative L.D.
iii) Short-axis, in-plane polarized transitions may have positive and negative L.D.: if the molecular shape is disc-like the transitions will have a positive L.D., whereas, if it is rod-like, their L.D. will be negative.

The detection and the assignment of out-of-plane polarized transitions is hence very direct and easy in disc-like-molecules like pyrimidine, pyrazine and pyridazine, where, as expected, the well isolated $n \to \pi^*$ out-of-plane and the $\pi \to \pi^*$ in-plane transitions show negative and positive L.D. respectively (Figure 2). The dichroic ratio $\Delta OD/OD$ of an out-of-plane transition, besides being opposite signed is also twice as intense as the dichroic ratio of an in-plane

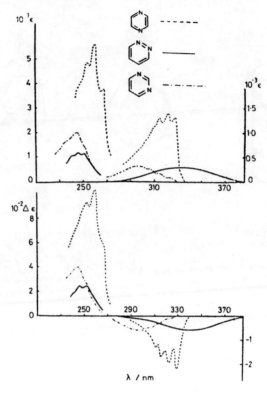

Fig. 2. Absorption (top) and linear dichroism (lower) spectra of aromatic diazines.

transition, therefore allowing the detection of transitions not revealable in isotropic spectra. The L.D. of 2,4,6 trimethyl pyridine N-oxide(2) (Figure 3) emphasizes this possibility; in fact at about 300 nm. a negative contribution is clearly apparent, by comparing the shapes of the anisotropic and isotropic spectra in this energy region; the presence of a $n \rightarrow \pi^*$ state (A_2), allowed by vibronic interaction, has been detected.

Long-axis-in-plane (positive L.D.) can be easily distinguished from the short-axis-in-plane transitions (negative L.D.) in rod-like molecules: the L.D. spectrum of 9,10-diazaphenantrene is a very clear example of this possibility: in fact the three bands present in the spectrum (Figure 4) have been previously assigned as $n \rightarrow \pi^*$ (395 nm.), $\pi \rightarrow \pi^*$ short-(350 nm.) and long- (300 nm.) axis polarized; our L.D. spectrum is in perfect agreement with these expectations.

The anisotropic absorption of plane polarized light (parallel and perpendicular to the optic axis of the sample) can be recorded by inserting and rotating a plane polarizer in a "normal" spectrophotometer(3) (static method) or by converting, in a dichrograph,

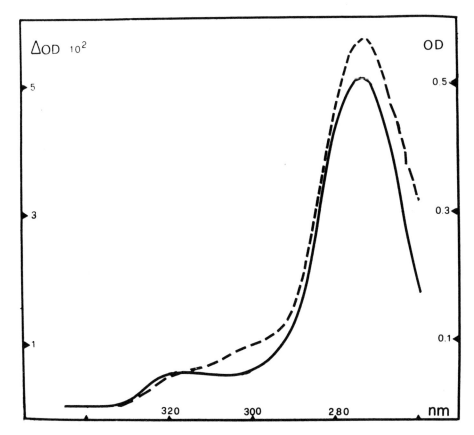

Fig. 3. Isotropic (dashed line) and linear dichroism (full line) spectra of 2,4,6-trimethylpyridine-N-oxide in the 260-350 nm region. The isotropic absorption was measured at 345°K. The optical density (OD) and the optical differences (ΔOD) are quoted to the right and to the left of the figure, respectively. Cell length 0.1 mm, conc. ~5×10^{-3} (mol dm^{-3}).

the right and left circularly polarized light into two perpendicularly modulated plane polarized components (lock-in method)(4). The lock-in technique, applied by us to the measurement of the L.D. of liquid crystalline samples allows a sensitivity at least 100 times higher than the static method.

The stability and reproducibility of the liquid crystal alignment has been improved by replacing the previously used d.c. electric field(3,5) by a sinusoidal a.c. one (1200 Hz)(1) and very recently by using a silane coating of the silica walls of the cell(6).

The use of the Merck bicyclohexyl eutectic mixture, (ZLI 1167) instead of the compensated nematic mixture of cholesteryl chloride

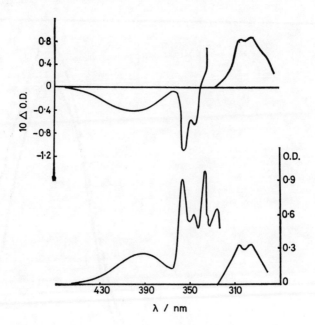

Fig. 4. Linear dichroism (top) and isotropic absorption (lower) spectra of 9,10-diazaphenantrene.

and laurate(1,3) allows now the recording of L.D. in the U.V. down to 200 nm, as shown by the spectrum of p-tolyl methyl sulphide where a short-axis-in-plane polarization has been detected in the 215 nm region(6). (Figure 5).

The usefulness of the L.D. data is connected to their relevance to the theoretical investigation of electronic states and to stereochemical studies carried out by Circular Dichroism, as this technique requires the knowledge of the polarization of the transitions under investigation in order to be interpreted.

In some particular cases the L.D. can give direct information about stereochemical problems: the uniaxial static distortion, induced by a liquid crystalline solvent in a guest cubic molecule like W (CO)$_6$, has been shown and studied by its L.D. spectrum(7).

3. INDUCED CHOLESTERIC MESOPHASES

When a chiral substance is dissolved in a nematic liquid crystal, a cholesteric mesophase is obtained(8). The cholesteric structure is characterized by its handedness (P-or M-helix) and pitch. Equal amounts of enantiomeric solutes of equal optical purity induce helical structures with identical pitch and opposite handedness

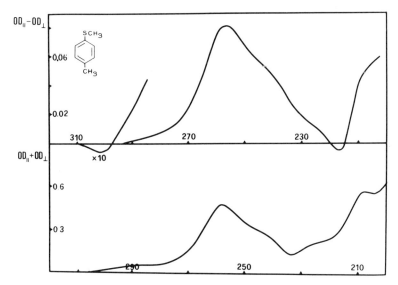

Fig. 5. Linear dichroism (top) and isotropic absorption (lower) spectra of p-tolyl methyl sulphide.

(Figure 6). Different substances show a different ability to twist the nematic phases. The "twisting power" of a chiral dopant can be defined as(9,10):

$$\beta = (p.c.r.)^{-1}$$

where p is the pitch (μm), c is the concentration (mol of solute/ mol of solution) and r the enantiomeric purity of the dopant. The parameter β, together with the sign (+) for P-helix or (-) for M-helix, characterizes the chiral solute in a similar way to a measurement of the rotatory power α. However, the physical origin of the two quantities is entirely different.

The origin of the optical rotation depends in fact on the interaction of light with molecules, while the twisting power originates from interactions between molecules of solute and solvent. Therefore, β should afford information on the chiral solute which is different and hopefully complementary to that given by the optical rotation.

Both the pitch and handedness of the cholesteric helices can be detected by chiroptical techniques, optical rotatory dispersion (O.R.D.) and circular dichroism (C.D.), by characterizing the selective reflection band. A right-handed helix (P-helix) originates a negative reflection Cotton-effect and vice versa (λ_o = np where λ_o is the wavelength of the reflection band, n the main refractive index and p the pitch).

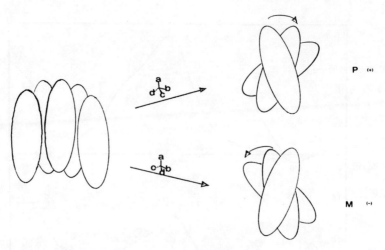

Fig. 6. Schematic picture of the structure of P and M cholesteric helices as obtained from a nematic liquid crystal.

Cholesteric mesophases induced by the addition of optically active non-nematogenic substances to nematic liquid crystals, are in most cases characterized by such long pitches that the corresponding reflection bands lie in the I.R. region(11,12), which is outside the range of commercial dicrographs. Nevertheless, also with standard dicrographs, it is possible to detect the chirality of the helices working at the absorption band of the liquid crystal chromophore(13). From the pratical chemist's point of view the study of the Grandjean-Cano disclinations by means of the polarizing microscope is more convenient, and both pitch and handedness can be measured at the same time(14,15,16).

The determination of the absolute configuration of a molecule is still a difficult task. The chemical correlation with a known compound and the anomalous X-ray diffraction are time consuming procedures. The chiroptical properties (O.R.D. and C.D.) can give correct informations in favourable cases only if the detailed electronic structures of the molecules are known. The usefulness of comparing the chirality of the induced cholesteric mesophases, in order to correlate the absolute configurations of the inducing molecules, relies on the expectation that molecules with similar chemical structures and with the same configuration, interact in a similar way with the nematic liquid crystal used as a solvent. It is fundamental for the competitivity of the method that the variations of the interactions between "host" and "guest", according to the molecular structure of the guest, are less dramatic than the variations observed in the chiroptical properties of the guest itself.

The first successful application of this idea was the correlation of the absolute configurations of a series of chiral secondary alcohols, containing an aromatic ring and having a single asymmetirc centre, using a qualitative comparison of the handedness of the helices induced in MBBA(17).

```
Ar - CH - Alk
     |
     OH
```

After the first experimental observations, a tentative explanation of the results was given in which the alcohols interact by means of a hydrogen bond with the nitrogen of the Schiff-base (MBBA), and the aromatic group (Ar) is aligned parallel to the nematic director. The steric bulk of the third substituent of the chiral carbon atom (Alk) prevents a nearby molecule of MBBA from lying parallel to the hydrogen-bonded one, and imposes a definite twist(13). The study was extended to the quantitative determination of the twisting power of a series of alcohols where both the polarity and the bulk of the alkyl group was varied(18). The results are shown in the Table.

With the exception of (2,2,2)-trifluoro-1-phenylethanol, which shows a very small twisting power of opposite handedness, all aromatic derivatives show a correlation between their absolute chirality (the R, S nomenclature is misleading owing to the different priority

Table 1. Dependence of the twisting power of the chiral aryl-alkyl carbinols on R and R' substituents and handedness of the cholesteric induced helices.

	Alcohol			Helix	Twisting power^a	Helix
n	R	R'	Absolute Configuration	(MBBA)	(MBBA)	(Phase IV)
1	CF_3	Phenyl	$\underline{S}^b$	P	0.2^c	P
2	Me	Phenyl	R	M	1.0^d	M
3	Et	Phenyl	R	M	1.3	M
4	Pr^i	Phenyl	R	M	1.4	M
5	Bu^t	Phenyl	R	M	1.5	M
6	Me	3-Pyridyl	R	M	1.0	M
7	Me	Mesityl	R	M	3.4	M
8	Me	1-Naphthyl	R	M	7.7	M
9	CF_3	1-Naphthyl	$\underline{S}^b$	M	8.5	M
10	CF_3	2-Naphthyl	$\underline{S}^b$	M	8.1	M

^{a}Values corrected to 100% optical purity of chiral alcohol. ^{b}(S)-trifluoromethylcarbinols are configurationally related to the corresponding (R)-methyl alcohols. ^{c}lit. (19); < 0.5. ^{d}lit. (19); 0.99.

of the CF_3 and the alkyl groups with respect to the aromatic moiety) and the handedness of the induced helices. As previously reported, the configuration corresponding to R-1-phenylethanol induces negative helices, and viceversa, the configuration corresponding to S-1-phenylethanol induces positive helices.

The values of the twisting power greatly increase by increasing the size of the aromatic moiety, and are less markedly, but clearly influenced, by increasing the bulk of the alkyl group when the aromatic moiety is the same. In derivatives 2 - 5 one observes a nice increase of β in passing from methyl to ethyl, isopropyl and t-butyl. In derivatives 8 and 9 the substitution of CH_3 with CF_3 causes a comparable relative increase of the twisting power.

The series of derivatives 2 - 9 seems to fulfill the requirement that the dominant "host-guest" interactions are of the same type in all members.

The variations of β with the structure are in agreement with the tentative model proposed.

In fact, large aromatic molecules are better orientated by liquid crystals than benzene and therefore β should increase in passing from Ar=phenyl to naphthyl. Similarly, an increase of the steric bulk of the alkyl group should increase the twist of the nearby molecules of MBBA or EBBA.

The deviating case of (2,2,2)-trifluoro-1-phenylethanol can be understood by considering a different interacting configuration with the Schiff bases, characterized by a double hydrogen bond(18).

A different approach to the determination of the absolute configuration was recently attempted and is based on measurements of the pitch and handedness of a large number of organic molecules by means of IR-ORD(19). As the results obtained did not show a correlation between the R-S Cahn-Ingold-Prelog configuration symbols (based on the atomic numbers) and the induced handedness, a different empirical system of priorities, allowing the correlation of a large number of derivatives was proposed. However, this scheme of priorities does not seem to be general and exceptions were found by the authors(19) and other workers(20). Generally speaking, it seems impossible to assign to a set of chemical groups a sequence of priorities which applies also to molecules having a very different structure(21); moreover, the specific chemical structure of the liquid crystalline solvent cannot be ignored. Before a confident use of the method for molecules with drastically different structures can be made, a great amount of work still has to be done in order to understand the physical basis of the twisting power. However, the technique seems to be promising and its range of applicability, in the worst case, is similar to that of the quasi racemate method(22), with the advantage of requiring only a few milligrams of only one enantiomer.

The fact that a low concentration of a chiral compound governs the molecular arrangement of the whole solution and that the consequent high order induces a pronounced effect in the characteristic physical parameters such as pitch and handedness, points out the possibility of "amplifying" a small dissymmetric characteristic of the solute.

Compounds whose chirality is due solely to isotopic substitution show very low optical rotation whose determination normally requires a considerable amount of meterial.

We naturally extended our research to this type of chiral molecule(23). The results obtained are shown in the Scheme 1.

		twisting power μm^{-1}
R		+ 0.070 ± 0.004
S		− 0.072 ± 0.003
R		− 0.064 ± 0.005
S		+ 0.062 ± 0.004
R		− 0.050 ± 0.003
S		+ 0.053 ± 0.004

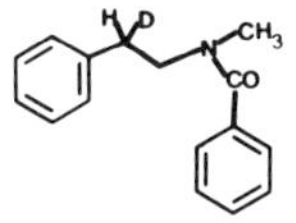

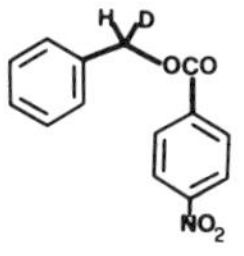

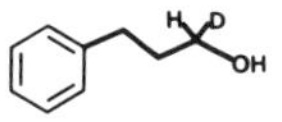

Scheme 1

The fact that well characterized values of β can be obtained for the compounds investigated by using only a drop of cholesteric solution for each measurement (altogether a few milligrams of chiral derivative), points out the possible practical use of the method.

4. LIQUID CRYSTALLINE ASYMMETRIC INDUCTION?

The possibility of performing a chemical reaction between preorientated molecules and therefore of influencing considerably the activation parameters is certainly an attractive one.

In particular, we were interested in the possibility of obtaining asymmetric induction by using cholesteric liquid crystals as solvents.

Several reports of success in this field are present in the literature(24,25,26), however a recent reinvestigation by Kagan and coworkers(27) has not confirmed these asymmetric inductions.

Almost at the same time of this reinvestigation we studied a set of bimolecular reactions(28): all the syntheses described above were unimolecular processes and bimolecular reactions are expected to be more sensitive to the alignment of the reactants imposed by the mesomorphic solvents.

These include the cycloaddition between a ketene and a Schiff base (reaction 1), two addition reactions to α,β-unsaturated ketones (reactions 2 and 3), and the condensation of two molecules of aldehyde to yield an oxirane derivative (reaction 4), see Scheme 2.

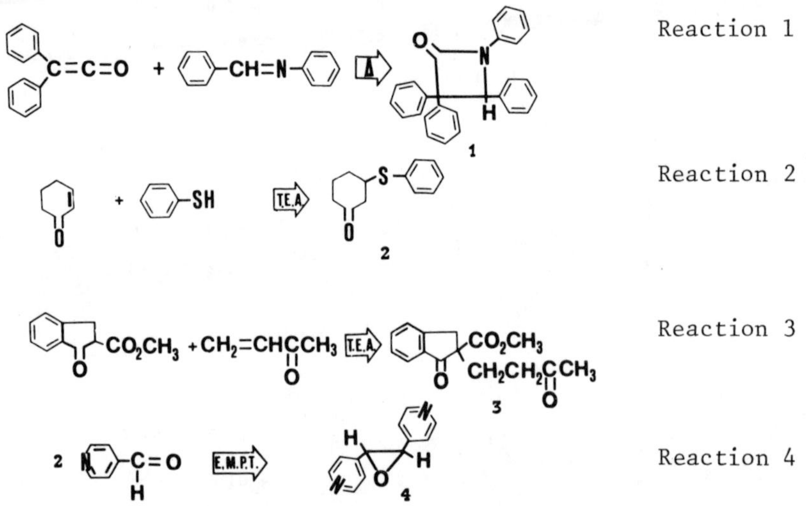

Reaction 1

Reaction 2

Reaction 3

Reaction 4

Scheme 2

The rotatory power of the β-lactame 1 in reaction 1 is not known, but is likely to be quite strong, due to the carbonyl chromophore surrounded by the dissymetrically disposed phenyl rings. Reactions 2 and 3 were specifically selected since they afford considerable enantiomeric excess in the presence of chiral catalysts and the optical rotatory power of products 2 and 3 are well known. Finally, the rotatory power of the oxirane 4 is particularly strong.

The reactions were carried out at room temperature in two different cholesteric mesophases with solute concentrations less than 5% in weight, in order not to disturb the anisotropic arrangement. The reaction products were separated properly from the liquid crystal and then carefully purified to eliminate any further contamination from the chiral phase. In all cases, the products did not show significant optical rotation, thus indicating the formation of racemic mixtures and the absence of asymmetric induction by cholesteric crystals.

However, recent reports in the literature seem to confirm the possibility of obtaining asymmetric induction, even if modest, by using cholesteric solvents (29,30).

As a general comment to results reported above it is reasonable to think that the macrostructural handedness of the mesophase can not influence significantly the stereochemistry of the reaction on account of the following points:

i) The pitch of the cholesteric helices is very large (normally 4000 A°) with respect to the molecular dimensions, and practically the molecules see in their neighbourhoods a nematic achiral structure;
ii) The orientation of the molecules, such as rod-shaped ones, with their long axis parallel to the local nematic director, is defined only with respect to their direction, and two orientations rotated by 180° are equally possible;
iii) There is considerable free rotation along the long molecular axis; each 'guest' molecule, therefore, shows different enantiotopic cross-section.

However the liquid crystalline local order can enhance and modify small solute-solvent interactions, which instead are not operative in an isotropic medium.

Acknowledgments. We thank C.N.R. (Rome) for financial support.

REFERENCES

1. G. Gottarelli, R. D. Peacock, and B. Samorì, J. Chem. Soc. Perkin, 2:1208 (1977).
2. G. L. Bendazzoli, G. Gottarelli, P. Palmieri, and B. Samorì, J. Chem. Phys. 67:2986 (1977).
3. E. Sackmann and H. Mohwald, J. Chem. Phys. 58:5407 (1973).
4. S. F. Mason and R. D. Peacock, Chem. Phys. Letters, 24:406 (1973).
5. R. Gale, R. D. Peacock, and B. Samorì, Chem. Phys. Letters, 37:430 (1976).

6. G. Gottarelli and B. Samorì, to be published.
7. B. Samorì, J. Phys. Chem. 83:375 (1979).
8. G. Friedel, Ann. Phys. Paris, 18:273 (1922).
9. H. Baessler and M. M. Labes, J. Chem. Phys. 52:631 (1970).
10. E. H. Korte, B. Schrader, and S. Baulek, J. Chem. Res. (M), 3001 (1978).
11. R. J. Dudley, S. F. Mason, and R. D. Peacock, J. Chem. Soc. Faraday 2, 997 (1975).
12. E. H. Korte, B. Schrader, and S. Baulek, J. Chem. Res. (S), 236 (1978).
13. G. Gottarelli, B. Samorì, and C. Stremmenos, Chem. Phys. Letters, 40:308 (1976).
14. J. P. Berthault, J. Billard, and J. Jacques, C. R. Acad. Sc. Paris, t. 284, Série C, 155 (1977).
15. G. Heppke and F. Oestreicher, Z. Naturforsch. 32a:899 (1977).
16. G. Heppke and F. Oestreicher, Mol. Cryst. Liq. Cryst. Letters, 41:245 (1978).
17. G. Gottarelli, B. Samorì, S. Marzocchi, and C. Stremmenos, Tetrahedron Letters, 1981 (1975).
18. G. Gottarelli, B. Samorì, C. Stremmenos, and G. Torre, Tetrahedron, (1980) in the press.
19. H. J. Krabbe, H. Heggemeier, B. Schrader, and E. H. Korte, J. Chem. Res. (S), 238 (1978); J. Chem. Res. (M), 3020 (1978).
20. G. Gottarelli, B. Samorì, U. Folli, and G. Torre, J. Phys. Colloq. (Orsay Fr.), C3:25 (1979).
21. H. Forster and F. Vogtle, Angew. Chem. Int. Ed. 16:429 (1977).
22. J. E. Ricci, Tetrahedron, 18:605 (1962).
23. G. Gottarelli, B. Samorì, C. Fuganti, and P. Grasselli, to be published.
24. F. D. Saeva, P. E. Sharpe, and G. R. Olin, J. Am. Chem. Soc. 97:204 (1975).
25. L. Verbit, T. R. Halbert, and R. B. Patterson, J. Org. Chem. 40:1649 (1975).
26. W. H. Pirkle and P. L. Rinaldi, J. Am. Chem. Soc. 99:3510 (1977).
27. C. Eskenazi, J. F. Nicoud, and H. B. Kagan, J. Org. Chem. 44:995 (1979).
28. S. Colonna, A. Dondoni, A. Medici, G. Gottarelli, and B. Samorì, Mol. Cryst. Liq. Cryst. 55:47 (1979).
29. M. Nakazaki, K. Yamamoto, and K. Fujiwara, Chem. Letters, 863 (1978).
30. P. Seuron and G. Solladie, J. Org. Chem. 45:715 (1980).

REACTIONS AND BEHAVIOUR OF ORGANIC ANIONS IN TWO-PHASE SYSTEMS

Mieczysław Makosza

Institute of Organic Chemistry
Polish Academy of Sciences
Warsaw, Poland

Abstract - Some specific features of the reactions of carbanions and carbenes in two-phase systems, along with reaction of these species at the phase boundary with aqueous NaOH are discussed.

Selection of a medium in which reacting compounds can come into the intimate contact required for a reaction has always been a general problem in organic synthesis. This goal is usually achieved by means of a solvent able to dissolve, at least partially, the reacting compounds.

In the case of reactions of anionic reagents with nonpolar organic compounds this is often a difficult task since sources of anions - corresponding sodium or potassium salts are soluble only in higly polar, preferentially protic, solvents. On the other hand these solvents are usually unsuitable for nonpolar compounds. Furthermore they interact strongly with anions affecting unfavourably the course of the reactions.

Corresponding tetraalkylammonium (TAA) salts ($R_4N^+Y^-$ or Q^+Y^-) provided that the cations contain more than 16 carbon atoms, have much better solubility patterns. Indeed salts $R_4N^+Y^-$ are well soluble in hydrocarbons, even when Y^- are such inorganic anions as F^-, Cl^- or CN^-. Thus Q^+Y^- could be very efficient sources of anionic reagents, not only assuring homogeneity in nonpolar solvents but also high activity of anions. The reason for the latter behaviour is that anions in nonpolar solvents are weakly solvated and that cation-anion interactions, in the case of tetraalkylammonium cations, are purely electrostatic; moreover the distance between positive and negative centres in Q^+Y^- are substantial due to the large size of the cations.

Unfortunately TAA salts cannot be widely applied as reagents in organic synthesis. They are much more expensive and much less available than the corresponding sodium or potassium salts and usually must be prepared before the use from commercial TAA halides via a laborious ion exchange. In the case of large scale preparations the TAA salts should be regenerated and reused, thus creating additional complications.

The most efficient new method of anion exchange in TAA salts, consists in the equilibration of their solutions in nonpolar solvents with an aqueous solution of the required sodium or potassium salts (1). On this basis convenient and simple methods of preparation of variety of TAA salts were developed(1,2).

$$Q^+X^-_{org} + Na^+Y^-_{aq} \rightleftharpoons Q^+Y^-_{org} + Na^+X^-_{aq}$$

The solution of a Q^+Y^- salt in a nonpolar solvent, which forms the organic phase in the ion exchange process, can be subjected directly to a reaction in which Y^- are consumed(1).

Consequently a process was developed where the required TAA salts are formed directly during the reaction and continuously regenerated via ion exchange with the aqueous phase. Since one TAA cation can transfer into the organic phase a great number of required anions via multiple ion exchange, the TAA salts can be used in much smaller than stoichiometric quantities, acting virtually as catalysts. The term phase-transfer catalysis is used for this type of processes(3). This principle is conveniently explained when the cyanation of an alkyl halide, dissolved in a hydrocarbon with aqueous solution of sodium cyanide, is considered(3).

When this two-phase system is stirred, the reaction does not occur since R-Cl and CN^- anions are located in different phases. The addition of a small amount of a lipophilic TAA chloride results in a rapid conversion of the alkyl chloride into the nitrile, via the continuous transfer of the cyanide anions to the organic phase and the chloride anions to the aqueous phase (see scheme in the next page).

This principle is quite general; a number of reactions, such as nucleophilic substitution, addition of anions, reduction and oxidation with anionic reagents etc. can be carried out very efficiently in this way(4).

$$R-Cl + \overset{+}{Q}\overset{-}{CN} \longrightarrow R-CN + \overset{+}{Q}\overset{-}{Cl}$$

$$\overset{+}{Q}\overset{-}{CN} \rightleftharpoons \overset{+}{Q}\overset{-}{Cl}$$

$$\overset{-}{Cl}\overset{+}{Na} \qquad \overset{-}{CN}\overset{+}{Na}$$

    The crucial feature of the phase transfer catalyzed reactions is that the ion exchange equilibrium should assure the transfer of reacting anions from the aqueous phase to the organic phase and the anions produced in the reaction in the opposite direction. Thus the catalytic process can occur anly when the reacting anions are more lipophilic than the anions produced during the reaction.

    It should be also pointed out that compounds able to complex sodium or potassium cations, giving them lipophilic properties (crown ethers, cryptands etc.), are also acting as catalysts in these reactions.

    A concept similar to the phase transfer catalysis is very efficient in the reactions of carbanions. These active intermediates are usually generated by the action of bases on CH acids

$$\rightarrow C-H + M^+B^- \xrightarrow{solv.} \rightarrow C^-M^+ + BH$$

    The selection of a base-solvent system is therefore essential in the reactions of carbanions. The low acidity of many CH acids requires that, as a rule, only strong bases: NaH, $NaNH_2$, t-BuOK etc., in strictly anhydrous organic solvents, can be employed for the generation of carbanions. The use of sodium hydroxide for generation of carbanions is rather limited because the equilibrium

$$\rightarrow C-H + NaOH \rightleftharpoons \rightarrow C^-Na^+ + H_2O$$

is shifted to the left due to the high acidity of water, and also because of its high hydrolytic activity. Both these limitations are eliminated when sodium hydroxide is employed in form of a concentrated aqueous solution and the generation and reactions of carbanions are carried out in two-phase systems in the presence of TAA salts or other sources of lipophilic cations(5). Here the acid-base equilibrium proceeds at the interface, producing carbanions in low concentration in the inactive form of sodium derivatives anchored at the phase boundary in a kind of adsorbed state. Subsequently these carbanions are continuously transferred into the organic phase as TAA derivatives, where they enter into the required reactions, thus shifting the acid base equilibrium to the right.

$$\geqslant\!\!C\text{-H} \rightleftharpoons \geqslant\!\!C^- \quad \geqslant\!\!C^- + Q^+Cl^- \rightleftharpoons \geqslant\!\!C^-Q^+$$

$$\geqslant\!\!C\text{-R} + Q^+Cl^- \longleftarrow \geqslant\!\!C^-Q^+ + R\text{-Cl}$$

NaOH Na$^+$ Na$^+$ Na$^+$Cl$^-$

This hyperbasic effect as well as the fact that the reacting compounds contact with aqueous alkali only at the interface are the most characteristic features of the catalytic two-phase generation of carbanions. These conditions are efficiently applicable for a variety of reactions of carbanions derived from the CH acids of pKa value up to 22.

Such important processes as alkylation of arylacetonitriles, cyclopentadiene hydrocarbons, aldehydes and ketones, esters, sulfones etc., condensation of carbanions with aldehydes and ketones, the Knoevenagel, Darzens, Michael and related reactions as well as many reactions involving sulfonium and phosphonium ylides have been successfully carried out under these conditions.

Another wide area of application of this two-phase system is the generation of dihalocarbenes, particularly dichloro- and dibromocarbenes via the α-elimination of hydrogen halide from the corresponding haloforms(5).

Generation of carbanions in the two-phase system occurs in the interfacial region between the organic and the aqueous phases. In presence of the catalyst the carbanions are continuously introduced into the organic phase where further reactions take place. In absence of the catalyst the carbanions are confined in the interfacial region, so they can react only with very active electrophiles.

Indeed aldehydes react easily with carbanions located in the interfacial region. For example the Darzens condensation of α-halonitriles of α-halosulfones with aldehydes (but not ketones) occurs when a benzene solution of these reagents is simply stirred with concentrated aqueous NaOH(6).

Specific situation of carbanions located at the interface can influence the steric course of the Darzens condensation. Thus the ratio of cis-trans isomers of 2,3-diphenylglycidonitrile produced in the reaction of phenylchloroacetonitrile with benzaldehyde in aqueous NaOH differs considerably <u>with</u> and <u>without</u> catalyst.

In this case the stereochemistry of the reaction is determined by the rate of cyclization of the isomeric intermediate chlorohydrines, so the decisive factor is the free energy of the transition states I and II.

PhCHCN + PhCH=O →(benzene, cat. / NaOH aq) [epoxide Ph/CN, H/Ph] + [epoxide Ph/CN, Ph/H]

cat. Et$_3$N$^+$CH$_2$PhCl$^-$ 90 %   10 %

no cat.   48 %   52 %

In the catalytic process TAA salts of chlorohydrines anions are dissolved in the organic phase, and for steric reasons I has lower energy than II; on the other hand at the interface the transition state II is additionally stabilized by the interaction of the negative poles of N, O and Cl with sodium cations located on one side of the molecule(6). (See Scheme 1).

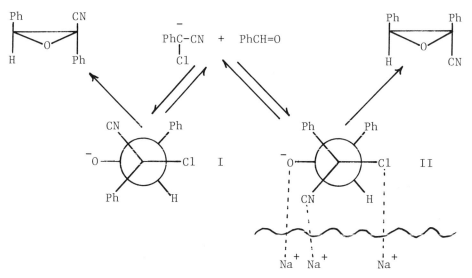

Scheme 1

Reactions of α-halonitriles with acrylonitrile derivatives also give cis-trans isomers of dicyanocyclopropanes in ratios depending as to whether the reaction proceeds at the interface or in the organic phase: the formation of cis dicyanocyclopropanes is favoured in the interfacial process. (See Scheme 2).

Also in these cases an additional stabilization of the transition state, leading to cis products by interaction of the negative centres with potassium cations located on the one side of the intermediate anion, is responsible for the preferential formation of the cis isomers.

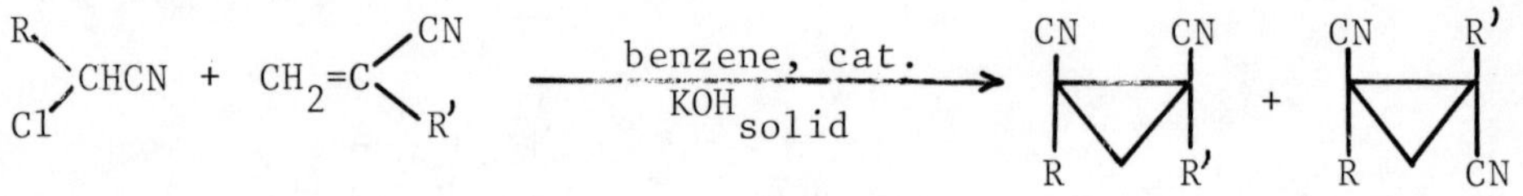

$$R\!\!>\!\!CHCN + CH_2=C\!\!<\!\!^{CN}_{R'} \xrightarrow[\text{KOH solid}]{\text{benzene, cat.}} \text{[cyclopropane with CN, CN, R, R']} + \text{[cyclopropane with CN, R', R, CN]}$$

R = Ph, R' = H	$Et_3\overset{+}{N}CH_2Ph\,Cl^-$	51%	49%
	no cat.	90%	10%
R = CH$_3$, R' = CH$_3$	$Et_3\overset{+}{N}CH_2Ph\,Cl^-$	49%	51%
	no cat.	76%	24%

Scheme 2

Upon analysis of numerous previous reports on the influence of solvents and counterions on the stereochemistry of these reactions, one can hardly find any consideration of the effects of the heterogeneity. On the other hand there are some apparent discrepancies in the previous results which can be reasonably explained by taking into account the fact that, when the systems are not clearly homogeneous, interfacial processes will favour the formation of cis isomers.

From many new applications of two-phase systems to reactions of carbanions some processes, connected with the direct introduction of a heteroatom to the carbanionic moiety, will be discussed.

Thioalkyl substituent (RS) is often introduced via the reaction of carbanions with alkyl disulfides. The catalytic two-phase system is unsuitable for this reaction since thiolate anions produced during the thioalkylation are highly extractible, so the catalysts do not operate. These difficulties are eliminated by the addition of an alkylating agent to the system, which reacts irreversibly with thiolate anions liberating the catalyst.

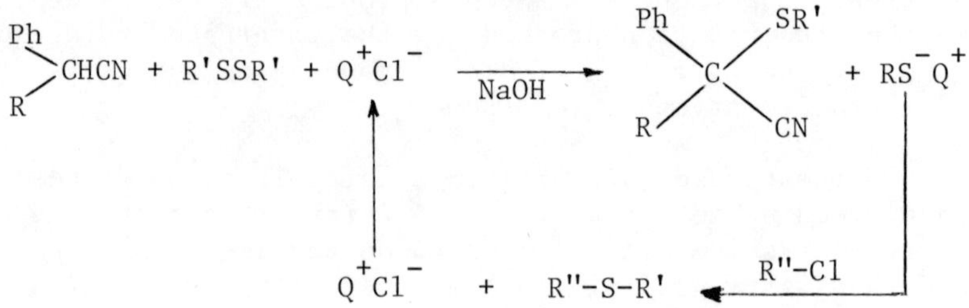

$$\text{Ph}\!\!>\!\!CHCN + R'SSR' + Q^+Cl^- \xrightarrow{\text{NaOH}} \text{Ph}\!\!>\!\!C\!\!<\!\!^{SR'}_{CN} + RS^-Q^+$$

$$Q^+Cl^- + R''\text{-}S\text{-}R' \xleftarrow{R''\text{-}Cl}$$

ORGANIC ANIONS IN TWO-PHASE SYSTEMS

Much more complicated is the hologenation of carbanions with polihalomethanes since the introduced halogen is both a good leaving and carbanion stabilizing group, so that further reactions often take place.

Indeed a reaction of phenylacetonitrile with CCl_4 leads to trans-dicyanostilbene via the initially formed phenylchloroacetonitrile(8). On the other hand phenylacetonitrile derivatives undergo simple halogenation

$$\underset{R}{\overset{Ph}{>}}CHCN + CCl_4 \xrightarrow[NaOH_{aq}]{Q^+Cl^-} \underset{R}{\overset{Ph}{>}}C\underset{Cl}{\overset{CN}{<}}$$

Even when R = OMe corresponding phenylchloromethoxyacetonitrile can be obtained in a resonable yield although this compound is very unstable toward hydrolysis(9).

Phenyldialkylaminoacetonitriles react with CCl_4 in a much more complicated way, giving 2-phenyl-2-dialkylamino-3,3,3-trichloropropionitriles, apparently according to the scheme reported below.

$$\underset{>N}{\overset{Ph}{>}}\bar{C}-CN + CCl_4 \longrightarrow \underset{>N}{\overset{Ph}{>}}C\underset{Cl}{\overset{CN}{<}} \longrightarrow \underset{\overset{+}{N}}{\overset{Ph}{\underset{\wedge}{\parallel}}}\overset{CN}{\underset{}{}}\,Cl^- \xrightarrow{CCl_3^-} \underset{>N}{\overset{Ph}{>}}C\underset{CCl_3}{\overset{CN}{<}}$$

Similar reactions occur with benzylidyne trichloride.

$$\underset{>N}{\overset{Ph}{>}}CHCN + PhCCl_3 \xrightarrow[NaOH_{aq}]{Q^+Cl^-} \underset{>N}{\overset{Ph}{>}}C\underset{CCl_2Ph}{\overset{CN}{<}}$$

As previously mentioned, phenylchloroacetonitrile, which is a versatile starting material in many reactions, cannot be obtained via reaction of phenylacetonitrile with CCl_4 in the two-phase system since, being a strong CH acid, is immediately deprotonated and undergoes further transformations. Under these conditions only a very small fraction of the reacting compounds is anionized so that it was possible to design a process in which phenylchloroacetonitrile does not react with itself but with a foreign electrophile.

The reaction of pehnylacetonitrile, CCl_4 and benzaldehyde thus gives 2,3-diphenylglycidonitrile (yield 80%!) according to the following scheme(9):

$$PhCH_2CN + CCl_4 \xrightarrow[NaOH]{Q^+Cl^-} Ph\underset{Cl}{\overset{}{C}}H-CN \longrightarrow Ph-\underset{Cl}{\overset{}{\bar{C}}}-CN \xrightarrow{PhCHO}$$

$$Ph-\underset{CN}{\overset{Cl}{C}}-\underset{O^-}{\overset{}{C}}HPh \longrightarrow Ph-\underset{CN}{\overset{}{C}}\underset{O}{\diagdown}\overset{}{\diagup}Ph$$

This is a general process for aromatic and some aliphatic aldehydes. Similar reactions occur between fluorene, CCl_4 and benzaldehyde.

The phenylchloroacetonitrile anion, generated in the reaction of phenylacetonitrile with CCl_4, can also be trapped by acrylonitrile. In this case, however, the main product is 1-phenyl-1,2-dicyano-2-chlorocyclopropane along with the expected phenyldicyanocyclopropane.

The pathway of the formation of these compounds is shown in the scheme 3(9).

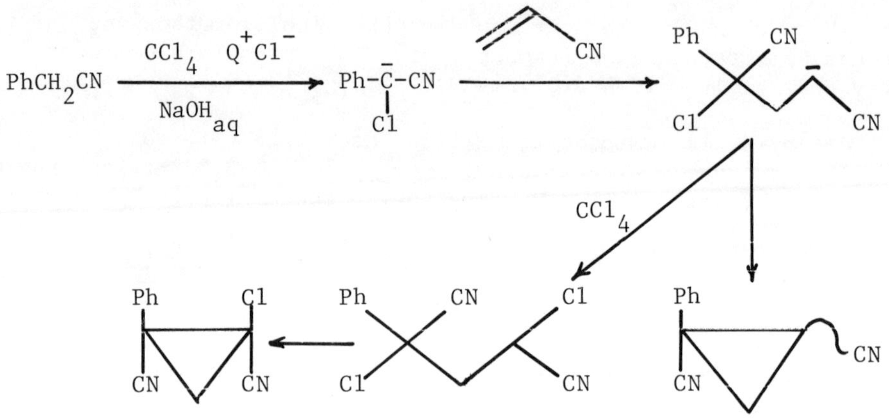

Scheme 3

Although the presented reactions involve a number of consecutive steps, they usually proceed with good yields of the corresponding oxiranes and cyclopropanes. Also isolation of the products is rather simple and therefore they offer considerable practical improvements in the synthesis of some complicated compounds.

Although heterogeneity of the catalytic two-phase system prevents alkaline hydrolysis of the reacting compounds, there are many cases in which aqueous sodium hydroxide cannot be used as a base. We have found that instead of NaOH, anhydrous sodium or potassium carbonates can be used for the generation of anions, even from rather weak CH acids. In these cases the system contains organic reactants in liquid phase, the catalyst (TAA salt or crown ether) dissolved in a nonpolar solvent and anhydrous alkali carbonate in solid phase. Here the acid-base equilibrium occurs on the surface of the solid phase, subsequently the carbanions are transformed from this adsorbed state into the organic phase in form of TAA salts. The carbonates are relatively mild bases so they can be used at relatively high temperature without decomposition of the starting materials and products.

This simple solid liquid system can be efficiently applied in many important reactions involving carbanions and other organic anions. It is particularly useful for alkylation of ethyl malonate, cyanoacetate and acetylacetate(10) e.g.

$$CH_2(COOEt)_2 + C_4H_9Br \xrightarrow[Na_2CO_3]{Q^+X^-} C_4H_9CH(COOEt)_2$$

These conditions are also very convenient for the N-alkylation of N-arylformamides and P-alkylation of dialkylphosphites. The latter process, known as the Michaelis-Becker reaction, is of particular practical value

$$R-Br + HOP(OEt)_2 \xrightarrow[K_2CO_3]{crown} R-\underset{O}{\overset{\|}{P}}(OEt)_2$$

Even such weak CH acids as phenylalkylacetonitriles can be deprotonated and acylated under these conditions (10).

$$\underset{R}{\overset{Ph}{>}}CHCN + ClCOOEt \xrightarrow[K_2CO_3]{crown} \underset{R}{\overset{Ph}{>}}C\underset{COOEt}{\overset{CN}{<}}$$

All of these processes cannot be carried out in the presence of aqueous alkali due to the rapid hydrolysis of the starting materials.

In the solid-liquid two-phase system it was possible to chlorinate secondary nitroalkanes with CCl_4.

$$\underset{CH_3}{\overset{CH_3}{>}}CHNO_2 + CCl_4 \xrightarrow[K_2CO_3]{Q^+Cl^-} \underset{CH_3}{\overset{CH_3}{>}}C\underset{Cl}{\overset{NO_2}{<}}$$

Thus the use of solid alkali carbonates considerably expands the area of application of the catalytic two-phase systems in organic synthesis.

This short presentation of the general concept and some new applications of the two-phase systems in organic synthesis and also some specific features of the interfacial processes is hoped to promote better understanding and wider applications of this approach to many synthetic problems.

This subject is thoroughly analyzed in many review and monographs(4,11).

REFERENCES

1. A. Brandström, "Preparative Ion Pair Extraction", AB Hassle, Lakemedel (1974).
2. M. Makosza and E. Białecka, Synth. Comm. 6:313 (1976).
3. C. M. Starks, J. Am. Chem. Soc. 93:195 (1971).
4. E. V. Dehmlow, S. S. Dehmlow, "Phase Transfer Catalysis" Verlag Chemie (1980); W. P. Weber, G. W. Gokel, "Phase Transfer Catalysis in Organic Synthesis", Springer Verlag (1977); M. Makosza, "Naked Anions - Phase Transfer. Modern Syntheitc Methods", ed., R. Schefold, Scweizerischer Chemiker Verband, Zurich (1976).
5. M. Makosza, Pure and Appl. Chem. 43:439 (1975).
6. A. Jonczyk, A. Kwast, and M. Makosza, J.C.S. Chem. Comm. 902 (1977).
7. A. Jonczyk, A. Kwast, and M. Makosza, Tetrah. Letters, 11 (1979).
8. M. Makosza, B. Serafinowa, and I. Gajos, Roczniki Chem, 43:671 (1969).
9. A. Jonczyk, A. Kwast, and M. Makosza, J. Org. Chem. 4:1192 (1979).
10. M. Fedorynski, K. Wojciechowki, Z. Matacz, and M. Makosza, J. Org. Chem. 43:4682 (1978).
11. W. E. Keller, "Compendium of Phase Transfer Reactions and Related Synthetic Methods", Fluka AG Buchs (1979).

SOLUTE-SOLVENT INTERACTIONS IN RING FORMATION

C. Galli, G. Illuminati, L. Mandolini, and B. Masci

Centro di Studi sui Meccanismi di Reazione del Consiglio
Nazionale delle Ricerche
Istituto di Chimica Organica dell'Università
00185 Roma, Italy

Abstract - Solute-solvent interactions may affect the cyclization of molecules in several ways including desolvation of short-chain substrates, interdependence of solvation and steric strains in the transition state of medium-ring formation, steric inhibition of solvation in special reaction series, and solvent-induced conformational changes. The effects are generally small and, particularly, do not seem to be responsible for the large rate-enhancement phenomena which are associated with the formation of common rings. Further studies are needed to obtain a more extensive knowledge of the possible modes of interaction and to provide broader quantitative bases in this field. The determination of the effective molarities for a given cyclization reaction over a spectrum of ring sizes has proved to be a valuable approach also in this connection.

The rate enhancements accompanying intramolecular reactions when compared to related intermolecular counterparts have withdrawn the attention of several investigators in recent years not only for the phenomenon per se but also because the factors involved have close connections with the mechanism of enzyme catalysis. Out of the possible factors involved in enhanced intramolecular reactivity, changes in solvation of the reacting species and/or transition state in going from the intermolecular over to the intramolecular process have been taken into account from time to time.

For a good understanding of the role of the solvent on the energetics of ring formation, we need a quantitative comparison of the effect of the solvent on both the intramolecular reaction leading to ring formation and the intermolecular reaction of an appropriate model system. It is also desirable to know how the free energy of activation is dissected into the enthalpy and entropy terms and how the reactivity change on transfer from one solvents (S_1) to another (S_2) obtains from the solvent activity coefficients ($^{S_1}\gamma^{S_2}$) for the intramolecular reactant, the intermolecular reactants and the related transition states(1).

Comparison of intra- vs. intermolecular reactivity is best effected by the determination of the effective molarity(2), EM, in each solvent, which is a k_{intra}/k_{inter} ratio in M units. The effect of transfer of the reacting systems from solvent S_1 to solvent S_2 on EM can be expressed by equation (1).

$$\log \frac{EM^{S_2}}{EM^{S_1}} = \log \frac{k_{intra}^{S_2}}{k_{intra}^{S_1}} - \log \frac{k_{inter}^{S_2}}{k_{inter}^{S_1}} =$$

$$= \log {}^{S_1}\gamma_{XY}^{S_2} - \log {}^{S_1}\gamma_{XY}^{S_2 \neq} - \log {}^{S_1}\gamma_{RX}^{S_2} - \log {}^{S_1}\gamma_{R'Y}^{S_2} + \log {}^{S_1}\gamma_{RXYR'}^{S_2 \neq}$$

(1)

In the lack of an analysis in terms of solvent activity coefficients, our discussion will be concerned with a limited number of available rate ratios and activation parameters.

Provided that solvent transfer does not give rise to change in mechanism, the $k_{intra}^{S_1}/k_{intra}^{S_2}$ ratio is expected to essentially reflect changes in solvation of the reactant and transition state. In contrast, structural effects such as steric strain and probability effects related to the encounters between ends of a connecting chain of a given length, which are known to play an important role in cyclization, are expected to cancel out in the above ratio. A clean-cut distinction between solvent and structural effects should however be treated with caution and interdependence between factors may not be negligible in all cases.

We may distinguish two different types of solvent interactions for the reaction of bifunctional molecules, i.e.,

(i) solvation at the reaction site involving the reacting ends and their interaction in the transition state,
(ii) solvation of the non reacting parts of the molecule affecting the conformation distribution of such a molecule.

Clearly both types of interactions affect the rate of reaction through the relative stabilities of reactant and transition state.

For widely different solvents in nucleophilic reactions involving anion-molecule interactions, $k_{intra}^{S_1}/k_{intra}^{S_2}$ may be very large(1). There is no doubt that solvation has an important role in such reactions. Now, it is of interest to understand whether effective molarities for ring closure reactions result not only from structural effects inherent to the formation of ring-shaped transition states but also, and to what an extent, they depend on solvation phenomena. The quasi-cyclic conformation of a short chain bifunctional molecule may involve solvent extrusion as compared to a pair of intermolecular reactants. Desolvated functional groups may interact more rapidly than more highly solvated ones. Part of the rate-enhancements as observed in cyclization reactions may originate in this way. Bruice and Turner(3) tested this hypothesis on 5-memebered cyclic anhydride formation, which is characterised by huge EM values, and found that solvation phenomena contribute very little to rate enhancements accompanying ring closure. For example, the intramolecular reaction of aryl hydrogen phthalates in water displays EM values in the order of 10^6 M and that of aryl monoesters of 3,6-endoxo-Δ^4-tetrahydrophthalic acid has EM values as large as 10^8 M in the same solvent (Table I). Despite such large effects, transfer from water

Table I. EM Data for Cyclic Anhydride Formation in Water and 1 M Water in Dimethylsulphoxide.[a]

EM, M 4.8×10^6 in H_2O

          2.6×10^6 in 1 M H_2O in DMSO

EM, M 1.3×10^8 in H_2O

          2.5×10^9 in 1 M H_2O in DMSO

Intermolecular model reaction $CH_3COOAr + CH_3CO_2^-$

[a] Data from ref. 3.

to 1 M H_2O in dimethylsulphoxide (DMSO) causes little change in the EM values. This behaviour suggests that there is little difference in solvation between transition states for the intra and intermolecular reactions and that similar extents of desolvation occur in going from reactant(s) to transition states in both reactions.

Similar views have been expressed by Dafforn and Koshland(4), who found that the rates of formation of some 5-membered ring lactones and thiolactones, although somewhat less spaced in sulpholane than in water, follow the same trends in the two basically different solvents (Table II).

The problem can also be approached by investigating the influence of solvent as a function of ring size along a uniform cyclization series. Rate data are available for the cyclization of $\underline{o}$-ω-bromoalkoxyphenoxides in 75% EtOH (v/v) and in 99% DMSO (v/v), which are two widely different media (equations 2 and 3)(5). The

$$\text{(2)}$$

$$\text{(3)}$$

comparison is extended over a broad spectrum of ring sizes, i.e., from 6 to 24. The reactivity increase in 99% DMSO relative to 75% EtOH involves factors from 5×10^3 to 1.5×10^4 for the cyclization reactions and is in the same order for the intermolecular model reaction. This large effect is an essential requirement for the search of solvation contributions to reactivity of some relevance as inherently bound to the intramolecular reaction.

The above-mentioned ring size range covers three groups of rings, i.e., common (n = 6), medium (n = 8-10), large (n = 14) and intermediate sizes in between that show borderline behaviour.

The general picture for the formation of the 6-membered ring is analogous to that found for 5-membered rings by Bruice and Turner(3), and by Dafforn and Koshland(4). The EM value is quite

SOLUTE-SOLVENT INTERACTIONS IN RING FORMATION

Table II. Thiolactone and Lactone Formation in Water and Sulpholane.[a]

	k_{rel} in H_2O	k_{rel} in sulpholane
THIOLACTONES		
HS–⋯–CONH$_2$	1.72	1.44
norbornyl-CH$_2$SH, CONH$_2$	1	1
norbornyl-HS, CONH$_2$	862	55.6
LACTONES		
HO–⋯–CONH$_2$	1	1
norbornyl-CH$_2$OH, CO$_2$H	83.5	48.5

[a] Data from ref. 4.

Table III. EM Data for the Formation of Catechol Polymethylene Ethers in 75% EtOH (E) and 99% DMSO (D).[a]

Ring size	EM^E, 25°C	EM^D, 25°C	EM^E/EM^D
Common ring region			
6	3,200	3,200	1.0
Medium ring region			
8	0.99	0.75	1.3
9	0.19	0.088	2.2
10	0.11	0.028	3.9
Large ring region			
14	0.016	0.026	0.62
16	0.023	0.027	0.74
24	0.070	0.041	1.7

[a] Data from ref. 5.

high and is very nearly the same whether the reactions take place in 75% EtOH or in 99% DMSO (Table III). We are inclined to think that such an insensitivity to solvent transfer does not result from a fortuitous compensation of opposing effects but, rather, it derives from the invariance of the solvation states in the intramolecular species as compared to those involved in the intermolecular reaction. As in the case of the 5-membered rings the large EM value exclusively arises from structural factors.

In the medium and large ring regions, most of the EM values (Table III) are less than 1 and the EM^E/EM^D ratios are somewhat greater than 1 in the former region and range from 0.85 to 1.6 in the latter. A first, general comment on this behaviour is that solvation effects do not contribute to a marked extent to the reactivity differences affecting the intramolecular reaction as compared to the intermolecular reaction. It is found that this is a general phenomenon encompassing common, medium, and large rings. In particular, in the large ring region not only structural contributions to reactivity cancel out in the EM value, but solvent contributions do so as well and the EM parameter accordingly tends to level off towards similar values whatever reaction is considered(2).

Minor effects can be noted for the medium rings, however. Even though small, they are significant since they consistently appear with the same trend ($EM^E/EM^D > 1$) along the sequence, n = 8, 9, 10, which is known to include the highest strained rings.

Solvation in the cyclic transition state leading to a medium-sized ring may be affected by the steric congestion typical of such ring, mainly resulting from transannular interactions and bond opposition forces. The data indicate that steric strains presumably destabilize the transition state by desolvation more effectively in 99% DMSO than in 75% EtOH, which is a stronger solvator for the species involved in this kind of reactions.

A more detailed speculation may be attempted in terms of changes in bond-breaking and bond-making contributions to the transition state stability in the two solvents. Transfer from 75% EtOH to 99% DMSO is expected to increase the nucleophilicity of the oxide anion and to decrease the leaving group ability of bromine. The two factors oppose to each other, with the net result that motion along the reaction coordinate (parallel effect) will be little affected. However, the indicated solvent change would favour a simultaneous decrease of both nucleophile-carbon and carbon-leaving group distances (perpendicular effect) and cause the transition state to be tighter in 99% DMSO than in 75% EtOH. This is illustrated by a More O'Ferrall(6) diagram (Figure 1). As a consequence the stability of the transition state should be reduced to a greater extent in 99% DMSO than in 75% EtOH by the steric congestion of the medium ring to be formed.

Unfortunately, the formation of catechol polymethylene ethers does not allow to investigate rings smaller than 6. Although for 5-membered ring formation there was found little influence of the possible desolvation of the reacting open chain as expected from Bruice and Turner's hypothesis(3), better tests should be sought by changing ring size (n) along a reaction series down to n = 3. For

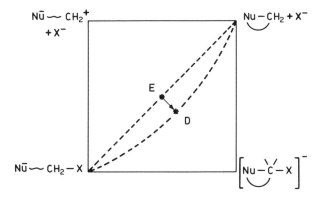

Fig. 1. Potential energy diagrams for an intramolecular S_N2 reaction of an anionic reactant showing the effect of changing solvent from protic to dipolar aprotic. E = 75% EtOH; D = 99% DMSO.

very short chains the solvation states for an intramolecular reaction should be distinctly different from those of its intermolecular counterpart (Figure 2). Lactone formation from ω-bromoalkanoate ions in 99% aqueous DMSO is a promising series of this kind as suggested by the entropies of activation(7,8). A partial plot of $\Delta S^{\neq}$ vs. n is shown (Figure 3). The entropy of activation for a bifunctional chain

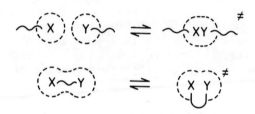

Fig. 2. Changes in solvation states accompanying cyclization of a very short bifunctional chain molecule as compared with the intermolecular model reaction.

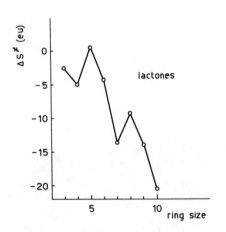

Fig. 3. $\Delta S^{\neq}$ plot vs. ring size for the lactonization of ω-bromo-alkanoate ions. Data from refs. 7 and 8 including a corrected value of +0.6 eu for ring size 5. Along this paper 1 eu = 4.187 J mol^{-1} K^{-1}

should consist of two main components, i.e., $\Delta S_{ir}^{\ne}$, due to the change of internal rotations around the single bonds along the chain and $\Delta S_{solv}^{\ne}$, due to the change in solvation on going from reactant to transition state. Total freezing of a single bond rotation causes a drop of about 4.5 e.u.(9) and cyclizations are accompanied by a general decrease in $\Delta S^{\ne}$ as n increases. The $\Delta S_{solv}^{\ne}$ component for a reaction involving an anionic nucleophile should be positive because the transition state should be less solvated than the reactant. Also, $\Delta S_{solv}^{\ne}$ should be constant as long as chains are long enough as to be unaffected by desolvation effects. If the hypothesis of desolvation of the functional ends of short chains is correct (Figure 2), the positive $\Delta S_{solv}^{\ne}$ contribution should decrease for small rings and the overall $\Delta S^{\ne}$ value be lower than expected. This may explain why in Figure 3 the $\Delta S^{\ne}$ values for n = 3 and 4 appear to be distinctly lower than those for n = 5 and 6.

Selected entropy data for the formation of small and common rings are reported in Table IV. Although limited in ring size range, they show for the very short chains an undeniable tendency to violate the expected trend toward more negative $\Delta S^{\ne}$ values with increasing chain length. Such anomalies were emphasized by Stirling(10) and find a possible explanation in the hypothesis of a reduced $\Delta S_{solv}^{\ne}$ contribution in the cyclization of a very short chains.

Solvent-induced conformational changes may also affect the tendency to chain cyclization. Bruice and Turner(3) found that succinate esters have intramolecular rate constants about 20-fold smaller in water than the phthalate esters do, whereas the rate constants for the two groups of esters are essentially the same in 1 M H_2O in DMSO. The observation was interpreted in terms of extended conformation of the succinate esters in water and predominant cisoid conformation in the H_2O - DMSO mixed solvent.

An extensive investigation of this problem was carried out by Winnik(11) in the intramolecular hydrogen transfer occurring on the decay of the excited triplet state of long-chain alkyl esters of 4-carboxy benzophenone (equation (4)). The rate constants, $k_{r,n}$, for this change, as obtained from flash photolysis experiments, showed to be affected by the solvent. The examined solvents included non polar solvents such as CCl_4, hydroxylic solvents such as AcOH and t-BuOH-H_2O and AcOH-H_2O mixtures, and the polar aprotic acetonitrile. The reaction occurs only if effectively ring-shaped conformations obtain to ensure contact between the alkane chain and the ketone carbonyl group, i.e., when the chain is no shorter than 9. Polar solvents increase the rate constant, $k_{r,n}$. The effects may arise from a change in the photoreactivity of the chromophore as well as from a change in the cyclization probability of the chain due to conformational dependence on solvent. The fact that significant solvent effects remain when a model intermolecular reaction is

Table IV. Entropy Data for Closure of Small and Common Rings.
1 eu = 4.187 J mol^{-1} K^{-1}

Reaction N°	Substrate	Solvent	Ring size n	$\Delta S^{\neq}$, eu
1[a]	$H_2N(CH_2)_{n-1}Cl$	50% dioxane	3	-15
			5	-13
2[a]	$C_6H_5NH(CH_2)_{n-1}Br$	60% EtOH	3	-11
			4	-11
			6	-17
3[a]	$C_6H_5NH(CH_2)_{n-1}Cl$	50% dioxane	3	-17
			5	-15
4[a]	$H_2NCH(C_6H_5)(CH_2)_{n-2}Cl$	50% dioxane	3	-7
			5	-10
5[b]	$p\text{-}CH_3C_6H_4S(CH_2)_{n-1}Cl$	80% EtOH (w/w)	3	-24
			5	-20
6[c]	$C_6H_5C(=O)(CH_2)_{n-2}Cl$	80% EtOH, Ag$^+$	4	-16.0
			5	-15.8
			6	-20.4
			7	-23.1
7[d]	$o\text{-}\bar{O}C_6H_4(CH_2)_{n-3}Br$	75% EtOH	5	+ 4.1
			6	+ 5.3
			7	- 3.2

[a] Data from R. Bird, A. C. Knipe, and C. J. M Stirling, J. Chem. Soc. Perkin TransII.:1215 (1973). [b] Data from R. Bird and C. J. M. Stirling J. Chem. Soc. Perkin TransII.:1221 (1973). [c] Data from D. J. Pasto and M. P. Serve, J. Am. Chem. Soc. 87:1515 (1965). Carbonyl oxygen participation is probably uncertain for compounds with n = 7, which show little anchimeric assistance. [d] Data from G. Illuminati, L. Mandolini, and B. Masci, J. Am. Chem. Soc. 97:4960 (1975). $\Delta S^{\neq}$ values up to ring size 10 are also available.

adopted and the effective molarity is determined therefrom does indeed indicate that the cyclization probably depends somewhat on solvent. The cyclization probability is enhanced in MeCN and 20% aqueous AcOH, whereas it is depressed in CCl_4. The overall rate changes involve factors close to 2.

Effects of desolvation in the transition state of cyclization reactions have been proposed to be responsible for rate depression (12). The intramolecular α-acylation of 2-ω-carboxyalkylthiophenes is shown to be appreciably retarded in the large ring region where strain effects are negligible. The EM values for this reaction are consistently found to be at least 10 times lower than those obtained for large ring formation which range between 0.01 and 0.05 M (Table V). The effect is attributed to disturbance of the solvation shell in the transition state by the crossing chain over

SOLUTE-SOLVENT INTERACTIONS IN RING FORMATION

Table V. Intramolecular Acylation of the Thiophene Nucleus.[a]

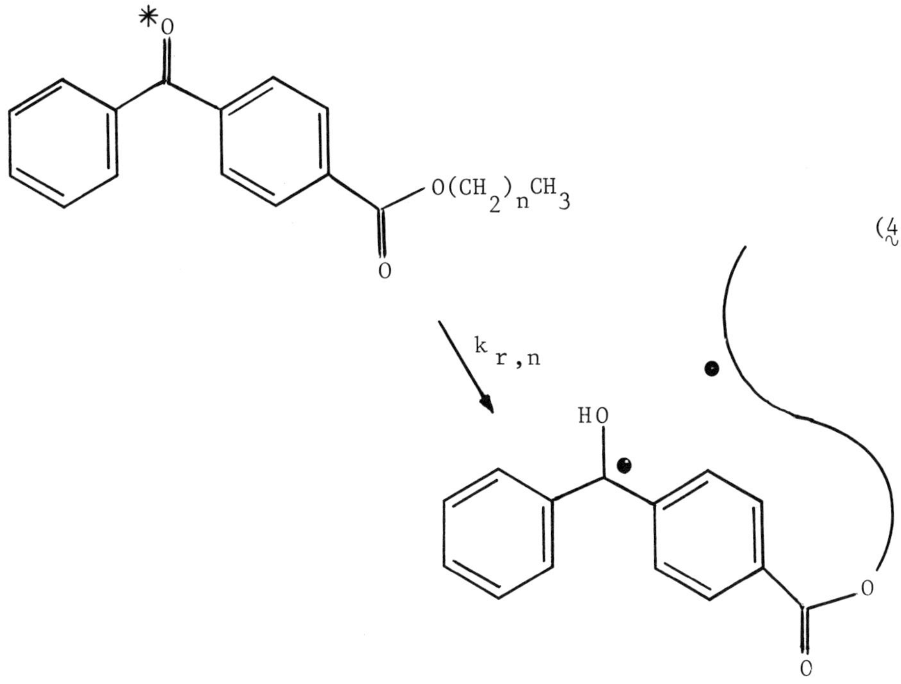

(4)

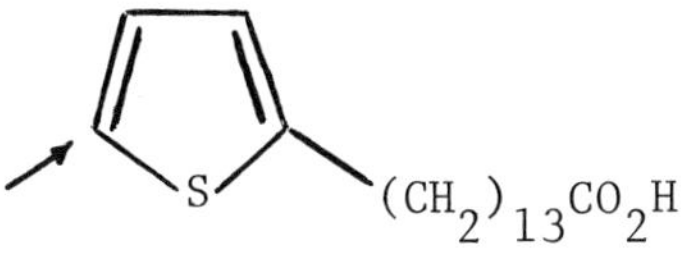

$k_{r,n}$

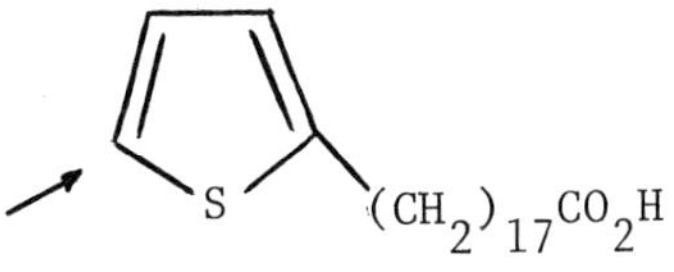

![thiophene-(CH$_2$)$_{13}$CO$_2$H] EM 1.5×10^{-3} M	![thiophene-(CH$_2$)$_{17}$CO$_2$H] EM 1.7×10^{-3} M

benzothiophene-(CH$_2$)$_{13}$CO$_2$H

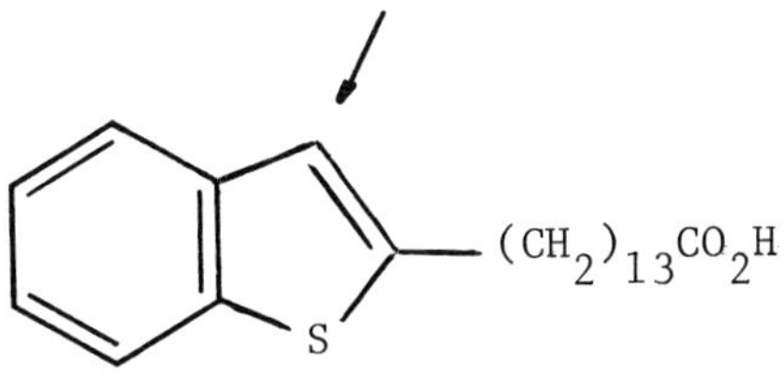

EM 1.1×10^{-2} M

[a] Cyclizations were carried out in CH$_3$CN at 50°C in the presence of $(CF_3CO)_2O-H_3PO_4$ (from ref. 12). The arrow indicates the position of attack.

the thiophene ring plane, as required by the commonly accepted geometry of the σ-adduct to be formed. This view is consistent with the finding that the intramolecular β-acylation of the 2-ω-carboxytridecyl benzothiophene, for which extensive desolvation is not expected, shows a "normal" EM value of 1.1×10^{-2} M.

In conclusion, although solvent effects as intrinsically bound to the cyclization phenomenon are generally small, they can arise from a number of situations including desolvation of a bifunctional short chain, interdependence of solvation and steric strains in the transition state of medium-ring formation, steric inhibition of solvation in special reaction series, and solvent-induced conformational changes. Some of the effects require further investigation for a more quantitative assessment of the energetic factors involved.

REFERENCES

1. G. Illuminati in "Techniques of Chemistry", M. R. J. Dack, ed., Vol. 8/2, ch. 12, p. 159, Wiley, New York (1976).
2. G. Illuminati, L. Mandolini, and B. Masci, J. Am. Chem. Soc. 99:6308 (1977).
3. T. C. Bruice and A. Turner, J. Am. Chem. Soc. 92:3422 (1970).
4. G. A. Dafforn and D. E. Koshland, Jr., J. Am. Chem. Soc. 99:7246 (1977).
5. A. Dalla Cort, G. Illuminati, L. Mandolini, and B. Masci, J. Chem. Soc. Perkin Trans. II, in press.
6. R. A. More O'Ferrall, J. Chem. Soc.(B), 274 (1970).
7. C. Galli, G. Illuminati, L. Mandolini, and P. Tamborra, J. Am. Chem. Soc. 99:2591 (1977).
8. L. Mandolini, J. Am. Chem. Soc. 100:550 (1978).
9. H. E. O'Neal and S. W. Benson, J. Chem. Eng. Data, 15:266 (1970).
10. C. J. M. Stirling, J. Chem. Ed. 50:844 (1973).
11. M. A. Winnik, Acc. Chem. Res. 10:173 (1977).
12. C. Galli, G. Illuminati, and L. Mandolini, J. Org. Chem. 45:311 (1980).

HYDROGEN ACIDITIES AND BRØNSTED RELATIONS

Andrew Streitwieser, Jr.

Department of Chemistry, University of California
Berkeley, California 94720, U.S.A.

Abstract - Hydrogen isotope exchange rates in methanolic sodium methoxide of hydrocarbons having delocalized carbanions give two separate linear Brønsted-type relations when log K values are plotted against the corresponding pK_{CsCHA} values, one of slope 0.36 at 45° applicable to fluorenes and indenes and another of slope 0.58 applicable to polyaryl methanes. Both calculations have been extended by measurements with additional compounds and the intersection of the two correlations has been characterized by phenylene derivatives. The results show that the proton transfer, even of extensively conjugated hydrocarbons, can show subtle variations between transition states and equilibrium carbanions.

1. INTRODUCTION

Carbon acids may be usefully dissected into the categories of Table I. Equilibrium acidity in solution is normally defined by equation (1), although in actual practice for many important classes of carbon acids, equation (2) is more appropriate.

Table I. Categories of Carbon Acids

Types of Acidity	Ion Type	Nature of Carbanion
Equilibrium	Ionic	Localized
Kinetic	Ion Pair	Delocalized

$$RH \xrightleftharpoons{K_a} R^- + H^+$$

$$pK(RH) = -\log K_a \qquad (1)$$

$$RH + B^- \xrightleftharpoons{K_{eq}} R^- + HB$$

$$pK(RH) = pK(HB) - \log K_{eq} \qquad (2)$$

In equation (2), B^- may be the lyate ion derived from solvent but in many determinations it is another carbon acid or indicator ion. The carbon acidity is then determined as an acidity difference between RH and BH. In solvent of lower polarity all of the ions are actually present as ion pairs. Such ion pairs are frequently further dissected into types such as "tight" or "loose", "contact" or "solvent-separated".

A number of reviews of carbon acidity have been published during the past decade (1-7). For more than two decades we have measured a number of equilibrium ion pair acidities using cesium cyclohexylamide in cyclohexylamine. These acidities, symbolized as pK_{CsCHA}, are defined by equation (3) in which 18.49 is the value for the pK of 9-phenylfluorene in aqueous solvents using the H_- method. A more complete discussion of this acidity scale with a summary of our pK_{CsCHA} values has recently been published(8).

$$RH + [\text{9-phenylfluorenyl}^- Cs^+] \xrightleftharpoons{K_{eq}} R^- Cs^+ + [\text{9-phenylfluorene}] \qquad (3)$$

$$pK_{CsCHA}(RH) = 18.49 - \log K_{eq}$$

Kinetic acidity is defined as the relative rate of proton abstraction by a base and is often measured by hydrogen isotope exchange. The linear free energy relationship often found between equilibrium and kinetic acidities for related groups of compounds is usually referred to as a Brønsted relationship (equation 4).

$$\log k_i/k_o = \beta(pK_o - pK_i) \qquad (4)$$

The proportionality constant β is usually, but not always, in the range from 0-1. This relationship recently has also been thoroughly reviewed(2,9). It is well known that the Brønsted relation can be valid for only a limited range. As the acid becomes weaker, the reprotonation of the anion becomes diffusion controlled and β approaches unity. As the acid becomes very strong the ionization of the acid approaches diffusion control and β approaches zero. For oxygen and nitrogen acids in aqueous solution the range between these two extremes is only a few pK units and the Brønsted equation applies to only a short range of acidity. An important requirement for such behavior is apparently that the acid be importantly involved in the hydrogen-bonded network of a hydroxylic solvent.

Carbon acids studied with respect to Brønsted relations include carbonyl compounds, nitroalkanes, sulfones and nitriles(2,9). Among the carbon acids the conjugated hydrocarbons constitute a unique class. They cover a broad range of pK values and the relative acidities depend almost wholly on delocalization of charge in the corresponding carbanions. For example, Table II summarizes the pK_{CsCHA} values of a number of hyrdrocarbons whose carbanions are expected to be approximately planar. For such systems, SCF-π MO calculations are applicable and the calculated π-energy differences between RH and R⁻ are also given. A plot of these quantities in Figure 1 shows that the π-electron energy differences alone provide an excellent correlation of acidity.

One consequence of such delocalization is that relative acidities of such hydrocarbons are remarkably insensitive to solvent and ion pairing. For example, Bordwell and his research group have determined the ionic acidities of a number of conjugated hydrocarbons in

Table II. SCF-π Calculations and Equilibrium Acidity[a]

Hydrocarbon	ΔE_π eV	pK_{CsCHA}[b]
Toluene	8.066	41.2
2-Methylnaphthalene	7.804	38.2
1-Methylnaphthalene	7.547	37.9
Diphenylmethane	6.398	(28.6)[c]
9-H-Benz[def]fluorene	5.796	22.9
10-H-Benz-[b]fluorene	5.788	23.7
Fluorene	5.775	23.0
Benzanthrene	5.699	21.43
6-H-Benz[cd]pyrene	5.625	19.91
7-H-Benz[c]fluorene	5.505	19.75
11-H-Benz[a]fluorene	5.421	20.35
Indene	5.419	19.9
Phenalene	5.277	18.49
Cyclopentadiene	5.132	16.25

[a] Adapted from ref. 10. [b] Statistically corrected; see ref. 8.
[c] Estimated for a hypothetical strain-free planar carbanion.

dimethyl sulfoxide (DMSO). Their pK_{DMSO} values are compared with our pK_{CsCHA} values in Table III and Figure 2. The excellent linear correlation with a slope close to unity demonstrates that structural changes affect the electrostatic interaction between the delocalized anionic charge and the large cesium cation within the contact ion pair to about the same degree as ionic solvation energies in DMSO.

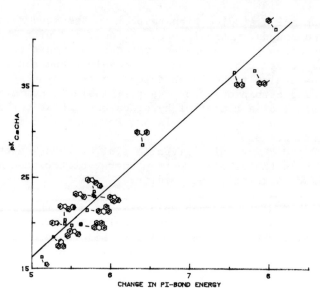

Fig. 1. Correlation of pK_{CsCHA} vs. SCF-π ΔE values. Data are summarized in Table II. The regression line is given by pK_{CsCHA} = -23.15 + 7.89 $ΔE_\pi$.

Table III. Comparison of pK_{CsCHA} and pK_{DMSO}

Compound	pK_{CsCHA} [a,b]	pK_{DMSO} [a,c]
9-Phenylfluorene	18.5	17.9
Indene	19.9	20.1
9-H-Benz|def|fluorene	22.9	22.4
9-Methylfluorene	22.3	22.3
Fluorene	23.0	22.9
10-H-Benz|b|fluorene	23.7	23.2
1,3,3-Triphenylpropene	26.6	25.9
9-Phenylxanthene	28.5	27.9
p-Biphenyldiphenylmethane	30.2	29.4
Triphenylmethane	31.45	30.6
Diphenylmethane	33.4	32.6

[a] per H. [b] Ref. 8. [c] Ref. 11.

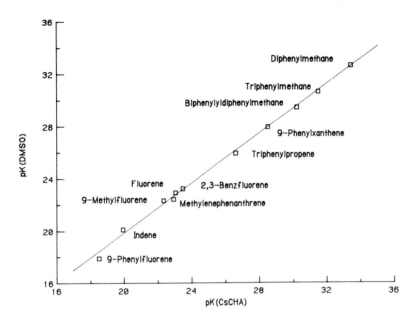

Fig. 2. Correlation of pK_{DMSO} vs. pK_{CsCHA}. The least squares regression line for the data in Table III: $pK_{DMSO} = 0.84 + 0.949 \times pK_{CsCHA}$.

Similar correlations apply to aqueous systems and to relative ion pair acidities in ethers(8).

2. FLUORENES AND INDENES

We have shown previously that fluorene and indene hydrocarbons give an excellent linear Brønsted correlation between pK_{CsCHA} and proton exchange rates with methanolic sodium methoxide(12). Although Brønsted correlations generally compare kinetic and equilibrium acidities in the same solvent, the general independence of the relative acidities of conjugated hydrocarbons to the medium provides adequate justification for the use of pK_{CsCHA} in this case.

Linearity was demonstrated over a range of 8 pK units. We have now extended the correlation by an additional 2 pK units. Of particular significance is the extension to 1,3-diphenylindene, DPI, whose pK_{CsCHA} is 13.6. The pK_{MeOH} of this compound has been measured directly(13); the value of 19.8 is within 3 units of that for the solvent methanol, 16.9(14). Curvature is frequently observed in Brønsted plots as the ΔpK between acid and base tends towards zero(2).

Some results with 1,3-diphenylindene-3-t are summarized in Table IV. The primary isotope effect, k_D/k_T = 2.3 at 25°, is essentially unchanged from that of fluorene which is less acidic by almost

Table IV. Kinetic Acidities of Hydrocarbons

Compound	pK_{CsCHA}	Temp °C	$k_2 T^a$ $M^{-1} s^{-1}$	K_D/k_T
1,3-Diphenylindene[b]	13.6	0	0.011	2.4
		25	.125	2.3
		45	$(.67)^c$	
2,7-Di-t-butylfuorene[d]	24.2	45	9.7×10^{-5}	
p-Methylbiphenyl	38.95	100	2.1×10^{-10}	
Toluene	41.2	100	$\sim 4 \times 10^{-11}$ [f]	
		178	3.5×10^{-7} [g]	1.05
6-H-Benz[cd]pyrene	19.9	45	3.0×10^{-3} [h]	
Benzanthrene	21.43	45	5.9×10^{-4} [i]	

[a] Second order rate constant for tritium exchange in $NaOCH_3/CH_3OH$.
[b] Unpublished results of C. C. Shen, J. R. Murdoch and J. T. Murphy.
[c] Extrapolated from results at other temperatures. [d] Unpublished results of C. C. Shen, [e] Ref. 18. [f] Ref. 20. [g] Ref. 19. [h] Unpublished results of F. Guibé. [i] Unpublished results of J. T. Murphy.

10 pK units. With the addition of 2,7-di-t-butylfluorene-9-t, tritium exchange rates now cover a range of over 10 pK units. The correlation is still linear over this range. This fact and the isotope effect comparisons strongly suggest that the mechanism for proton transfer and the structure of the transition state in methanolic $NaOCH_3$ are essentially the same over the entire range even as the system approaches the pK of methanol itself. The extended correlation is shown in Figure 3.

This correlation can also be used to estimate pK values of related hydrocarbons whose exchange rates have been measured. For example, Hine and Knight(15) have recently determined the rate of proton abstraction from 6,6-dimethylfulvene with sodium methoxide in CH_3ONa at 35°. Using known or estimated values for $k(CH_3OD)/k(CH_3OH)$, primary isotope effects and activation energy(12), we estimate tritium exchange at 45° in CH_3OH to have $\log k_2 = -4.0$ which corresponds to a $pK_{CsCHA} = 24.7$.

3. POLYARYLMETHANES

In our earlier work we also discovered the remarkable fact that polyarylmethanes give a linear Brønsted correlation between kinetic acidity in methanolic sodium methoxide and pK_{CsCHA} which is different from that given by the fluorenes and indenes(16). Triphenylmethane and related hydrocarbons are less reactive than the fluorenes and most of the kinetics were run at temperatures around 100°. The Brønsted slope determined from five compounds is 0.46 at 100° (Figure 4) and

HYDROGEN ACIDITIES AND BRØNSTED RELATIONS

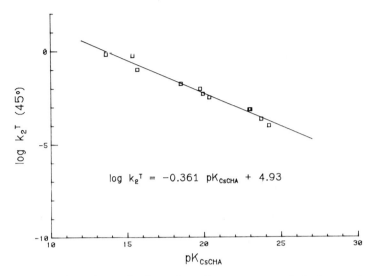

Fig. 3. Extended Brønsted plot for fluorenes and indenes from data of Table IV and ref. 12. The far left point is 1,3-diphenyl-indene; the far right is 2,7-di-t-butylfluorene.

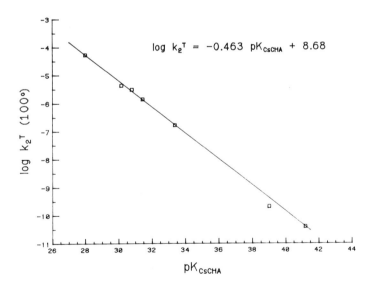

Fig. 4. Brønsted plot for arylmethanes at 100°. The least squares line is that of the first five points (ref. 16). The last two are p-methylbiphenyl and toluene (Table IV).

extrapolates to 0.58 at 45° for comparison with the value of 0.36 for fluorenes.

We are interested in extending the correlation to higher pKs. p-Methylbiphenyl has pK_{CsCHA} = 38.95(17). In 0.5M NaOCH$_3$ in methanol containing CH$_3$OT the hydrocarbon picks up tritium slowly but linearly at 100°. Material recovered after three months showed 0.06% reaction (Table IV). This number has been corrected for the 21% incorporation of tritium into the rings established by oxydation to p-biphenylcarboxylic acid(18).

Toluene is less reactive still. At 178° in methanolic sodium methoxide the kinetics of loss of isotope from a mixture of toluene-α-d and α-t gives k_D/k_T = 1.05(19). Only approximate rates are available for incorporation of tritium at 100° (Table IV)(20). The results for p-methylbiphenyl and toluene provide the two additional points at high pK shown in Figure 4. Note that these points were not used to determine the least squares correlation, yet they fit very well. The correlation appears to be linear over a range of 13 pK units.

The unchanging slope even for toluene may, however, be an artifact. The extrapolated k_2^H for toluene at 25° $\simeq 10^{-16}$ M^{-1} s^{-1}. If k_{-2} is of diffusion-controlled magnitude, $\sim 10^{10}$ M^{-1} s^{-1}, K_{eq} for equation (2) is about 10^{-26} or, since pK_{MeOH} (MeOH) = 17, pK_{MeOH} (toluene) $\simeq$ 43. This number is close to the value of pK_{CsCHA} = 41.2 but there are two problems. One is that a Brønsted slope of <1 implies that the back reaction is <u>not</u> diffusion controlled. The other problem is that from the pK measurements of diphenylindene, pK_{MeOH} of conjugated hydrocarbons are 6-7 units higher than the corresponding pK_{CsCHA}. However, a pK_{MeOH} of about 47 for toluene would require a back reaction of benzyl anion with methanol that is faster than a diffusion rate. The most probable resolution of this dilemma is that benzyl anion is more strongly solvated by hydrogen bonding in methanol than more delocalized carbanions and that the pK difference, pK_{MeOH} - pK_{CsCHA} for toluene is less than the 6-7 appropriate for other hydrocarbons Thus, in Figure 4, a number somewhat <u>less</u> than the actual pK_{CsCHA} for toluene would probably be more appropriate and, as a result, the Brønsted slope at toluene is actually closer to unity.

4. PHENALENES

The question remains: why do two such closely related groups of compounds produce such markedly different Brønsted plots? Fluorenes and indenes are derivatives of cyclopentadiene and one wonders if the magic properties of 4n + 2 ring systems are responsible. To test this point we examined some phenalene hydrocarbons that are relatively acidic but are not cyclopentadienes.

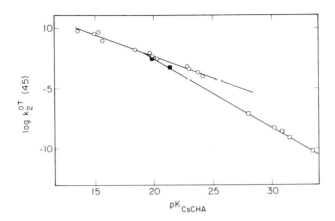

Fig. 5. Combined Brønsted plot including the phenalene derivatives (black squares).

At 45° the polyarylmethane correlation intersects that of the fluorenes at $pK_{CsCHA} = 19.2$, $\log k_2^T = -2.0$. The phenalene hydrocarbons have pK_{CsCHA} in this region(10). Some values and kinetic results are summarized in Table IV.

A combined Brønsted correlation is shown in Figure 5. The phenalene derivatives do indeed form an extension of the polyarylmethane correlation. The fact that benzanthrene in particular is significantly less reactive than a fluorene of comparable acidity suggests that the cyclopentadienyl moiety is more efficient at delocalizing negative charge. Accordingly, relatively little reorganization is required for effective delocalization of charge on proton transfer. That is, the transition state is reached at a relatively early stage in proton transfer. For the polyarylmethanes more reorganization is required and the transition state may be described as late. This analysis is also consistent with the observed patterns of primary isotope effects. The analysis also suggests that for these conjugated hydrocarbons the Brønsted β is a valid measure of charge delocalized at the transition state compared to the equilibrium carbanion.

Acknowledgment. This research was supported in the past by USPH NIH grant no. GM12855 and more recently by NSF grant no. CHE79-10812. I am also indebted to past and present members of my research group for the experimental work cited in this paper.

REFERENCES

1. J. R. Jones, "The Ionization of Carbon Acids", Academic Press, New York (1973); see also Quart. Rev. Chem. Soc. 25:365 (1971); Prog. Phys. Org. Chem. 9:241 (1972).
2. R. P. Bell, "The Proton in Chemistry", 2nd ed., Cornell University Press, Ithaca, N. Y. (1973).
3. E. Buncel, "Carbanions. Mechanistic and Isotopic Aspects", Elsevier, Amsterdam (1975).
4. O. A. Reutov, I. P. Beletskaya, and K. P. Butin, "CH-Acids", Pergamon Press, Oxford (1978); see also O. A. Reutov, K. P. Butin and I. P. Beletskaya, Russian Chem. Revs. 43:1 (1974).
5. M. Schlosser, "Polare Organometalle", Springer-Verlag, Berlin (1973).
6. M. Szwarc, A. Streitwieser, and P. C. Mowery, in "Ions and Ion Pairs in Organic Reactions", M. Szwarc, ed., Vol. 2, Ch. 2, Wiley Interscience, New York (1974).
7. F. G. Bordwell, Pure and Appl. Chem. 49:963 (1977).
8. A. Streitwieser, Jr., E. Juaristi, and L. L. Nebenzahl, in "Comprehensive Carbanion Chemistry", E. Buncel and T. Durst, eds., Elsevier Publishing Co., Amsterdam, in press; an earlier review for Japanese readers is available in J. Synth. Org. Chem. Japan, 33:797 (1975).
9. F. Hibbert in "Comprehensive Chemical Kinetics", C. H. Bamford and C. F. H. Tipper, eds., Vol. 8, Elsevier Publishing Co., Amsterdam (1977).
10. A. Streitwieser, Jr., J. M. Word, F. Guibé, and J. S. Wright, Tetrahedron Letters, submitted.
11. M. S. Matthews, J. E. Bares, J. E. Bartmess, F. G. Bordwell, F. J. Cornforth, G. E. Drucker, Z. Margolin, R. J. McCallum, G. J. McCollum, and N. R. Vanier, J. Am. Chem. Soc. 97:7006 (1975), and references cited therein.
12. A. Streitwieser, Jr., W. Hollyhead, A. Pudjaatmaka, P. H. Owens, T. Kruger, P. Rubenstein, R. McQuarrie, M. Brokaw, W. Chu, and H. M. Niemeyer, J. Am. Chem. Soc. 93:5088 (1971).
13. A. Streitwieser, Jr., C. J. Chang, and A. T. Young, J. Am. Chem. Soc. 94:4888 (1972).
14. J. Koskikallio, Suom. Kemistilehti B, 30:111 (1957).
15. J. Hine and D. B. Knight, J. Org. Chem. 45:991 (1980).
16. A. Streitwieser, Jr., W. Hollyhead, G. Sonnichsen, A. Pudjaatmaka, C. J. Chang, and T. Kruger, J. Am. Chem. Soc. 93:5096 (1971).
17. A. Streitwieser, Jr. and F. Guibé, J. Am. Chem. Soc. 100:4532 (1978).
18. Unpublished experiments of J. Cambray.
19. Unpublished results of T. A. Keevil.
20. Unpublished experiments of A. P. Marchand and J. S. Wright.

SOLVENT EFFECTS ON SOME NUCLEOPHILIC SUBSTITUTIONS

Michael H. Abraham

Department of Chemistry, University of Surrey
Guildford, U. K.

Abstract - The effects of hydroxylic solvents on $\Delta G^{\ddagger}$ values for the solvolysis of t-butyl chloride have been dissected into initial-state and transition-state contributions. The $\Delta G°_t(Tr)$ values, after correction for the "size" or "cavity" effect, are well correlated with values for the Me_4N^+ Cl^- ion-pair and for α-aminoacids. Similar calculations have been carried out for other alkyl halides and for S_N2 reaction of triethylamine with ethyl iodide. Dissections of $\Delta H^{\ddagger}$ and $\Delta S^{\ddagger}$ for these S_N1 and S_N2 reactions into initial-state effects are also reported.

1. INTRODUCTION

The first systematic investigation on the influence of solvent on reaction rates was reported by Menschutkin(1) as long ago as 1890. Quite soon after this study, chemists began to consider whether or not solvent influences on reaction rates were connected with the effect of solvents on the reactants (i.e. with initial-state effects). However, a careful and extensive investigation by Von Halban(2) in 1913 showed conclusively that for the reaction of trimethylamine with p-nitrobenzyl chloride, solvent effects on the reactants could not account quantitatively for the overall influence of solvent on the reaction rate constant. Little further progress was made on these lines until the advent of transition state theory, when it then became clear that in principle it was possible to dissect the influence of solvent on rate constants into initial-state and transition-state contributions(3-5).

Thus Grunwald and Winstein(6), and then Winstein and Fainberg (7), used the vapour pressure data of Olsen and co-workers(8,9) to obtain the effect of aqueous alcoholic solvents on the Gibbs energy of t-butyl chloride, and combined these results with kinetic measurements to deduce the solvent effect on the transition state. Some years later, Arnett and co-workers(10-14) carried out an analysis in terms of enthalphy (and hence entropy) for the solvolysis of t-butyl chloride in the aqueous alcoholic solvents used before, and Rudakov and Tretyakov(15) extended the Gibbs energy measurements to a series of pure solvents. Bimolecular reactions were also examined, Haberfield and co-workers(16-18) carrying out analyses in terms of enthalpy on Menschutkin reactions, and Eckert and co-workers(19-21) discussing the Diels-Alder reaction. Many other studies were also reported(22-30), so that by the early 1970's the above work together with that of Parker(31) on bimolecular reactions involving ionic species had led to an entirely new method for the examination of transition states.

The relevant equation for the dissection of solvent effects in terms of the Gibbs energy may be deduced from the Pronsted-Bjerrum equation 1 for the reaction of Q and R through a transition state Tr(3-5,32,33). Here, k_o is the rate constant in a standard solvent and k is the rate constant in any other solvent. The terms γ_x, γ_R and γ_{Tr} are primary medium activity coefficients. Equation 1 can be recast as equation 2, and the latter simplified by defining the

$$k/k_o = \gamma_Q \cdot \gamma_R / \gamma_{Tr} \tag{1}$$

quantity ΔG_t^o, the standard free energy of transfer of a given species

$$RT\ln \gamma_{Tr} = RT\ln \gamma_Q + RT\ln \gamma_R - RT\ln k/k_o \tag{2}$$

from a standard solvent A to another solvent B. The change in the

$$\Delta G_t^o(Tr) = \Delta G_t^o(Q) + \Delta G_t^o(R) + \delta \Delta G^\ddagger \tag{3}$$

activation free energy, $\Delta G^\ddagger = \Delta G_B^\ddagger - \Delta G_A^\ddagger$ is given by $\delta \Delta G^\ddagger = -RT\ln (k_B/k_A)$. Alternatively, equation 3 can be deduced(34) by considering the solvent effect on the quasi-equilibrium 4.

$$Q + R \rightleftharpoons Tr \tag{4}$$

The problem of dissection of solvent effects thus reduces to the determination of rate constants in a series of solvents to obtain the $\delta \Delta G^\ddagger$ term, and the determination of values of ΔG_t^o for the reactants by standard thermodynamic measurements. This paper is restricted to reactions in which the reactants are nonelectrolytes. For such compounds that are not too involatile, ΔG_t^o values can be obtained through measurements of Raoult's Law activity coefficients of the solute at infinite dilution in a series of solvents, or through the corresponding Henry's Law constants, equation 5 and 6.

If a reactant is solid at the temperature in question, ΔG_t^o values are often obtained through solubility measurements, via equation 7

$$\gamma^\infty = \left(\frac{P}{P^o X}\right)_{X \to 0} \qquad K^H = \left(\frac{P}{X}\right)_{X \to 0} \qquad (5)$$

$$\Delta G_t^o = RT\ln(\gamma_B^\infty/\gamma_A^\infty) = RT\ln(K_B^H/K_A^H) \qquad (6)$$

where S_B and S_A are the solubilities in the respective solvents. Equation 7 is only valid if the same solid is in equilibrium with solvents A and B (i.e. if no solid solvates are formed) and rigorously should be applied only to sparingly soluble species; however, in the case of nonelectrolytes the latter condition is not too critical.

$$\Delta G_t^o = -RT\ln(S_B/S_A) \qquad (7)$$

Values of ΔG_t^o may also be obtained by partition measurements of the solute between pairs of immiscible solvent, but this method of course is restricted in terms of solvents. In all investigations of ΔG_t^o values, it is necessary to specify the concentration units employed, and to ensure that in the basic equations 2 and 3 the concentration units of the rate constant (for second-order constants) are the same as those used in the calculation of the ΔG_t^o values. In the present paper, the concentration units in solution will be mol fractions, and all second-order rate constants will be corrected to units of (mol fraction)$^{-1}$s^{-1} before calculation of the $\delta\Delta G^{\ddagger}$ values (34).

2. THE SOLVOLYSIS OF t-BUTYL CHLORIDE

If only one reactant needs to be considered, equation 3 reduces to equation 8, where RX signifies an alkyl halide. Values of

Table I. Calculation of $\Delta G_t^o(Tr)$ for Solvolysis of t-Butyl chloride[a]

Solvent	γ^∞	ΔG_t^o(t-BuCl)	$\delta\Delta G^{\ddagger}$	ΔG_t^o(Tr)
Water	2.2 x 10^4	4.57	-6.22	-1.65
MeOH	10.0	0	0	0
EtOH	6.40	-0.26	1.32	1.06
1-PrOH	5.59	-0.34	1.68	1.34
1-BuOH	4.62	-0.46	1.94	1.48
2-PrOH	6.04	-0.30	2.24	1.94
t-BuOH	4.60	-0.46	2.96	2.50

[a] ΔG values in kcal mol^{-1}[†], mol fraction scale, from refs. 29 and 34.

[†] along this paper 1 cal = 4.187 J

ΔG_t^o(t-BuCl) have been deduced from γ^∞ values obtained by the technique

$$\Delta G_t^o(Tr) = \Delta G_t^o(RX) + \delta \Delta G^\ddagger \qquad (8)$$

of gas chromatographic head-space analysis, and in Table I are all the relevant results needed to calculate $\Delta G_t^o(Tr)$ for the solvolysis of t-butyl chloride; methanol is taken as the reference solvent, A (29,34). The ΔG_t^o(t-BuCl) values are not at all exceptional. It is known that for nonpolar or slightly polar solutes there is a very large solvent effect on transfer to or from water, or aqueous organic mixtures but that for transfers between nonaqueous solvents, ΔG_t^o values are generally quite small(34-36). In Table II are given ΔG_t^o values for a number of nonpolar and other solutes to illustrate the trends; the ΔG_t^o values for t-butyl chloride are quite close to those for n-pentane, a nonpolar solute of almost the same size as t-butyl chloride. However, the values of ΔG_t^o for the transition state, Table I, do not resemble those for nonpolar or slightly polar solutes (nitromethane has a dipole moment of 3.5 D); clearly more polar species need to be examined as models for the transition state. Some years ago(24,29), the ion-pair $Me_4N^+Cl^-$ was suggested as a suitable model, and more recently, ΔG_t^o values for the transition state have been compared to those for zwitterionic α-aminoacids(35). The relevant ΔG_t^o values are collected in Table III.

Table II. Values of ΔG_t^o for Solutes, kcal mol^{-1} on the mol fraction scale[a]

Solute: V(ml mol^{-1}): Solvent	Ethane 55	Pentane 116	Octane 164 ΔG_t^o	MeCOEt 90	MeNO$_2$ 54	t-BuCl 111
Water	2.83	4.90	7.17	1.40	1.04	4.57
MeOH	0	0	0	0	0	0
EtOH	-0.30	-0.56	-0.65	0.03	0.16	-0.26
1-PrOH	-0.46	-0.82	-0.95	0.05	0.32	-0.34
1-BuOH	-0.59	-0.99[b]	-1.13	-0.03	0.30	-0.46
2-PrOH	-0.44[b]	-0.66[b]	-0.98	-0.08	0.33	-0.30
t-BuOH	-0.73[b]	-1.01[b]	-1.25	-	0.32	-0.46
PC	-0.05	0.13	0.33	-0.22	-1.10	-0.22
DMSO	0.25	0.50	0.63	0.01	-1.27	0.08
MeCN	-0.11	-0.17	0.07	-0.31	-1.07	-0.39
MeCOMe	-0.53	-0.96	-0.92	-0.49	-1.02	-0.83
Benzene	-0.79	-1.51	-2.01	-0.36	-0.25	-1.06
Hexane	-1.22	-1.91	-2.35	0.40	1.11	-1.00

[a] From refs. 34-36. [b] Estimated values, ref. 37

Table III. Comparison of ΔG_t^o values for the t-Butyl Chloride Solvolysis Transition State with Values for Polar Species(34, 35,37)

Solute V(ml mol^{-1}) Solvent	Glycine 57	α-Amino caproic acid 122	α-Phenylalanine 135 ΔG_t^o	Me$_4$N$^+$Cl$^-$ 107	Tr 116
Water	-3.41	-0.90	-0.93	-3.0	-1.65
MeOH	0	0	0	0	0
EtOH	1.22	1.03	0.86	1.5	1.06
1-PrOH	1.39		1.29	1.9	1.34
1-BuOH	1.65	1.44	1.32	2.2	1.48
2-PrOH	1.52		1.53	2.3	1.94
t-BuOH	1.74		1.88	3.9	2.50

Plots of $\Delta G_t^o(Tr)$ against ΔG_t^o (Me$_4$NCl) or ΔG_t^o(α-Phenylalanine) yield quite good straight lines with slopes of 0.624 and 1.394 respectively, so that, perhaps surprisingly, the transition state responds to change of solvent to a larger extent than does the α-amino-acid. A plot of $\Delta G_t^o(Tr)$ against ΔG_t^o(Glycine) does not give a very good straight line, however. One major reason for these effects can be seen by inspection of the results in Table II. For nonpolar solutes such as ethane, pentane, and octane, there are substantial changes of ΔG_t^o with change of solvent from methanol to the less polar alcohols, and very large changes on transfer from methanol to water. Furthermore, these "nonpolar" effects depend on the size of the solute, becoming larger as the solute size increases. Thus if ΔG_t^o values are compared between species of different size, there will be a contribution, possibly quite large, due to the "nonpolar" or "size" effect. Many workers have recognised such effects(38,39), and, indeed, it has been shown(28) that ΔG_t^o values for transfer from methanol to water can usefully be broken down into a neutral or non-electrostatic contribution, ΔG_N, and an electrostatic contribution, ΔG_E, equation 9, before comparisons are made between different species. In equation 9, the ΔG_N term is calculated from data on ΔG_t^o

$$\Delta G_t^o = \Delta G_N + \Delta G_E \tag{9}$$

values for nonpolar solutes, and the ΔG_E term obtained by difference. Quite recently, equation 9 has been applied to the ΔG_t^o values given in Table III to obtain the corresponding ΔG_E quantities for all species. Another approach(37) is to consider the solution of a solute to involve firstly creation of a cavity in the solvent, followed by insertion of the solute into the correctly-sized cavity and subsequent interaction of the solute with the surrounding solvent. In order to probe solute-solvent interactions, the work needed to create the cavity must be subtracted out, for example through equation 10.

Table IV. Calculation of the ΔG_E and ΔG_{INT} Terms for the Transition State[a]

Solvent	ΔG_t^o	ΔG_N	ΔG_E	ΔG_t^o	ΔG_{CAV}	ΔG_{INT}
Water	-1.65	4.78	-6.43	-1.65	3.86	-5.51
MeOH	0	0	0	0	0	0
EtOH	1.06	-0.47	1.53	1.06	-0.13	1.19
1-PrOH	1.34	-0.75	2.09	1.34	-0.18	1.52
1-BuOH	1.48	-0.99	2.47	1.48	0.14	1.34
2-PrOH	1.94	-0.66	2.60	1.94	-0.55	2.49
t-BuOH	2.50	-1.01	3.51	2.50	-0.49	2.99

[a] In kcal mol^{-1}. The ΔG_{CAV} term is that calculated by SPT.

Two methods have been used to calculate the cavity term(37), the

$$\Delta G_t^o = \Delta G_{CAV} + \Delta G_{INT} \qquad (10)$$

scaled-particle theory of Pierotti (SPT method)(40) and the theory of Halicioglu and Sinanoglu as modified by Reisse and Moura Ramos (SRMR method)(41), see Table IV. If ΔG_E or ΔG_{CAV} for the transition state are plotted against values of ΔG_E or ΔG_{CAV} for the solutes shown in Table III, excellent straight lines are obtained, not only for Me$_4\overset{+}{N}$Cl$^-$ and α-phenylalanine but also for glycine as well. Details of the slopes, m, of the obtained straight lines are in Table V. After correction via equation 9 or equation 10, the values for all three α-aminoacids are reasonably constant. The value of the slopes, m, of the plots of transition-state ΔG_E or ΔG_{INT} quantities against those for the Me$_4\overset{+}{N}$Cl$^-$ ion-pair are denoted(27) as Z values, the average Z value (Table V) being 0.84; this is very close to the value of 0.85 obtained previously with rather less data(27). It is tempting to equate the Z value with the extent of charge separation in the transition state and thus with the extent of progress along the

Table V. Plots of Transition-state values against those for Model Solutes(37)

Function plotted : Model Solute	ΔG_t^o	ΔG_E m[a]	ΔG_{INT}(SPT)	ΔG_{INT}(SRMR)
Glycine	≈ 0.71	1.13	0.97	1.24
α-Aminocaproic acid	1.03	1.06	1.11	1.03
α-Phenylanine	1.39	1.01	1.11	0.98
Me$_4$N$^+$Cl$^-$	0.62	0.83	0.79	0.90

[a] Slope of the straight lines obtained.

SOLVENT EFFECTS ON SOME NUCLEOPHILIC SUBSTITUTIONS 347

reaction coordinate, α. Indeed, the present value of $Z = 0.84$ is near to that of 0.8 units deduced by Clarke and Taft(43) from a study of salt effects. Albery and Kreevoy(42) have recently suggested that Z is related to α via the equation $\alpha = \sqrt{Z}$, but for the solvolysis of t-butyl chloride and also for a number of Menschutkin reactions (see later), use of Z itself correlates better with other measures of α than does $\sqrt{Z}$.

If ΔG_E or ΔG_{INT} values for the transition state are plotted against those calculated for the ($Me_4N^+ + Cl^-$) pair of dissociated ions, the obtained m value is only about 0.5, thus demonstrating that the pair of dissociated ions does not resemble the transition state. In the reaction scheme for solvolysis, equation 11 with RX = t-BuCl, it can therefore be deduced that the step k_1 is the determining process, so that $k_1^{obs} = k_1$ and the transition state lies between the t-BuCl initial state and the first-formed $t\text{-Bu}^+Cl^-$ ion-pair.

$$t\text{-BuCl} \underset{k_{-1}}{\overset{k_1}{\rightleftarrows}} t\text{-Bu}^+Cl^- \longrightarrow t\text{-Bu}^+ + Cl^- \longrightarrow \text{Product} \qquad (11)$$
$$\searrow \text{Product}$$

The values of m (Table V) using α-aminoacids as model solutes are quite high, for glycine m averages 1.11, so that if the aminoacids had a charge separation of unity, this would imply an unrealistically high value of Z for the transition state. Pople and co-workers(44), however, have shown that in the gas phase, the separation of charge on zwitterionic glycine is less than unity, and if in solution the effective charge separation is perhaps around 0.7-0.8 units, then it becomes clear why glycine and other α-aminoacids are good models for the t-butyl chloride transition state (after correction for the size effect).

Several years ago, initial-state and transition-state contributions were reported for the solvolysis of t-butyl chloride in a-queous alcoholic solvents(6,7). In these solvents, the size effect is very critical and unless suitably sized model solutes are available or unless the size effect can be corrected, quantitative deductions on the nature of the transition state cannot be made, see ref. 45.

Dissection of solvent effects on $\Delta H^{\ddagger}$ and $\Delta S^{\ddagger}$ into initial-state and transition-state contribution have also been carried out(46), Tables VI and VII, but in the absence of ΔH_t^o and ΔS_t^o values for suitable model solutes, little quantitative deductions can be made. Comparison of transition-state values for those of the dissociated pair of ions, ($Me_4N^+ + Cl^-$), shows again that the pair of ions is not at all a good model for the transition state.

In the cited work(46), values of $\Delta H_t^o(t\text{-BuCl})$ were obtained

Table VI. Calculation of $\Delta H_t^o(Tr)$ for the Solvolysis of t-Butyl Chloride[a]

Solvent	ΔH_s^o(t-BuCl)	ΔH_t^o(t-BuCl)	$\delta \Delta H^{\ddagger}$	$\Delta H_t^o(Tr)$
Water	–	–1.7[b]	–1.7	–3.4
MeOH	0.36	0	0	0
EtOH	0.34	0.0	1.0	1.0
1-PrOH	0.43	0.1	0.9	1.0
1-BuOH	0.53	0.2	–1.5	–1.3
2-PrOH	0.84	0.5	–0.6	–0.1

[a] Values in kcal mol^{-1} at 298 K from ref. 46. [b] Estimated value.

through measurement of the enthalpy of solution at infinite dilution, ΔH_s^o, of t-BuCl in a series of alcohols, equation 12. The values of

$$\Delta H_t^o = \Delta H_{s,B}^o - \Delta H_{s,A}^o \qquad (12)$$

$\Delta H_t^o(Tr)$ were then obtained through an equation similar to equation 8; the ΔS_t^o values were then calculated from the corresponding values of G and H.

$$\Delta H_t^o(Tr) = \Delta H_t^o(\text{t-BuCl}) + \delta \Delta H^{\ddagger} \qquad (13)$$

3. THE SOLVOLYSIS OF ALKYL HALIDES

Solvent effects of aqueous alcoholic mixtures on $\Delta G^{\ddagger}$ have also been directed into initial-state and transition-state contributions for a few other alkyl halides(34), but no quantitative conclusions were reached except for solvolyses in methanol-water mixtures when equation 9 was applied in order to obtain ΔG_E values for the transition states(28,34). It is now possible to extend this type of

Table VII. Calculation of $\Delta S_t^o(Tr)$ for the Solvolysis of t-Butyl Chloride[a]

Solvent	ΔS_t^o(t-BuCl)	$\delta \Delta S^{\ddagger}$	$\Delta S_t^o(Tr)$
Water	–21[b]	15	–6
MeOH	0	0	0
EtOH	0.8	–1	0
1-PrOH	1.4	–3	–2
1-BuOH	2.1	–11	–9
2-PrOH	2.6	–10	–7

[a] Values in cal K^{-1} mol^{-1} at 298 K from ref. 46. [b] Estimated value.

Table VIII. Values of $\Delta G_t^o(Tr)$ and $\Delta G_E(Tr)$ for Some Solvolyses Transition States[a]

RX : Solvent	Ph_2CHCl	PhCHMeCl	$PhCH_2Cl$	n-PrCl	
		$\Delta G_t^o(Tr)$			
Water	2.21	−0.48	1.88	1.18	
MeOH	0	0	0	0	
EtOH	1.3	1.2	0.70		
2-PrOH	2.6				
		$\Delta G_E(Tr)$			$\Delta G_E(Me_4\overset{+}{N}\overset{-}{Cl})$
Water	−5.48	−6.30	−3.12	−2.68	−7.40
MeOH	0	0	0	0	0
EtOH	1.92	1.74	1.19		1.94
2-PrOH	3.56				2.93

[a] Values in Kcal mol^{-1} at 298 K, this work

analysis to other solvents; details of the $\Delta G_t^o(Tr)$ and $\Delta G_E(Tr)$ values thus obtained are in Table VIII. Comparison of the values of $\Delta G_E(Tr)$ with those of ΔG_E for the $Me_4\overset{+}{N}\overset{-}{Cl}$ ion-pair allows, again, estimates of Z. These are in Table IX, together with Z values obtained previously(28,34) for the methanol-water system. The general gradation of Z in the sense primary-alkyl halide < sec - alkyl halide < tert-alkyl halide is in accord with recent work suggesting that the extent of nucleophilic solvent participation decreases gradually along the same series(47,48). The greater the extent of nucleophilic participation, the less carbonium ion character is there in the transition state, and the "earlier" is the transition state.

4. THE MENSCHUTKIN REACTION

In view of the very large body of work reported on the Menschutkin reaction of amines with alkyl halides, there have been very few investigations in which solvent influences have been dissected into initial-state and transition-state contributions, as regards the parameters G, H, and S. (see ref. 34). The exception is a rather large scale investigation of the triethylamine/ethyl iodide reaction (35,49-52). The relevant equation needed for a dissection of solvent influences in terms of the Gibbs energy is equation 14; similar equations may be set up for the enthalpy or entropy function. Values of

$$\Delta G_t^o(Tr) = \Delta G_t^o(Et_3N) + \Delta G_t^o(EtI) + \delta\Delta G^\ddagger \qquad (14)$$

the solvent effect on the reactants, $\Delta G_t^o(Et_3N)$ and $\Delta G_t^o(EtI)$ have been obtained through gas chromatographic determinations of Raoult's Law activity coefficients(50-52), equation 5, and details of the analysis for a series of hydroxylic solvents are in Table X. As is

Table IX. Calculated Z Values for Solvolyses Transition States

RX	Z(This work)[a]	RX	Z(28,34)
PhCHMeCl	0.86	t-BuCl	0.84
Ph$_2$CHCl	0.84	t-BuBr	0.82
t-BuCl	0.83	t-BuI	0.77
iso-PrCl	0.58[b]		
PhCH$_2$Cl	0.45	iso-PrBr	0.51
n-PrCl	0.36	PhCH$_2$Cl	0.40
		n-PrCl	0.35
		EtBr	0.31
		MeI	0.31
		MeBr	0.28
		n-BuBr	0.27

[a] From results in Table VII. [b] From data in water and ethanol only.

nearly always the case for rather nonpolar solutes, values of ΔG_t^o for Et$_3$N and EtI are quite small for transfer between the alcohols, but large for transfer to water. It is interesting to compare the ΔG_t^o(Tr) values in Table X with values for the t-butyl chloride transition state (Table I) and for the various solutes collected in Table II. Quite clearly, the transition state in reaction 14 is rather nonpolar. The ΔG_t^o(Tr) values may be compared with those for the product ion-pair, Et$_4$N$^+$I$^-$, both directly and after calculation of ΔG_E or ΔG_{INT} values. The relevant ΔG_E values, calculated from the observed values of ΔG_t^o via equation 9 are collected in Table XI. A plot of ΔG_E(Tr) against ΔG_E(Et$_4$NI) has a correlation constant of 0.994 and a slope m of 0.47, thus indicating that the charge sepa-

Table X. Calculation of ΔG_t^o(Tr) in the Et$_3$N/EtI Reaction[a]

Solvent	$\delta\Delta G^{\ddagger}$	ΔG_t^o		
		Et$_3$N	EtI	Tr
Water	-1.45	2.51	3.34	4.40
MeOH	0	0	0	0
EtOH	0.34	0.15	-0.27	0.22
1-PrOH	0.71	-0.05	-0.42	0.24
1-BuOH	0.90	-0.09	-0.55	0.26
2-PrOH	0.49	0.23	-0.43	0.29
t-BuOH	0.41	0.36	-0.44	0.33

[a] Values in kcal mol^{-1} at 298K, on the mol fraction scale, from refs. 35,49-52.

Table XI. Values of ΔG_E for the Et_3N/EtI Transition State and Model Solutes[a]

Solvent	$\Delta G_E(Tr)$	$\Delta G_E(Et_4\overset{+-}{NI})$	$\Delta G_E(Et_4\overset{+}{N} + I^-)$
Water	-3.86	-7.81	-9.01
MeOH	0	0	0
EtOH	0.86	1.67	2.75
1-PrOH	1.21	2.36	4.15
1-BuOH	1.45	2.77	5.25
2-PrOH	1.32	3.04	5.40
t-BuOH	1.65	4.55	9.13

[a] Values in kcal mol^{-1} at 298K, this work.

ration in the transition state is not large, and that the transition state is quite early. The pair of dissociated ions, $(Et_4N^+ + I^-)$ is a very poor model for the transition state, and quite incorrect conclusions could be drawn from comparisons of the transition state with this particular species. It is, after all, the ion-pair that is the product in the elementary reaction 15.

$$Et_3N + EtI \longrightarrow [Et_3N\text{--}Et\text{--}I] \longrightarrow Et_4\overset{+-}{NI} \qquad (15)$$

A similar dissection to that shown in Table X has been carried out for 32 aprotic solvents(51). Without giving all the details, suffice it to say that the solvent effect on the transition state is again much less than on the product ion-pair. Unlike the case of alcohols and water, the nonelectrostatic or ΔG_N term for transfer between aprotic solvents is quite small, so that a plot of $\Delta G_E(Tr)$ against $\Delta G_E(Et_4\overset{+-}{NI})$ leads to similar conclusions as does that of $\Delta G_t^o(Tr)$ against $\Delta G_t^o(Et_4\overset{+-}{NI})$, discussed before(51). Details of the various estimates of Z, from the slope of the plots, are in Table XII. All these values suggest that the transition state is early;

Table XII. Estimates of the Z value for the Et_3N/EtI Transition State[a]

Function plotted	Aprotic solvents	Alcohols and water
$\Delta G_t^o(Tr)$ vs $\Delta G_t^o(Et_4\overset{+-}{NI})$	0.37	-
$\Delta G_E^t(Tr)$ vs $\Delta G_E^t(Et_4\overset{+-}{NI})$	0.46	0.47
$\delta\Delta G$ vs $\delta\Delta G^o$	0.45[b]	-

[a] Ref. 51 and this work. [b] $\delta\Delta G^o$ refers to the change in the standard free energy for reaction 15 with change in solvent.

Table XIII. Transfer values for the Et_3N/EtI Transition State[a]

Solvent	ΔG_t^o	ΔH_t^o	ΔS_t^o
Hexane	4.3	4.1	-1
Ethyl acetate	0.7	-0.5	-4
Acetone	0	0	0
Acetonitrile	-0.1	1.2	4
DMSO	-0.3	3.0	11
$CHCl_3$	-1.6	-4.4	-9

[a] ΔG_t^o and ΔH_t^o in kcal mol^{-1}, ΔS_t^o in cal K^{-1} mol^{-1} at 298K on the mol fraction scale, ref. 52.

compare the S_N2 solvolysis transition states listed in Table IX, where e.g. Z for the n-PrCl solvolysis transition state is 0.36 units.

Very recently, a dissection of $\delta\Delta H^{\ddagger}$ and $\delta\Delta S^{\ddagger}$ values for reaction 15 has been completed(52). Unfortunately, since values of ΔH_t^o and ΔS_t^o for the product ion-pair are not available, it is not possible to obtain the equivalent of Z values from any enthalpy or entropy plots. However, dissections in terms of enthalpy or entropy can be used to test for specific solvation of the transition state in particular solvents. In Table XIII are details of the values of ΔG_t^o, ΔH_t^o, and ΔS_t^o for the Et_3N/EtI transition state; the "normal" solvent acetone has been taken as the reference solvent.

It is clear that there is some specific interaction with chloroform leading to a lowering of Gibbs energy and enthalpy; the entropy of the transition state is more negative than expected, again indicating some type of associative interaction.

5. CONCLUSION

The dissection of solvent influences on rate constants into initial-state and transition-state contributions, especially when combined with a knowledge of solvent effects on the thermodynamic properties of suitable model solutes, is a powerful method for the examination of transition states. In favorable cases with electrically neutral transition states, it is possible to estimate the degree of charge separation in the transition state and its position along the reaction coordinate. Dissections in terms of enthalpy and entropy are useful also in probing specific transition-state/solvent interactions.

REFERENCES

1. N. A. Menschutkin, J. Phys. Chem. 6:41 (1890).
2. H. von Halban, J. Phys. Chem. 84:129 (1913).
3. W. F. K. Wynne-Jones and H. Eyring, J. Chem. Phys. 3:492 (1935).
4. K. J. Laidler and H. Eyring, Ann. New York Acad. Sci. 39:303 (1939-40).
5. S. Glasstone, K. J. Laidler, and H. Eyring, "The Theory of Rate Processes", McGraw-Hill, New York (1941).
6. E. Grunwald and S. Winstein, J. Am. Chem. Soc. 70:846 (1948).
7. S. Winstein and A. H. Fainberg, J. Am. Chem. Soc. 79:5937 (1957).
8. A. R. Olsen and R. S. Halford, J. Am. Chem. Soc. 59:2644 (1937).
9. A. R. Olsen, W. C. Reubsamen, and W. E. Clifford, J. Am. Chem. Soc. 76:5255 (1954).
10. E. M. Arnett, W. G. Bentrude, and P. McC. Duggleby, J. Am. Chem. Soc. 87:2048 (1965).
11. E. M. Arnett, P. McC. Duggleby, and J. J. Burke, J. Am. Chem. Soc. 85:1350 (1963).
12. E. M. Arnett, W. G. Bentrude, J. J. Burke, and P. McC. Duggleby, J. Am. Chem. Soc. 87:1541 (1965).
13. E. M. Arnett and D. R. McKelvey, Record Chem. Progress, 26:185 (1965).
14. E. M. Arnett, in "Physicochemical Processes in Mixed Aqueous Solvents", ed., F. Franks, Heinemann, London (1967).
15. E. S. Rudakov and V. P. Tretyakov, Reakts. Spos. Organ. Soedin. Tartusk. Gos. Univ. 2:142 (1965).
16. P. Haberfield, A. Neudelman, A. Bloom, R. Romm, H. Ginsberg, and P. Steinherz, J.C.S. Chem. Comm. 194 (1968).
17. P. Haberfield, A. Nudelman, A. Bloom, R. Romm, and H. Ginsberg, J. Org. Chem. 36:1792 (1971).
18. P. Harberfield, J. Am. Chem. Soc. 93:2091 (1971).
19. C. A. Eckert and K. F. Wong, Trans. Faraday Soc. 66:2313 (1970).
20. R. A. Grieger and C. A. Eckert, J. Am. Chem. Soc. 92:7149 (1970).
21. R. A. Grieger and C. A. Eckert, Ind. Eng. Chem. Fundamentals, 10:369 (1971).
22. M. H. Abraham, J.C.S. Chem. Comm. 1307 (1969).
23. M. H. Abraham, F. Behbahany, M. J. Hogart, R. J. Irving, and G. F. Johnston, J.C.S. Chem. Comm. 117 (1969).
24. M. H. Abraham, Tetrahedron Letters, 5233 (1970).
25. M. H. Abraham, R. J. Irving, and G. F. Johnston, J.C.S. (A), 199 (1970).
26. M. H. Abraham, J.C.S. (B), 299 (1971).
27. M. H. Abraham, J.C.S. (A), 1061 (1971).
28. M. H. Abraham and G. F. Johnston, J.C.S. (A), 1610 (1971).
29. M. H. Abraham, J.C.S. Perkin Trans. II, 1343 (1972).

30. M. H. Abraham and F. J. Dorrell, J.C.S. Perkin Trans. II, 444 (1973).
31. A. J. Parker, Chem. Rev. 69:1 (1969).
32. J. M. Brønsted, J. Phys. Chem. 102:169 (1922).
33. N. Bjerrum, J. Phys. Chem. 108:82 (1924).
34. M. H. Abraham, Prog. Phys. Org. Chem. 1:11 (1974).
35. M. H. Abraham and P. L. Grellier, J.C.S. Perkin Trans. II, 1856 (1975).
36. M. H. Abraham, J. Am. Chem. Soc. 101:5477 (1979).
37. M. H. Abraham, A. Nasehzadeh, J. Reisse, and J. J. Moura Ramos, J.C.S. Perkin Trans. II, (1980) in the press.
38. N. Bjerrum and E. Jozefowicz, J. Phys. Chem. 159A:194 (1932).
39. M. Alfenaar and C. L. Ligny, Rec. Trav. Chim. 86:929 (1967).
40. R. A. Pierotti, Chem. Rev. 76:717 (1976).
41. J. J. Moura Ramos, M. Lemmers, R. Ottinger, M.-L. Stein, and J. Reisse, J. Chem. Reserach (S), 56 (1977); (M) 0658 (1977).
42. W. J. Albery and M. M. Kreevoy, Adv. Phys. Org. Chem. 16:87 (1978).
43. G. A. Clarke and R. W. Taft, J. Am. Chem. Soc. 84:2295 (1962).
44. Y.-C. Tse, M. D. Newton, S. Vishveshwara, and J. A. Pople, J. Am. Chem. Soc. 100:4329 (1978).
45. M. H. Abraham, D. H. Buisson, and R. A. Schulz, J.C.S. Chem. Comm. 693 (1975).
46. M. H. Abraham, J.C.S. Perkin Trans. II, 1028 (1977).
47. P. E. Peterson, D. W. Vidrine, F. J. Waller, P. M. Henrichs, S. Magaha, and B. Stevens, J. Am. Chem. Soc. 99:7969 (1977).
48. T. W. Bentley and P.v. R. Schleyer, Adv. Phys. Org. Chem. 14:1 (1977).
49. M. H. Abraham and P. L. Grellier, J.C.S. Perkin Trans. II, 623 (1975).
50. M. H. Abraham, P. L. Grellier, and J. Mana, J. Chem. Thermodynamics, 6:1175 (1974).
51. M. H. Abraham and P. L. Grellier, J.C.S. Perkin Trans. II, 1735 (1976).
52. M. H. Abraham and A. Nasehzadeh, unpublished work.

INITIAL STATE AND TRANSITION STATE SOLVENT EFFECTS: REACTIONS IN

PROTIC AND DIPOLAR APROTIC MEDIA

E. Buncel
Department of Chemistry, Queen's University
Kingston, Ontario, Canada, K7L 3N6

and E. A. Symons
Physical Chemistry Branch
Chalk River Nuclear Laboratories
Atomic Energy of Canada Limited
Chalk River, Ontario, Canada, K0J 1J0

Abstract - This work is concerned with the origin of effects on rate equilibrium processes when reactions are performed in dipolar aprotic media such as dimethyl sulfoxide and dimethylformamide compared to typical hydroxylic media. An example of processes studied in our laboratory is the competition between proton abstraction and σ-complex (Meisenheimer adduct) formation in the interaction of nitroaromatic compounds with basic systems (e.g. 1,3-dinitrobenzene in $DMF-D_2O-NaOD$). Other reactions studied are the base catalyzed isotopic exchange of D_2 in aqueous DMSO mixtures, and also isotopic exchange of fluoroform. It has been found possible in certain cases to dissect the initial state and transition state contributions to the reaction rates from the kinetically measured enthalpies of activation and the thermodynamically evaluated enthalpies of transfer of the reactants. This procedure affords insight into transition state properties and behaviour.

1. INTRODUCTION

In the past, transition state theory(1) was typically used in a qualitative manner to explain the effect of medium changes on reaction rates. This approach reached its quintessence in the systematic prediction by Hughes and Ingold(2) of the effect of increased solvent polarity on the rates of various types of nucleophilic substitution reactions. Their classification of S_N2 and S_N1 reactions by charge type, and the prediction of how an increase in solvent polarity affects the rate on the basis of the change in charge distribution in the transition state relative to the reactants, has since been reproduced in most texts dealing with structure and reactivity in organic chemistry(3).

Using the dielectric constant (ε) of the medium as a measure of solvent polarity, a number of expressions were derived relating ε to some kinetic parameter via transition state theory, e.g. the Laidle-Eyring equation(4), and the Christiansen-Scatchard equation (5). These relationships, however, were found to have only limited applicability.

As an alternative to a macroscopic parameter such as ε, a number of empirical measures of solvent polarity have been proposed (6). Since these parameters are based upon microscopic model processes (both chemical and physical), they act as probes of solute solvent interactions such as hydrogen-bonding, charge-dipole, dipole-dipole, and dipole-induced dipole. These empirical parameters have found extensive usage(7), and new empirical parameters continue to be proposed, even during the past few years(8-10). A generalized solvent polarity scale(11) has been found to correlate satisfactory with the most widely used of the empirical polarity scales.

2. REACTIONS IN DIPOLAR APROTIC MEDIA - PROTON TRANSFER vs. σ-COMPLEX FORMATION.

The present study, concerned with dipolar aprotic media, had its origin in the 1960's, with the discovery that certain solvents such as dimethyl sulfoxide (DMSO) dimethylformamide (DMF) and hexamethylphosphortriamide (HMPA), had the property of accelerating enormously certain reactions relative to protic solvents such as alcohols(12-14). These reactions included nucleophilic substitutions, both aliphatic and aromatic, as well as proton abstraction processes proceeding by carbanion intermediates.

Our own work with dipolar aprotic media began with study of the interaction between nitroaromatic compounds and bases(15). Depending on the nature of the nitroaromatic and the base, the following types of interaction are possible(15-22).

(i) A charge transfer interaction, giving rise to a donor-acceptor type complex.
(ii) Electron transfer, yielding a radical anion.
(iii) Proton abstraction, giving an aryl carbanion or a benzylic anion.
(iv) Covalent addition to a ring carbon, yielding a Meisenheimer type complex, also known as a σ-complex.
(v) Substitution of a displaceable group, resulting in an overall nucleophilic aromatic substitution, a process believed to occur via σ-complex intermediates(23,24).

It will be apparent that in many systems more than one of these interactions may be possible(25,26); in the particular studies to

INITIAL STATE AND TRANSITION STATE SOLVENT EFFECTS 357

be described we have found that σ-complex formation effectively competes with proton abstraction(27-30).

The processes of proton abstraction and σ-complex formation in the interaction of OH⁻ with 1,3-dinitrobenzene (DNB) are shown in Scheme 1. σ-Complex formation is the predominant process, as shown by the characteristic electronic absorption spectrum of this species. However, a small concentration of the aryl carbanion is also formed and can be detected by the occurrence of deuterium exchange (followed by IR or NMR spectroscopy) when the process is carried out in deuterated medium.

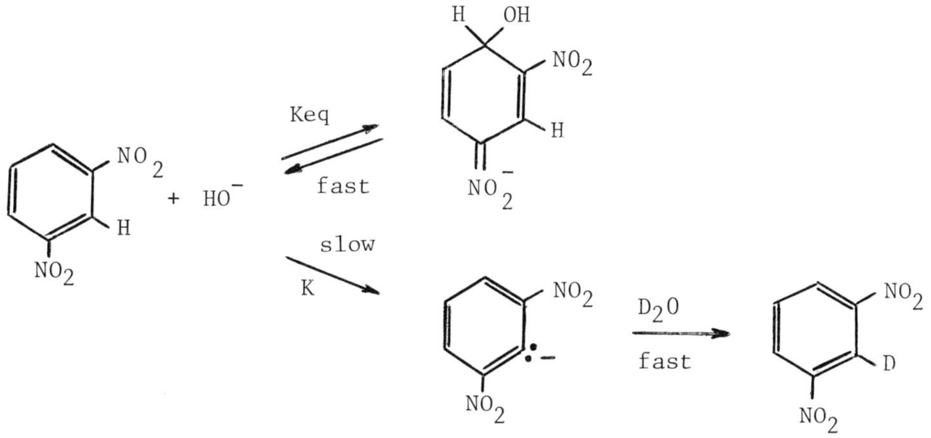

Scheme 1

The results for the equilibrium formation of σ-complex, and the rate constant for deuterium exchange of DNB (0.48M), in DMF-D_2O mixtures containing deuteroxide ion (0.004M), are shown in Figure 1. It is seen that σ-complex formation, though relatively slight in aqueous media, increases fairly sharply with DMF content, and under the conditions of exchange, with DNB in large excess, ∼ 99% complexing of OD⁻ occurs in ∼ 70 mole % DMF. The observed rate constant for deuterium exchange also increases at first but reaches a maximum at ∼ 70 mole % DMF and then decreases.

Corresponding results for 1,3,5-trinitrobenzene (TNB) have shown that in this case σ-complex formation increases more steeply with DMF content, and under the conditions of exchange, again with TNB in excess, ∼ 99% the base will already be complexed at ∼ 20 mole % DMF. Interestingly, k_{obs} for exchange was found to decrease with increasing DMF content over the entire region studied, which is unexpected since base catalyzed proton exchange is generally enhanced in dipolar aprotic media, as we have seen.

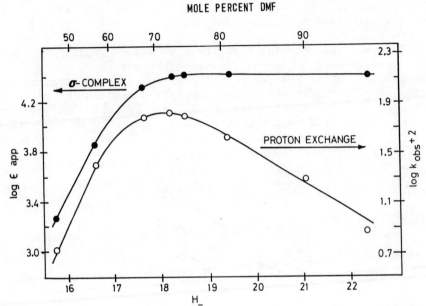

Figure 1. Competition between hydrogen exchange and σ-complex formation for m-dinitrobenzene in the $OD^-/D_2O/DMF$ system.

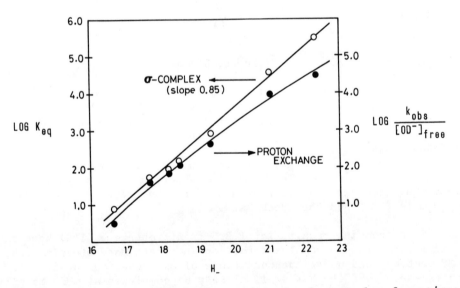

Figure 2. Dependence of hydrogen exchange and σ-complex formation on solution basicity for m-dinitrobenzene in the $OD^-/D_2O/DMF$ system.

However, the principles governing the TNB and DNB systems must be analogous, the apparent differences arising at the quantitative level. That is, in the TNB case, predominant (say 99%) σ-complex formation has set in at a much lower DMF composition. Therefore, if one were to extrapolate the curve for exchange to lower DMF content, one would expect to observe an initial rise in k_{obs}, as in the DNB system. Hence, the results obtained in both systems are in accord with the σ-complex being an unreactive form of the substrate towards deuterium exchange.

The origin of the decreasing exchange rate with increasing DMF content is shown in an alternative fashion in Figure 2, which is applicable to the DNB case. Plotting log $k_{obs}/[OD^-]_{free}$ (in effect log k_2, the second order rate constant for exchange) vs. H_-, one obtains an initial linear portion followed by a downward curving plot. However, the plot of log K_{eq} vs. H_- remains linear over the whole range of medium composition. It is steeper dependence of σ-complex formation on medium basicity, compared to proton exchange, which is thus the underlying reason for the decreasing exchange rate in media of high DMF content.

Enhancement of rate processes in dipolar aprotic media was initially attributed principally to desolvation of the anionic species (12a). Thus an anion such as OH^- will be strongly solvated in a protic solvent by hydrogen bonding, but such interactions are absent in a dipolar aprotic solvent like DMF. If a qualitative potential energy diagram showing the medium effect on these processes were desired, then the initial state would be indicated as being considerably destabilized, on the assumption that the neutral nitroaromatic molecule would not be appreciably affected by solvent change.

What conjecture should one make concerning transition states and intermediates, for example in proton abstraction and σ-complex formation? Predictions of this type can be risky as has already been illustrated(7). The transfer function approach(31,32), does provide a way of gleaning insight into transition state behaviour, through dissection of measured solvent effects into initial state and transition state contributions.

3. THERMODYNAMIC TRANSFER FUCNTIONS: INITIAL STATE AND TRANSITION STATE.

In Figure 3 is represented the superposition of the potential energy-reaction coordinate diagrams for a given reaction occurring in two solvents. One can define, according to this figure, the terms δH_{tr}^R and δH_{tr}^T as the enthalpies of transfer of the reactants and of the transition state between the two solvents.

It is seen that the enthalpy of transfer of the transition state on going from a standard solvent O to a solvent S is given by the expression

$$\delta H_{tr}^{T} = \delta H_{tr}^{R} + (\Delta H_{s}^{\ddagger} - \Delta H_{o}^{\ddagger})$$
$$= \delta H_{tr}^{R} + \delta \Delta H^{\ddagger} \qquad (1)$$

Thus δH_{tr}^{T} can be evaluated from the measurable transfer enthalpies of stable solute species and kinetic activation parameters. The required transfer enthalpies can be obtained from measurements of heats of solution, or by application of Henry's law for gaseous solutes. An analogous expression to that in eq. 1 can readily be derived for the equilibrium situation. Moreover, corresponding expressions would be applicable to free energies and entropies of transfer, as well.

The dissection of medium effects on reaction rates into initial state and transition state contributions leads to the development of a systematic approach to classify reaction types. The resulting classification is shown in Table I, which presents all possible outcomes of medium change: rate acceleration, rate retardation and no change. Thus we have classified reaction types as balanced, re-inforced, positive or negative, initial state or transition state controlled, ans so on. Not all of these situations have been observed so far and it would be intriguing to design appropriate systems which would complete this classification. The studies to be described forthwith serve to illustrate some of these reaction types.

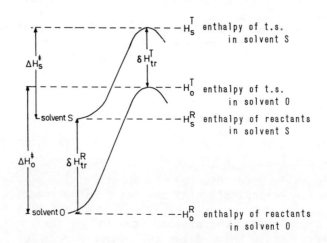

Fig. 3. Relationship between enthalpies of transfer of initial state and transition state for a reaction occurring in two different media.

Table I. Transfer free energies of reactants (δG_{tr}^R), of transition states (δG_{tr}^T) and solvent effects on reaction rates. Classification of reaction types.

Case	δG_{tr}^R	δG_{tr}^T	Effect on Rate[a]	Reaction type
1	−	−	+, 0, or −	Balanced
2	+	−	+	Positevely reinforced
3	0	−	+	Positive transition state control
4	−	0	−	Negative initial state control
5	+	0	+	Positive initial state control
6	0	0	0	Solvent independent
7	−	+	−	Negatively reinforced
8	+	+	+, 0, or −	Balanced
9	0	+	−	Negative transition state control

[a] The plus sign refers to rate acceleration, the minus sign to rate retardation, and zero to no effect.

4. ILLUSTRATION OF TRANSFER FUNCTION APPROACH FOR VARIOUS REACTION TYPES

In Figure 4 are illustrated several reaction types, all of which have been reported by Haberfield and co-workers(33) (see also 12b,34,35). The figures are intended to be schematic, only relative energy levels being implied, and are based on measurements of enthalpies of transfer of reactants and enthalpies of activation.

In A we have two neutral reactant molecules which are not appreciably solvated, hence will not be appreciably affected by the change in medium from protic to dipolar aprotic, so that the initial state is coincident in the two media. However, the polarizable transition state will be more solvated in a dipolar aprotic medium such as DMF, so there should normally be a rate enhancement on changing to a dipolar aprotic medium (δH_{tr}^T is negative). This is the Menschutkin type reaction which is discussed in detail by Professor Abraham(36), who has observed cases in which the realationship of the coincident initial states in the two media does not hold(31). However this relationship does hold for this particular reaction and we selected it so as to contrast with the other reaction types shown in this Figure, in which the initial state is not coincident in the two media. It may perhaps be emphasized that this is the only case, of the systems illustrated here, in which changes in the transition state solvation are solely responsible for the medium effect on the reaction rate. In terms of the classification in Table I we can describe this as a "positive transition state control" reaction type.

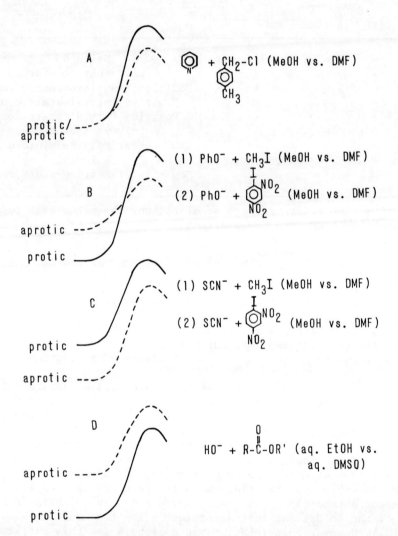

Fig. 4. Free energy profiles for several nucleophilic reactions occurring in protic and dipolar aprotic media.

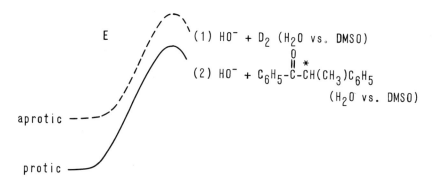

Fig. 5. Free energy profiles for several proton transfer processes

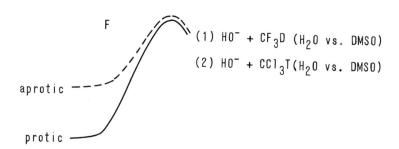

Fig. 6. Enthalpies of transfer of reactants and transition state for the exchange reaction $D_2 + H_2O \rightarrow HD + DOH$ in the $OH^-/H_2O/DMSO$ system.

In B we have S_N2 and S_NAr processes in which one reactant, phenoxide ion, is destabilized by the dipolar aprotic medium. However, the transition state is stabilized since the charge is now more widely dispersed, especially when the polarizable iodine is involved.

In C we have again S_N2 and S_NAr reactions but the charge on the reactant anion is dispersed on polarizable atoms so that the initial state is now stabilized, as well as the transition state.

In D we have the case of base-induced ester hydrolysis, a reaction known to occur via a tetrahedral intermediate. The transition state, which will resemble the tetrahedral intermediate, will have charge concentrated on the two oxygen atoms, and such a species is expected to be destabilized in a dipolar aprotic medium relative to the protic solvent.

In C and D are situations in which the effects on reactants and on the transition state are balancing, so that the magnitude of the rate enhancement is dependent on the relative stabilization in case of C, or the relative destabilization in case of D, of the initial state and the transition state. Hence C and D could in principle lead to retardation in a dipolar aprotic medium relative to protic. On the other hand, B is a case in which the two effects reinforce one another so that there should always be a rate enhancement on changing the medium. Our classification denotes the systems C and D as balanced reaction types, and B as a positively re-inforced reaction type.

5. ISOTOPIC EXCHANGE OF D_2 IN AQUEOUS DMSO

The base-catalyzed exchange of molecular deuterium in the DMSO-water system has yielded data of interest(37-39). At constant $[OH^-]$, the rate constant for exchange as function of mole % DMSO increases gently at first, and then more steeply in the media of high DMSO content. The log k vs. H_- plot has an initial slope of ~ 0.2, increasing to $\sim 0.3-0.4$ in the high DMSO range. The slope of ~ 0.3 in the H_- plot is one of the smallest observed in a proton transfer process. Of course, the small slope only reflects the relatively small overall increase in rate over the range of medium composition, about 4 powers of 10. It is interesting that a slope of 0.9 in the log k vs. H_- plot would have corresponded to an overall rate increase of more than 10 powers of 10.

The question is, how does one explain such a low dependence of rate on medium basicity in this system? The transfer function method has shed some light on this problem.

The thermodynamic parameters for the D_2-OH^- exchange reaction have been determined for the entire range of DMSO-water compositions. The enthalpy of transfer term for D_2 is quite small relative to that for OH^-, which is extremely large. Thus, the δH_{tr}^R term for the reactants reaches ca. 80 kJ mol^{-1} in nearly pure DMSO (Figure 6). The enthalpy of activation for the reaction starts at 100 kJ mol^{-1} in aqueous medium, passes through a shallow minimum at about 60 mole % DMSO, and then increases again. The difference in enthalpy of activation between H_2O and the DMSO media, when combined with H_{tr}^R according to eq. 1, yields the enthalpy of transfer of the transition state. The H_{tr}^T term is positive throughout the medium composition (Figure 6), indicating that the transition state is extensively desolvated, or destabilized, on going from an aqueous medium to DMSO-containing media.

What are the mechanistic implications of these results? The main conclusion of the enthalpies of transfer data is that destabilization of OH^- is largely retained in the transition state. This means that the rate-determining transition state retains negative charge on an electronegative atom, since that would lead to desolvation in DMSO rich media. Of the various mechanisms that have been considered for this deceptively simple reaction(37-40), we can probably exclude the simplest mechanism for exchange, as given in eq. 2, namely rate-determining abstraction of a proton to give a deuteride ion which would be rapidly discharged by reaction with solvent. Here the negative charge in the transition state (1) resides in part on oxygen and in part on deuterium, and since hydride (or deuteride) ion is considered to be a soft polarizable base, it would have only small solvation requirements, so that the δH_{tr}^T term should be small, contrary to what is found.

$$HO^- + D\text{-}D \xrightarrow{slow} HOD + D^- \qquad (2a)$$

$$D^- + HOH \xrightarrow{fast} DH + OH^- \qquad (2b)$$

$$\left| \overset{\delta-}{H} \ldots \overset{\delta+}{D} \ldots \overset{\delta-}{D} \right|^-$$

$$\underline{1}$$

In the mechanism shown in eq. 3, formation of free deuteride ion is avoided by participation of a water molecule in the rate-determining step, providing electrophilic assistance for removal of D^- and thereby avoiding the formation of a high energy intermediate. Negative charge is here retained on electronegative oxygens, so that the transition state will have appreciable solvation requirements. This is in agreement with the observation of a large positive transition state enthalpy of transfer term δH_{tr}^T.

$$HO^- + D\text{-}D + HOH \longrightarrow HOD + DH + OH^- \qquad (3)$$

$$\left| \overset{\delta-}{HO} \ldots \overset{\delta+}{D} \ldots \overset{\delta-}{D} \ldots \overset{\delta+}{H} \ldots \overset{\delta-}{OH} \right|^-$$

$$\underline{2}$$

6. IONIZATION OF CARBON ACIDS

Another area to which the thermodynamic transfer function approach has been applied is that of ionization of carbon acids. One such example is the racemization of D-α-methyl-α-phenylacetophenone (MPA) in hydroxide/water/DMSO mixtures, where heats of solution of reactant species have been combined with previously reported kinetic data(41). In Figure 7 are shown the enthalpy of transfer functions for the individual and combined reactants, and the enthalpies of activation for this system. The resulting calculated δH_{tr}^{T} has a somewhat surprising dependence on solvent composition over the range accessible experimentally. The transition state is generally less well solvated as the DMSO content increases, but this solvent effect is only half as great as on the hydroxide ion desolvation. The apparent maximum and minimum for δH_{tr}^{T} at ca. 37 and 43% DMSO, respectively, seen in Figure 7 may be real; however, these features could perhaps be artifacts that would result if the δH_{tr}^{MPA} datum corresponding to 43.3% DMSO were low by $\sim$ 4 kJ mol^{-1}. Within experimental error, a fairly smooth increase in δH_{tr}^{T} would then be obtained. This result would be in accord with a transition state structure in which negative charge is largely localized on the two oxygens, as expected on the basis of the generally accepted mechanism of enolisation.

Another case of the combination of old and new data involves hydrogen isotope exchange with chloroform (as CCl$_3$T), again in basic water/DMSO mixtures(42). The rapidly increasing rate of loss of tritium limited the accessible solvent range to 0-24 % DMSO. The δH_{tr} data are plotted in Figure 8. Here, solvation of the transition state (believed to be product-like, $[HO^-..H^+....^-CCl_3]$)(42), is initially enhanced in 5-15% DMSO. Thus as the DMSO content of the solvent is raised, increased transition state solvation is responsible at first for the observed rate acceleration. However, decreased hydroxide ion solvation becomes the major factor above about 15% DMSO.

It has been possible to obtain useful information over a much wider range of water/DMSO composition with a similar but less reactive hydrocarbon acid, namely fluoroform (CF$_3$H, CF$_3$D)(43). As for chloroform, a product-like transition state is expected. Enthalpy of transfer data are shown in Figure 9. The general similarity between Figures 8 and 9 is not surprising; the minimum near 10% DMSO (mainly the result of the δH_{tr}^{-OH} dip at this point(34)) is about -7 kJ mol^{-1} for $[HO^-..D^+...^-CF_3]$ compared to -10 kJ mol^{-1} for $[HO^-..T^+...^-CCl_3]$. In fact, over most of the 25% DMSO region available for comparison, the fluoroform transition state is slightly less solvated by 2-5 kJ mol^{-1}.

These haloform systems are apparently the first examples reported where relatively little change in transition state solvation occurs with significant change in solvent character. As a result,

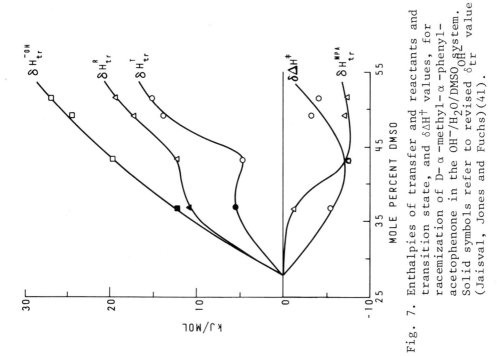

Fig. 7. Enthalpies of transfer and reactants and transition state, and $\delta\Delta H^{\ddagger}$ values, for racemization of D-α-methyl-α-phenyl-acetophenone in the OH$^-$/H$_2$O/DMSO system. Solid symbols refer to revised δ_{tr} value (Jaisval, Jones and Fuchs)(41).

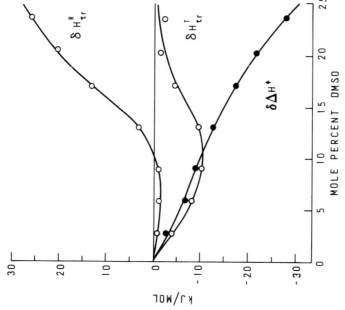

Fig. 8. Enthalpies of transfer of reactants and transition state, and $\delta\Delta H^{\ddagger}$ values for exchange reaction $CCl_3T + H_2O \rightarrow CCl_3J + TOH$ in the OH$^-$/H$_2$O/DMSO System. (Jones and Fuchs)(42).

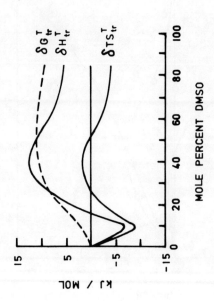

Fig. 10. Enthalpy, entropy, and free energy of transfer functions for the transition state of the isotopic exchange reaction $CF_3D + H_2O \to CF_3H + DOH$ in the $OH^-/H_2O/DMSO$ system.

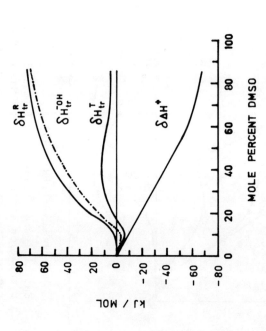

Fig. 9. Enthalpies of transfer of reactants and transition state, and $\delta\Delta H^{\ddagger}$ values, for exchange reaction $CF_3D + H_2O \to CF_3H + DOH$ in the $OH^-/H_2O/DMSO$ system.

above ca. 10% DMSO, the observed large increase in isotope exchange reactivity arises solely from desolvation of the reactants, and more specifically of the hydroxide ion. This is shown pictorially in Figure 5, which is a complement to Figure 4. It will be readily apparent that one has, for hydrogen exchange involving fluoroform and chloroform in solvents rich in DMSO, according to Table I, a case of <u>positive initial state control</u>.

For the case of fluoroform exchange the transfer function technique has been taken one stage further, in that δG_{tr}^T and δTS_{tr}^T values have been estimated as a function of solvent composition(43); these are given in Figure 10. The δG_{tr}^{-OH} data used were those of Villermaux and Delpuech(44), as found in Table XXIV or ref. 22. Incorporation of $\Delta G^{\ddagger}$ and $\Delta S^{\ddagger}$ values for CF_3D exchange(43), and $\delta G_{tr}^{CF_3H}$ values estimated from CF_3H solubilities, gave δG_{tr}^T, the δTS_{tr}^T data being obtained from the equation $\delta G_{tr}^T = \delta H_{tr}^T - \delta TS_{tr}^T$. Above 30% DMSO, δG_{tr}^T remains almost constant, with changes in δH_{tr}^T and δTS_{tr}^T balancing each other. Between 0 and 30 % DMSO it is the enthalpy term that contributes more to the increase in δG_{tr}^T to its constant level.

The present transfer function results on isotope exchange in the haloform systems are of interest in relation to the proposal that these represent cases of encounter-controlled reprotonation of the carbanion conjugate bases(45).

7. CONCLUSIONS

A potential problem that has not yet been broached is the possibility that the transition state does not remain constant as the medium is changed. Thus it is conceivable that there are two or more pathways form reactants to products and that the preferred path may vary as the solvent is changed, in which case δG_{tr}^T would be measuring the difference in free energies of two different species. The possibility of changing transition states is one that could in principle be detected through kinetic isotope effects(46) or through synergistic and antagonistic effects(32,47).

It should be mentioned that related principles have been applied to acid catalyzed processes(48,49) and to excited state processes (50,51).

Finally, it is pointed out that methods are currently being explored to predict δG_{tr}^T values(10,36), which would enable one ultimately to predict reaction rates in different media.

<u>Acknowledgment</u>. Financial support from the Natural Sciences and Engineering Research Council of Canada in the form of an operating grant (to E. B.) is gratefully acknowledged. We also wish to

thank our co-workers named in the references herein for their contributions to this work.

REFERENCES

1. M. G. Evans and M. Polanyi, Trans. Faraday Soc. 34:11 (1938).
2. E. D. Hughes and C. K. Inogld, J. Chem. Soc. 244 (1935).
3. J. March, "Advanced Organic Chemistry. Reactions, Mechanisms and Structure", 2nd edn., Table IX, p. 332, McGraw Hill, New York (1977).
4. K. J. Laidler and H. Eyring, Ann. N. Y. Acad. Sci. 39:303 (1940).
5. E. S. Amis, "Solvent Effects on Reaction Rates and Mechanisms", p. 20, Academic Press, New York (1966).
6. C. Reichardt, "Solvent Effects in Organic Chemistry", Verlag Chemie, Weinheim (1979); Angew. Chem. Int. Ed. 18:98 (1979).
7. E. M. Kosower, "An Introduction to Physical Chemistry", Wiley, New York (1968).
8. T. M. Krygowski and W. R. Fawcett, J. Am. Chem. Soc. 97:2143 (1975).
9. A. Arcoria, V. Librando, E. Maccarone, G. Musumarra, and G. A. Tomaselli, Tetrahedron, 33:105 (1977).
10. A. J. Parker, U. Mayer, R. Schmid, and V. Gutmann, J. Org. Chem. 43:1843 (1978).
11. J. L. Abboud, M. J. Kamlet, and R. W. Taft, J. Am. Chem. Soc. 99:8325 (1977).
12. (a) A. J. Parker, Quart. Rev. 16:163 (1962); (b) A. J. Parker, Chem. Rev. 69:1 (1969).
13. J. J. Delpuech, Bull. Soc. Chim. France, 1624 (1966).
14. C. D. Ritchie, in "Solute-Solvent Interactions", eds., J. F. Coetze and C. D. Ritchie, Marcel Dekker, New York (1969).
15. E. Buncel, A. R. Norris, and K. E. Russell, Quart. Rev. 22:123 (1968).
16. M. R. Crampton, Adv. Phys. Org. Chem. 7:211 (1969).
17. F. Pietra, Quart. Rev. 23:504 (1969).
18. M. J. Strauss, Chem. Rev. 70:667 (1970).
19. G. Doddi, G. Illuminati, and F. Stegel, J. Org. Chem. 36:1918 (1971).
20. C. F. Bernasconi, MTP Int. Rev. Sci. Org. Chem., Ser. One, 3:33 (1973).
21. F. Terrier, A. P. Chatrousse, C. Paulmier, and R. Schaal, J. Org. Chem. 40:2911 (1975).
22. E. Buncel and H. Wilson, Adv. Phys. Org. Chem. 14:133 (1977).
23. J. A. Orvik and J. F. Bunnett, J. Am. Chem. Soc. 92:2417 (1970).
24. J. Miller, "Aromatic Nucleophilic Substitution", Elsevier, Amsterdam (1968).
25. M. R. Crampton and V. Gold, J. Chem. Soc.(B), 498 (1966).

26. I. R. Bellobono and G. Sala, J. Chem. Soc. Perkin Tans. II, 2:169 (1972).
27. E. Buncel and E. A. Symons, Can. J. Chem. 44:771 (1966); J. Org. Chem. 38:1201 (1973).
28. E. A. Symons and E. Buncel, Can. J. Chem. 50:1729 (1972).
29. E. Buncel and A. W. Zabel, J. Am. Chem. Soc. 89:3082 (1967) and unpublished work.
30. E. Buncel, J. A. Elvidge, J. R. Jones, and K. T. Walkin, J. Chem. Res. in press.
31. M. H. Abraham, Prog. Phys. Org. Chem. 11:2 (1974).
32. E. Buncel and H. Wilson, Acc. Chem. Res. 12:42 (1979).
33. P. Haberfield, J. Am. Chem. Soc. 93:2091 (1971).
34. R. Fuchs, C. P. Hagan, and R. F. Rodewald, J. Phys. Chem. 78:1509 (1974).
35. B. G. Cox, Ann. Repts. Chem. Soc. (A), 70:249 (1973).
36. M. H. Abraham, Abstracts, 5th International Conference on Solute-Solvent Interactions, Florence, Italy (1980).
37. E. A. Symons and E. Buncel, J. Am. Chem. Soc. 94:3641 (1972); Can. J. Chem. 51:1673 (1973).
38. E. Buncel and E. A. Symons, J. Am. Chem. Soc. 98:656 (1976).
39. E. Buncel, R. A. More O'Ferrall, and E. A. Symons, J. Am. Chem. Soc. 100:1084 (1978).
40. C. D. Ritchie and J. F. King, J. Am. Chem. Soc. 90:833 (1968).
41. D. K. Jaiswal, J. R. Jones, and R. Fuchs, J. Chem. Soc. Perkin II, 102 (1976).
42. J. R. Jones and R. Fuchs, Can. J. Chem. 55:99 (1977).
43. E. A. Symons and M. J. Clermont, unpublished work, submitted to J. Am. Chem. Soc. (1980).
44. S. Villermaux and J. J. Delpuech, Bull. Soc. Chim. France, 2534 (1974).
45. A. J. Kresge, Acc. Chem. Res. 8:354 (1975).
46. K. C. Westaway, Can. J. Chem. 56:2691 (1978).
47. R. M. Pollack and M. Brault, J. Am. Chem. Soc. 98:247 (1976).
48. G. Scorrano, Abstracts, 5th International Conference on Solute-Solvent Interactions, Florence, Italy (1980).
49. K. Yates and T. A. Modro, Acc. Chem. Res. 11:190 (1978).
50. P. Haberfield, M. S. Lux, and D. Rosen, J. Am. Chem. Soc. 99:6828 (1977).
51: P. Haberfield, M. S. Lux, I. Jasser, and D. Rosen, J. Am. Chem. Soc. 101:645 (1979).

SOLVATION ENERGIES IN ACID CATALYZED PROCESSES

Gianfranco Scorrano

Istituto di Chimica Organica
Centro CNR Meccanismi di Reazioni Organiche
Via Marzolo 1, 35100 Padova, Italy

Abstract - The modern methods of abstracting mechanistic information from kinetic data of acid catalyzed reaction are described and the shortcomings of the older methods indicated.

The interest of organic chemists in the interpretation of acid catalyzed reactions dates back to the very early measurements of reaction rates(1). The most widely known and applied theory is the one proposed more than 40 years ago by Zucker and Hammett(2) and later reviewed by Paul and Long(3). According to this theory we could reveal whether the reaction mechanism is of the A-1 or A-2 type by checking whether the increase in logarithm of the experimental reaction rate depends on the first power of H_o or log c_{H^+}, respectively. Probably because of its deceptive simplicity, this criterium has found very wide applications. Although Hammett himself has pointed out(4) on the shortcomings of this criterium there are few researchers who still apply the Zucker-Hammett hypothesis. Let's quickly discuss the reasons why this criterium is not valid. The rate expression for the A-1 mechanism, depicted in equation 1 and 2, is reported in equation 3

$$S + H^+ \underset{K_{SH^+}}{\rightleftharpoons} SH^+ \tag{1}$$

$$SH^+ \xrightarrow{\text{slow}} \neq \longrightarrow \text{products} + H^+ \tag{2}$$

$$v = k_o [SH^+] f_{SH^+}/f_{\neq} \tag{3}$$

where the bracketed term represents the concentration, and f's are the activity coefficients of the protonated substrate and of the transition state ($\neq$). By substituting for $[SH^+]$ the value obtained through the equilibrium constant (K_{SH^+}), and the rearranging we obtain equation 4, which becomes equation 5 after introducing

$$\log \frac{v}{[S]} = \log \frac{k_o}{K_{SH^+}} + \log a_{H^+} + \log f_S/f_{\neq} \qquad (4)$$

$$\log k_\psi = \log \frac{k_o}{K_{SH^+}} - H_o + \log \frac{f_S}{f_{\neq}} \frac{f_{BH^+}}{f_B} \qquad (5)$$

the acidity function H_o [by definition(2,3) $H_o = -\log a_{H^+} f_B/f_{BH^+}$, where a_{H^+} is the activity of the proton, and f_B, f_{BH^+} are the activity coefficients of free and protonated nitroanilines].

It follows that the Zucker-Hammett hypothesis, requiring that $\log k_\psi + H_o =$ constant, holds only if

$$\frac{f_S}{f_{\neq}} = \frac{f_B}{f_{BH^+}} \qquad (6)$$

in all the acidity range.

Of course equation 6 cannot be directly checked. However, there are many examples which show equation 6 does not hold even when we are comparing the activity coefficient ratios of two bases of different structure. For instance, if we compare the acidity functions defined through primary (H_o)(5) and tertiary (H_o''')(6) nitroanilines we found that their difference ($H_o - H_o'''$), i.e. the difference between the activity coefficients [$H - H_o''' = \log (f_{B'''}/f_{B'''H^+}) - \log (f_B/f_{BH^+})$], is 0.22, 1.16 and 2.22 in 10, 50 and 90% H_2SO_4. If equation 6 does not hold even when the comparison is between two very similar bases, there is little doubt of its validity when S and B are structurally very different.

A very similar discussion for the A-2 reactions shows that the Zucker-Hammett hypothesis (i.e. $\log k_\psi - \log c_{H^+} =$ constant) holds only in the very unlikely case where $\log(f_{H^+} f_S a_{H_2O}/f_{\neq}) = 0$.

The experimental failure of the Zucker-Hammett hypothesis led other authors to propose alternative equations even before its theoretical weakness was completely clear. Bunnett proposed(7) that the amount of variation in the experimental rate constant which cannot be accounted for by the acidity function H_o should depend on variations in the activity of water and defined, according to equation 7, a parameter

$$\log k_\psi + H_o = w \log a_{H_2O} + \text{constant} \qquad (7)$$

w linked to the role played by water in the transition state. Later on equation 7 was modified into equation 8 by Yates(8), who recognized the fact that each family of compounds follows its own function

$$\log k_\psi + H_X = r \log a_{H_2O} + \text{constant} \qquad (8)$$

In this case r should give the reaction order in water molecules. In some cases, as will be shown later, unreliable w and r parameters have been obtained. Moreover, the theoretical foundament of 7 and 8 appears doubtful.

If we consider more carefully the definition of the A-1 process (equation 1 and 2) and its general expression for the rate equation (equation 3), we may easily realize (see equation 9, where $|S|_S$ stands for the stoicheiometric concentration of the substrate) that

$$\log \frac{v}{[S]_S} = \log k_\psi - \log \frac{[SH^+]}{[S]_S} = \log k_o + \log \frac{f_{SH^+}}{f_{\neq}} \qquad (9)$$

once the experimental pseudo first order rate constant (k_ψ) is corrected for the amount of protonated substrate ($[SH^+]/[S]$), we must obtain a constant value, unless the $f_{SH^+}/f_{\neq}$ term changes with acidity. This is always the case and the more recent treatments have been aimed to revealing the relationships between these changes and the reaction mechanisms. Although this approach gives less direct answers than, for instance, the Zucker-Hammett's one we must face the fact (see equation 9) that the only parameter which really changes when acid catalyzed reactions are studied at different acid concentrations is the $f_{SH^+}/f_{\neq}$ ratio.

Two main approaches have been developed to evaluate and interpret this ratio: the first by Bunnett and Olsen(9), lather extended by ourselves(10), and the second by Yates and his coworkers(11).

Yates proposes to evaluate the transition state activity coefficient and how it changes with acidity. Comparison of its behavior with that of charged species of known structure would give information on the structure of the transition state. Of course the activity coefficient of the transition state cannot be directly measured. However, we may rearrange equation 3 into equation 10 which becomes equation 11, if $[S] = [S]_S$ (if the substrate is not a weak base the expression must be modified by taking into account the protonation fraction which can be evaluated by knowing the pK and the acidity function followed by the substrate) from which equation 12 is obtained:

$$v = k_o [SH^+] f_{SH^+}/f_{\neq} = \frac{k_o}{K_{SH^+}} [S] \frac{f_S}{f_{\neq}} a_{H^+} \qquad (10)$$

$$\frac{v}{[S]_S} = k_\psi = \frac{k_o}{K_{SH^+}} \frac{f_S}{f_{\neq}} a_{H^+} \qquad (11)$$

$$\log f_{\neq} = \log k_o - \log K_{SH^+} - \log k_\psi + \log f_S + \log a_{H^+} \qquad (12)$$

All terms in the right hand part of equation 12 are either constants

(k_0, K_{SH^+}) or easily measured (k_ψ, f_s) as a function of the medium acidity but for the proton activity. A proton activity scale has been however defined, relative to the tetraethylammonium ion as standard, in the following way(12). The acidity function H_{GF} has been defined electrometrically by following the ferrocene oxidation as a function of the medium acidity. We may then express the proton activity as

$$\log a_{H^+} = -H_{GF} - \log f_{Fec^+} + \log f_{Fec} \qquad (13)$$

The ferrocene activity coefficient (f_{Fec}) can be obtained by distribution measurements up to 34,3% H_2SO_4 and extrapolated at higher acid concentrations. The evaluation of the activity coefficient for the ferrocinium ion (f_{Fec^+}) requires the extrathermodynamic assumption, summarized in equation 14, which also shows that it can be obtained from the relative solubilities of ferrocinium pentacianopropenide (Fec^+ PCP^-) and tetraethylammonium pentacianopropenide (TEA^+ PCP^-).

$$f^*_{Fec^+} = f_{Fec^+}/f_{TE\ A^+} = \frac{f_{Fec^+} f_{PCP^-}}{f_{TEA^+} f_{PCP^-}} = \left[f_{\pm(Fec^+PCP^-)} / f_{\pm(TEA^+PCP^-)} \right]^2 \qquad (14)$$

The (relative) activity coefficient of the ferrocinium ion has been measured up to 70% H_2SO_4 and, in this range, we may evaluate also the $a^*_{H^+}$ term, and, therefore, the (relative) activity coefficient of the transition state (see equation 12). In the following figure, reproduced from ref. 11, the activity coefficients of some cationic molecules, which can be taken as models for transition states of hydrolysis reactions, are compared with the transition states activity coefficients for the hydrolysis of esters.

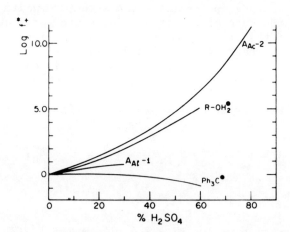

Fig. 1. Activity coefficients of cationic molecules and transition states of ester hydrolysis (ref. 11).

The different behavior of the different models allows an easy differentiation among the reaction mechanisms.

This treatment has, however, some limitations. We have already mentioned that the equations discussed hold for weak bases, and that the proton activity scale has been directly measured in a limited range of acidity. Moreover, the comparison of curved plots is not always straightforward.

We therefore prefer to deal with the problem of defining the $f_{SH}/f_{\neq}$ ratio (equation 9) in a different way(10). From studies of the protonation equilibria of weak bases, it has been shown that two acidity functions are linearly related according to equation 15

$$H_S + \log c_{H^+} = \log \frac{f_{SH^+}}{f_S f_{H^+}} = (1-\phi_e)(H_o + \log c_{H^+}) \quad (15)$$

If we write a similar relationship for the acidity function $H_{\neq}$ described by the equilibrium $S + H^+ \rightleftharpoons \neq$ and then substract $H_{\neq}$ to H_S we obtain

$$H_{\neq} + \log c_{H^+} = \log f_{\neq}/f_S f_{H^+} = (1-\phi_{\neq})(H_o + \log C_{H^+}) \quad (16)$$

$$H_S - H_{\neq} = \log f_{SH^+}/f_{\neq} = (\phi_{\neq} - \phi_e)(H_o + \log c_{H^+}) \quad (17)$$

which upon substitution in equation 9 gives

$$\log k_{\psi} - \log \frac{[SH^+]}{[S]_S} = \log k_o + (\phi_{\neq} - \phi_e)(H_o + \log c_{H^+}) \quad (18)$$

The $\log [SH^+]/[S]$ term is equal to zero when the substrate is completely protonated ($[S]_S = [SH^+]$, strong base) or can be evaluated for moderately weak bases, or for very weak bases ($[S]_S = [S]$) may be computed from the definition of pK_{SH^+} and equation 15 $[\log [SH^+]/[S] = \phi_e (H_o \log c_{H^+}) + p K_{SH^+} - H_o]$. Substitution of these values into equation 18 leads to the three general equations 18a, b, and c to be applied when studying reactions of strongly, moderately, or weakly basic substrates respectively

$$\log k_{\psi} = \log k_o + (\phi_{\neq} - \phi_e)(H_o + \log c_{H^+}) \quad (18a)$$

$$\log k_{\psi} - \log \frac{[SH^+]}{[S]_S k_o} = \log k_o + (\phi_{\neq} - \phi_e)(H_o + \log c_{H^+}) \quad (18b)$$

$$\log k_{\psi} + H_o = \log \frac{1}{K_{SH^+}} + \phi_{\neq}(H_o + \log c_{H^+}) \quad (18c)$$

It follows that, by plotting the appropriate function against ($H_o + \log c_{H^+}$) we may obtain straight lines whose slope will be related, as shown later, to the mechanism.

The slope parameter will be, depending on the equation used, either $\phi_{\neq}$ of ($\phi_{\neq} - \phi_e$). Their meaning may be assesed by rearranging equations 15 and 16. We obtain:

$$\log f_{\neq} - \log f_{SH^+} = (\Phi_{\neq} - \Phi_e)(\log f_{H^+} - \log \frac{f_{BH^+}}{f_B}) \quad (19)$$

$$\log f_{H^+} - \log \frac{f_{\neq}}{f_S} = (1 - \Phi_{\neq})(\log f_{H^+} - \log \frac{f_{BH^+}}{f_B}) \quad (20)$$

We know from acidity functions(13), protonation equilibria(14), and activity(15) measurements that the activity coefficients difference in the right hand part of equations 19 and 20 is always positive. This implies that a positive ($\Phi_{\neq} - \Phi_e$) slope parameter will be obtained when $f_{\neq}$ changes with acidity more rapidly than f_{SH^+} and a positive $\Phi_{\neq}$ when the ratio $f_{\neq}/f_S$ changes more rapidly than f_{BH^+}/f_B. The activity coefficients here defined represent the free energy of transferring the indicated species from the reference aqueous standard state to the concentrated acid solutions, and therefore their energy of solvation. Positive slope parameters are hence associated with reactions occurring through transition states which require large solvation energies. How could this be translated into information on the transition state structure?

We may obtain some hint by considering the large collection of slope parameters obtained through protonation equilibria (Table I). Here we see that the positive values are associated with the protonation of species leading to oxonium ions. These ions have larger solvation energies than ammonium or sulfonium ions since the positive charge is more localized on the small oxygen atom, therefore giving stronger interaction with the solvent water. On the contrary, carbonium ions, where the positive charge is highly delocalized, have negative slope parameters thus indicating weak interaction with the solvent.

Table I. Φ Values for Several Bases in Sulfuric Acid

Acidity function	Base	Φ
H_R	Triarylcarbinol triarylcarbonium ion	-1.20 to -1.59
H_I	Indoles	-0.26 to -0.46 and -0.67 to 0.85
H_C	Azulenes	-0.70
H_o'''	Tertiary anilines	-0.33 to 0.48
H_T	Thioamides	-0.36
H_S	Dialkyl sulfides	-0.26 to -0.29
H_o	Primary anilines	0
H_A	Amides	+0.42 to +0.55
H_{ROR}	Dialkyl Ethers	+0.75 to +0.82
H_{ROR}	Alcohols	+0.85
H_{H_2O}	Water	+1.00

In conclusion, the sign and the magnitude of the slope parameter is related to the nature of the atom bearing the positive charge and its ability to disperse the positive charge internally or externally through solvation.

We may now illustrate the merit of this treatment by considering few examples.

We'll first consider the carbon-oxygen bond breaking which occurs upon protonation of t-butyl esters or ether. The Yates r treatment gives clearly unreasonable "orders" in water of -8.9 and -3.15. The ($\Phi_{\neq} - \Phi_e$) slope parameter, instead, is negative in both cases, as expected for reactions going from an oxonium ion to a transition state which resembles a much less solvated carbonium ion. The fact that the magnitude of the slopes is not the same derives from the different degree of carbonium ion character of the transition states, as well as from the difference in activity coefficients of the protonated substrates. In other terms, even for reactions which belong to the same mechanistic category, we have no right to expect slopes of the same magnitude, although certainly we expect to find the same sign.

The second example we wish to discuss is a classical of acid catalyzed reactions whose interpretation has shown the weakness of both the Zucker-Hammett hypothesis and the Bunnett w treatment. In fact, the lactonization(16) of γ-hydroxybutyric acid follows h_o and gives w = 2.21 and 2.23 in $HClO_4$ and HCl, respectively, whereas the hydrolysis of γ-butyrolactone(16) follows c_{H^+} and gives w = 8.50 and 6.11, respectively. In other terms, the transition states for the two reactions are classified into two categories, with the water molecule being implied as nucleophile only in the latter reaction. This is clearly untenable, since both the forward and the reverse reaction must have a common transition state. The $\Phi_{\neq}$ parameters are positive (0.23 and 0.31 for the lactonization and 1.08 and 0.88 for the hydrolysis in $HClO_4$ and HCl, respectively) meaning that for both reactions the $f_{\neq}/f_S$ term increases with acidity faster than f_{BH^+}/f_B, and hence that the transition state has a fairly large solvation energy. Since both reactions have a common transition state, i.e.

Table II. Acid Catalyzed Carbon-Oxygen Bond Breaking

$R-O-Bu-t+H^+ \rightleftharpoons$	$R-\overset{+}{\underset{H}{O}}-Bu-t$	$\longrightarrow tBu^+$	+	ROH

R =	CH_3CO			CH_3
r	-8.9			-3.15
$(\Phi_{\neq} - \Phi_e)$	-1.23			-0.87

equal $f_{\neq}$'s, the larger $\phi_{\neq}$ value for the hydrolysis must be associated with a smaller f_S term for the lactone than for the acid, as indeed found from solubility measurements in aqueous solutions of sodium perchlorate and sodium chloride(17).

A final case deals, in some more detail, with the acid catalyzed hydration of ethylenes and acetylenes. The reactivity ratio between two pairs of similarly substituted alkenes and alkynes is very much near unity(18), contrary to what expected from the large difference in stability of carbonium ions and vinyl cations. One explanation maintains that the two intermediates ions $\underline{1}$ and $\underline{2}$ are

$$\text{C=C} + H_3O^+ \longrightarrow H-\overset{|}{C}-\overset{+}{C} + H_2O \longrightarrow \text{products}$$
$$\underline{1}$$

$$-C\equiv C- + H_3O^+ \longrightarrow \underset{H}{\text{C=C}}^+ + H_2O \longrightarrow \text{products}$$
$$\underline{2}$$

differently solvated(19), and that the greater stabilization through solvation of $\underline{2}$ could overcome its intrinsic lower stability as compared with $\underline{1}$.

We have compared a series of acetylenes and ethylenes(18) and always found very similar $\phi_{\neq}$ parameters which indicate that the solvation requirements of the two reactions are very similar. Yates has reached similar conclusions(11b) form the evaluation of the transition states activity coefficients, although his computation from our data shows a difference of about 4 kJ mol^{-1} in the $f_{\neq}$ term. This difference must be, however, somewhat overestimated. In fact, the plot of log $a_{H^+}^*$ - log k_ψ = log $f_{\neq}$ / $f_S k_o$ vs. the medium acidity (see Figure 2) runs parallel for styrene and phenylacetylene in all

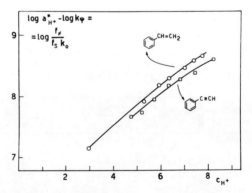

Fig. 2. Activity coefficients for the hydration of styrene and phenylacetylene.

the range measured. Since the experimental f_S values for the two compounds are equal(18), we must conclude that there is really no change in the activity coefficients of the transition states due to changes in medium acidity.

The substituent effect on the hydration of styrenes and acetylenes is described, according to our treatment(18), by the general equation $\rho = \rho + \dfrac{\phi_S^{\neq} - \phi_4^{\neq}}{\sigma^+}$ (H_o + log $[H^+]$) which for styrenes becomes $\rho = -3.0$ (±0.1) + 0.3 (±0.08) (H_o + log $[H^+]$) which shows that the ρ values depend on the acidity, becoming larger as the acidity increases.

This conclusion has been challenged by Johnson(20) who in a recent publication reported ρ values independent of the acid concentration. However, since Johnson has used the same set of data we have used, the discrepancy must be due to the different ways used to compute the experimental data. Actually, Johnson did correlate reaction rates with acidity in a fashion very similar to the one required by the Zucker-Hammett hypothesis. In fact he did extrapolate the experimental data from plots of log k_ψ vs. H_c, the acidity function believed to describe the transition state of the hydration reaction. We may settle this difference by looking at the experimental data reported by Coussemant(21) (see Figure 3) which have

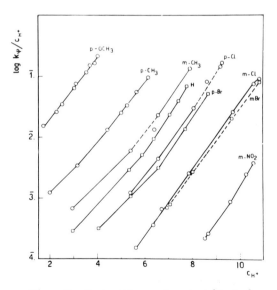

Fig. 3. Hydration of substituted styrenes (ref. 21)

been the base for both computations. Being the overlap relatively good, we may compute the ρ values at several acidities without any extrapolation. This has been done at four acidities and as shown in Figure 4 the ρ values do increase with acidity, as described by our equation and contrary to what reported by Johnson. While this last example shows the shortcomings of using incorrect equations to correlate rates with acidity, we like to stress again what we believe is the important message to be taken from this work. It is impossible to extract direct mechanistic conclusions from the dependence of rates on acidity as attempted in previous treatments (i.e.; Zucker-Hammett, Bunnett's w, Yates r). However knowledge of the transition state solvation requirements, made possible by the treatments here discussed, and the comparison with the behavior of model species, allows indirect, but realistic, conclusion on the structure of the transition state which can be easily translated in mechanistic categories.

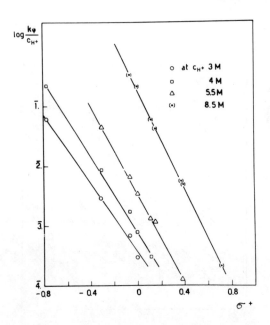

Fig. 4. Substituents effect on the hydration of styrenes at several acidities.

REFERENCES

1. B. Holberg, *Ber*, 45:2997 (1912)
2. L. Zucker and L. P. Hammett, *J. Am. Chem. Soc.* 61:2791 (1939).
3. M. A. Paul and F. A. Long, *Chem. Rev.* 57:1 (1957); F. A. Long and M. A. Paul, *ibid.* 57:953 (1057).
4. L. P. Hammett, "Physical Organic Chemistry", Mc Graw Hill, New York (1970).
5. C. D. Johnson, A. R. Katritzky, and S. A. Shapiro, *J. Am. Chem. Soc.* 91:6654 (1969).
6. E. M. Arnett and G. W. Mach, *J. Am. Chem. Soc.* 86:2671 (1964).
7. J. F. Bunnett, *J. Am. Chem. Soc.* 83:4956, 4968, 4973, 4978 (1961).
8. K. Yates and R. A. Mc Clelland, *J. Am. Chem. Soc.* 89:2686 (1967).
9. J. F. Bunnett and F. P. Olsen, *Can. J. Chem.* 44:1917 (1966).
10. V. Lucchini, G. Modena, G. Scorrano, and U. Tonellato, *J. Am. Chem. Soc.* 99:3387 (1977).
11. (a) R. A. Mc Clelland, T. A. Modro, M. F. Goldman, and K. Yates, *J. Am. Chem. Soc.* 97:5223 (1975); (b) K. Yates and T. A. Modro, *Acc. Chem. Res.* 11:190 (1978).
12. T. A. Modro, K. Yates, and J. Janata, *J. Am. Chem. Soc.* 97:1492 (1975).
13. E. M. Arnett and G. Scorrano, *Adv. Phys. Org. Chem.* 13:83 (1976).
14. P. Bonvicini, A. Levi, V. Lucchini, G. Modena, and G. Scorrano, *J. Am. Chem. Soc.* 95:5960 (1973); A. Levi, G. Modena, and G. Scorrano, *ibid.* 96:6585 (1974).
15. K. Yates and R. A. McClelland, *Progr. Phys. Org. Chem.* 11:323 (1974).
16. F. A. Long, F. B. Dunkle, and W. F. McDevit, *J. Phys. Chem.* 55:829 (1951).
17. F. A. Long, W. F. McDevit, and F. B. Dunkle, *J. Phys. Chem.* 55:813 (1951).
18. G. Modena, F. Rivetti, G. Scorrano, and U. Tonellato, *J. Am. Chem. Soc.* 99:3392 (1977).
19. K. Yates, G. H. Schmid, T. W. Regulski, D. G. Garratt, H. W. Leung, and R. McDonald, *J. Am. Chem. Soc.* 95:160 (1973).
20. C. C. Greig, C. D. Johnson, S. Rose, and P. G. Taylor, *J. Org. Chem.* 44:745 (1979).
21. J. P. Durand, M. Davidson, M. Hellin, and F. Coussemant, *Bull. Soc. Chim. France*, 43 (1966).

INDEX

Absolute Configuration, 303
Acetylcholine receptor, 231, 240, 244
Acid-base group constants, 139, 142
Acid-base microconstants, 139, 142
ACTH acid-base properties, 144, 145
Acid catalyzed reactions, 373
Acidity functions, 379
Adamson's rules, 91
Ammine Complexes, 67, 74-76
Amphiphile structure, 210
Apamin, 231, 234, 235-9
Aqua complexes, 67, 71, 73-74
Asymmetric Electrolytes, 48-51
Asymmetric Induction, 305, 307
Asymmetric molten salts, 60

Bilayers, 209-11, 213-4, 217-8
Bis(acetylacetonato)copper(II), 82, 85-86
Bis(dithiocarbamato)copper(II), 86
Bis(dithiophosphato)copper(II), 87-88
BPTI Acid-base equilibria, 145
β-bromoethylnaphtalene, 225-229

Brønsted relations, 331-3, 335-7
t-butyl chloride, 343-5, 347

^{43}Ca NMR, 192-194, 197-201, 204
^{111,113}Cd NMR, 191, 194, 201-204
^{59}Co NMR, 185
Calmodulin complexes, 201
Carbon acidity, 332
Carbon dioxide hydration, 256, 258, 260, 265, 271
Carbonic anhydrase, 253
  anionic inhibition, 261, 267
  models, 224
Carbanions, 311-2, 314
Cavity, 181-3, 222, 345
  formation, 27
Chelation, 178-9
Chirality, 300-3, 305
Cholesterol, 217
Chromium(III) complexes, 94, 97-99
Cyclization, 319-22, 327-9
Cyclodestrins, 222
Clusters, 176
Cobalt(III) complexes, 100-103, 105-8
Cobalt(II) complexes, 118, 162 165
  ESR, 115
  dioxygen carriers, 162, 165
  as catalysts, 170
Contact shift, 131
Copper(II) complexes, 81
Corticotropin, 144

Dance-Miller parameters, 134
Debye-Hückel Theory, 42, 43

385

Deuterium exchange, 105, 235
Dioxygen carriers, 161
Dipolar shift, 131

Electrolyte solutions, 41, 61
  osmotic coefficients, 46
  activity coefficients, 46
Electrophilic catalysis, 84
Enzyme catalysis, 319
Ethylamine complexes, 72, 76-78

Garrick theory, 73
Glycine, 346-7
Guanosine, 180

Hammett equation, 277, 284-5, 287-288
Hill plots, 177
H-isotope exchange, 331-2
Hole structure of liquids, 7
Host-guest interactions, 221
Hydration thermodynamics, 13, 73-4
Hydrophobic hydration, 24
Hydrophobic interaction, 149, 154-7
Hydrophobic recognition, 223

I_d substitution, 105
Inclusion complexes, 221
Ionic atmosphere, 42
Ionic fluids, 41
Ionic hydration, 5,6
Ion pair, 184-7, 332, 351
Iridium(III) complexes, 100, 103
Iron(III) complexes, 108-113
Iron(III) dithiocarbamates, 129

Lactones, 322-3, 326
Linear dichroism, 296, 298, 299
Liquid crystals, 212, 214, 295, 299, 300, 307
Liquid, structure of, 1

McMillan-Mayer theory, 42, 43
Membranes, 209-11, 213-4, 217-8
Metal deactivators, 87
^{25}Mg NMR, 192-194
Michaelis-Menten parameters, 264, 265
Molten salts, 55-60

^{23}Na NMR, 181, 183, 194
Nitric oxide carriers, 163
Non-electrolytes, 13

Ornstein-Zernicke equation, 44

Perchloric acid, 4, 5
Percus-Yervic terms, 53
Phase transfer, 184, 310
Phospholipase A_2 complexes, 205
Photochemical rearrangement, 91
Photosubstitution reactions,
Piezosolvatochromism, 283
Poisson-Boltzmann Equation, 43
Polypeptides protonation, 143
  complexes, 147
Proton abstraction, 218, 332, 336, 357, 359, 364-5
  exchange, 335, 357-8, 336
  transfer, 336, 356
Pyridinium-N-phenoxide betaines, 283

Quadrupole moments, 176

Redox catalysts, 84
Relaxation rate, 194
Rhodium(III) complexes, 100-3
Ring stacking interactions, 149, 152, 157

Solvation shell, 175
Solvato-chromism, 277-282
Solvent assisted dissociation, 107

Solvent polarity, 275, 276
Solvolysis of
   t-butylchloride, 343-5, 347
   β-bromoethylnapthalene, 224, 228
Spin density, 122-3
Spin polarization, 125
Symmetric Electrolytes, 46-50

Taft equation, 286
Ternary complexes, 149
Tetraalkylammonium salts, 309, 316
Thermosolvato-chromism, 282
Thiolactones, 322-3, 326
Tolman's parameter, 127
Troponin-C complexes, 197
Two-phase systems, 311-2, 314, 316-7

Vesicles, 210, 213-4

Water structure, 2

Zucker-Hammett theory, 373-5, 379